T0094011

Fundamentals of Artificial Intelligence

Problem Solving and Automated Reasoning

About the Author

MIROSLAV KUBAT is an associate professor emeritus of Electrical and Computer Engineering at the University of Miami. He taught artificial intelligence and related courses for more than a quarter century. He has published more than a hundred peer-reviewed papers and is the author of the commercially successful textbook *An Introduction to Machine Learning*.

Fundamentals of Artificial Intelligence

Problem Solving and Automated Reasoning

Miroslav Kubat

New York Chicago San Francisco Athens
London Madrid Mexico City Milan
New Delhi Singapore Sydney Toronto

Fundamentals of Artificial Intelligence: Problem Solving and Automated Reasoning

1 2 3 4 5 6 7 8 9 LCR 28 27 26 25 24 23

Library of Congress Control Number: 2022052800

ISBN 978-1-260-46778-9
MHID 1-260-46778-3

Sponsoring Editor
Lara Zoble

Editorial Supervisor
Patty Mon

Project Manager
Tasneem Kauser

Acquisitions Coordinator
Caitlin Cromley-Linn

Copy Editor
Nupur Kakkar

Proofreader
Rup Narayan

Indexer
Miroslav Kubat

Production Supervisor
Lynn M. Messina

Composition
KnowledgeWorks Global Ltd.

Illustration
KnowledgeWorks Global Ltd.

Art Director, Cover
Jeff Weeks

To my late wife, Verunka

Contents

Preface

Psychologists and philosophers always had a hard time trying to put a finger on the essence of intelligence—whether natural or artificial. No wonder. The phenomenon is elusive and resents being squeezed into definitions, characteristics, and descriptions. Modern technologist, suspicious of abstract arguments, subscribes to a pragmatic viewpoint: Artificial intelligence (AI) is a discipline that studies algorithms, data structures, and perhaps even mathematical theorems, that allow the computer to do justice to domains where traditional programming struggles.

AI was not born yesterday. First ideas popped up when World War II was still in vivid memory. What followed was a long period of hectic developments, full of bold promises, painful disappointments, staggering discoveries, unexpected hurdles, genuine revolutions, and countless twists and turns. Every now and then, a fascinating novelty grabbed the limelight, self-promoting itself as the ultimate answer to all challenges and conundrums—only to give way to another panacea that replaced it in a few years' time. Gradually, though, concentrated efforts of generations of scholars bore fruit. AI's impact on our lives is hard to ignore. Growing interest from well-paying industries boosted its popularity among undergraduates, and entry-level courses are now in high demand.

How to answer this demand? What should an introductory course look like? What topics should it cover? What, come to this, constitutes the very essence of AI? These are the questions that a conscientious educator has to deal with. Attempts to answer them have undergone stormy evolution.

Early AI pioneers had sweeping ambitions. Apart from reasoning and problem solving, they also embraced computer vision, natural language processing, machine learning, even robotics. In due time, however, some of these fields have strayed. They are taught in their own specialized courses; they rely on their own textbooks and employ techniques that no longer have much to do with their parent discipline. Sure enough, vision, language understanding, and ability to learn *are* genuine aspects of any intelligent behavior. However, they have long since abandoned their AI cradle and become largely independent. Old textbooks also made it a point to cover dedicated programming languages, habitually devoting a few chapters to *Prolog* and *Lisp*. Today, inclusion of these languages in an introductory text is no longer imperative. They are useful, but not vital.

What, then, *should* be included? In this author's view, the foundations of AI are delineated by two questions. First, how to write programs capable of solving difficult problems? Second, how to endow the computer with the ability to reason, and perhaps even to argue? In regards to problem solving, classical AI relies on techniques to *search* through the space of potential solutions. To these, modernity has added biologically inspired alternatives such as the *genetic algorithm* and the now-so-fashionable *swarm intelligence*.

The second strand, automated reasoning, rests on knowledge representation and inference techniques. Most of these take advantage of first-order logic, but logic alone is not enough. Our human thinking easily copes with information that is unreliable, tentative, incomplete, and sometimes downright misleading. AI scientists asked themselves how to achieve something similar. Their answers rely on the theory of probability, fuzzy-set theory, and some other paradigms. All these efforts culminated in the advent of *expert systems*, software packages meant to emulate the thought processes of human experts.

There has never been any shortage of teaching texts for AI; each generation produced quite a few. The best known are impressive both in their scope and in their authors' erudition. And yet, some of them look rather like manuals or encyclopedias. Learned and scholarly though there are, they do not target beginners. They offer invaluable material to specialists interested in advanced areas, but they cannot be recommended to readers who want an easy-to-digest introduction. Besides, they tend to be too bulky and heavy to carry around the campus.

It was with these thoughts in mind that this author embarked on his project. His intention was to write a cohesive text that would present the foundations of AI in an easy-to-read and reasonably sized volume. He wanted to make early chapters prepare the soil for later ones, to provide motivation for each new topic. He sought to offer practical advice, even at the cost of sacrificing theoretical depth. No one has ever become an AI expert by just mastering the algorithms. Rather, one has to know which of them to employ under what circumstances, how to combine them, how to tweak them to suit the specific needs of realistic applications. All this, too, the author hoped to deliver.

Last but not least, there is more to AI than what's on the surface. Each algorithm has its story, historical context, a specific need that once motivated its conception. All this deserves to be known, too. Appreciation for the bigger picture helps develop fondness for the discipline—thus making its study more enjoyable. This is why the author slipped-in contextual information wherever possible. He felt the field deserved it. After all, we are dealing here with nothing less than the birth, growth, and maturity of one of humanity's greatest technological achievements.

Miroslav Kubat
Miami, FL, U.S.A.
June 2022

Acknowledgment

I nearly abandoned the project. The whole adventure seemed far-fetched, presumptuous. Did the world really need yet another AI textbook? And was I the one destined to deliver it? Paralyzed by doubts, I dawdled, dragged my feet. One memorable evening, I thought I had enough and I told my wife I was giving up.

She would have none of it. Browsing through what I had written so far, she said this would be not just a good book, but a great one! In her view, the manuscript covered precisely the topics needed by modern AI classes. The chapters were structurally balanced and never lost their focus. She liked the way the problems were presented and their solutions explained. As for the pictures, they were nicer and more informative than in any educational text she had seen before. She went on and on.

You would have to see her shining eyes. She was so excited, so passionate—and successful! Under the onslaught of her enthusiasm, my writer's block melted, then vanished altogether. I got back to work, knowing that I owed it to her.

This is what one devoted fan can do for you. Oh my, was I lucky!

We both knew the cruel truth: She would never hold in her hands the neatly bound volume she so much desired me to complete. Her diagnosis did not leave the tiniest shred of hope.

Tomorrow, I will send the manuscript to the publisher. In a few months, the product will hit the market. Over the coming years, instructors will be choosing it for their courses, and many students will read it. Alas, I will never have a chance to tell my lifelong companion that we made it, in the end. She is no longer around.

It is to her memory that I dedicate these *Foundations of AI*. It is thanks to her that I wrote them, forever grateful for her shining eyes, for her faith in my work, for her encouragements, for her unwavering love.

How I wish you were here, Verunka. Your departure has made the world such an empty place.

CHAPTER 1
Core AI: Problem Solving and Automated Reasoning

The core of artificial intelligence (AI) consists of two major areas. The first focuses on *problem solving*: the studies of techniques that help the computer address difficult tasks. For some of these tasks, available algorithms are either too complicated or prohibitively expensive; for others, they are not known at all. The second area, *automated reasoning*, wants to impart to computers our human ability to benefit from expert knowledge, to answer sophisticated queries, and to infer previously unknown consequences of this knowledge.

To set the stage for the rest of the book, this chapter briefly outlines the history of AI and offers more information about the two fundamental tasks.

1.1 Early Milestones

The history of a scientific discipline helps us appreciate its underlying principles, immediate goals, and long-term ambitions. Let us take a cursory look at how and why AI was born, and let us examine the expectations and hopes cherished by its early trailblazers.

1.1.1 Limits of Number Crunching

The last salvos of World War II had barely faded away when the first digital computers were officially introduced. The experience was overwhelming; humanity gasped in awe. "Just fancy what a machine capable of hundreds of arithmetic operations per second can do for us!" Engineers, scientists, mathematicians, statisticians, and just about everybody else realized that the new invention would usher in a brand new world, a future rife with unheard-of feats and staggering accomplishments.

Imagination ran high. In no time, a group of visionaries started to look beyond mere number crunching. They wanted more. They predicted that computers would one day play chess; that they would be endowed with the ability to see, to listen, to read; that they would even converse with their users. Machines would tackle tasks that had so far been the exclusive domain of experts and specialists. In short, computers would exhibit *intelligence*.

All this was no empty philosophizing, nor was it castles in the air. These people meant business. As early as in 1949, Claude Shannon explored the possibilities of chess programming.[1] Other ideas soon followed suit.

1.1.2 AI Is Born

In 1956, an impressive array of brilliant minds convened at Dartmouth College. The scholars had diverse backgrounds, but they all shared one great ambition: to pave the way for the birth of intelligent machines. What would it take to make the dream a reality? What algorithms were needed, what data structures, what programming languages, what hardware? What concrete tasks were the machines to be applied to? Questions of this kind then motivated generations of scholars.

This is when AI was born. The infant scarcely deserved the proud name of science, but as the cliché goes, even the longest march begins with a single step. The significance of this particular step is hard to overstate. The Dartmouth College gathering did nothing less than inaugurate the era of AI.

1.1.3 Early Strategy: Search Algorithms

It is one thing to agree on the goals; it is another to know how to achieve them. What was needed was a strategy, a unifying principle to lean on. Finding it was not an easy undertaking. Fruitful paradigms are only uncovered after a fair amount of effort, and only when all the jigsaw puzzle's pieces are at hand—which they were not! Available hardware was lamentably slow and unreliable, programming techniques were in diapers, practical experience virtually nonexistent.

The problem-solving part of AI received attention earlier than automated reasoning. One idea for dealing with it employed *search* through the space of potential solutions. An early triumph came in the mid-1960s with the introduction of a computer program known as General Problem Solver.[2] The product was hailed as the first successful attempt to address problems that appeared to require intelligence. By today's standards, these problems were still simple. What matters is that their solutions were discovered by *search* techniques.

1.1.4 Early Wisdom: Computers Need Knowledge

It did not take long for the scientific community to figure out that search was not enough. Another critical ingredient was needed. To wit, one had to find ways of conveying to the machine some version of the background knowledge available to humans. Besides, one needed algorithms to facilitate automated reasoning with this knowledge.

Adequate methods were developed and systematically explored—and their importance was duly appreciated. In the 1990s AI was dominated by studies of *knowledge representation* and *reasoning*. Most efforts sought to benefit from the diverse variations of first-order logic, but other approaches were explored, too, among them *frames* and *semantic networks*.

---

[1] More famous is the journal version of this paper: Shannon (1950).

[2] The program was referred to by its initials, GPS, the same acronym that we today use for the global positioning system.

1.1.5 Programming Languages

Appreciating the need for better tools, scientists embarked on studies of programming languages that would facilitate the implementation of the newly developed techniques. Quite a few were proposed, two of them legendary: *Prolog*, which was good at conveying knowledge in the form of logical statements, and *Lisp*, which was an early example of what came to be known as *function-oriented programming*. With these, writing AI programs was much easier than, say, in FORTRAN.

The two languages happily coexisted, with no consensus regarding the superiority of the one or the other. Preferences went largely along geographical lines. *Prolog* dominated in Europe because it had been developed mainly in France and the U.K. By contrast, Americans tended to prefer *Lisp* because it was associated with M.I.T. In the 1980s and 1990s, AI research was colored by friendly competition and rivalry.

1.1.6 Textbooks: Many Different Topics

Alongside advances in problem solving and reasoning, breakthroughs were reported in computer vision, natural language processing, and machine learning. Early AI textbooks almost always devoted at least one chapter to each of these disciplines. This was only natural because, in those days, all these fields largely relied on the techniques typical of early AI research: search and the use of knowledge.

Besides, the relative novelty of *Prolog* and *Lisp*, and the practical benefits they brought to AI, made it imperative to cover these languages in some detail.

1.1.7 Twenty-First Century Perspective

Over the decades, however, vision, speech, and learning started to draw their ideas and inspiration from elsewhere. So fruitful were these borrowings that the three disciplines gradually abandoned their AI cradle and acquired lives of their own. The development has its parallels. Ancient philosophy sought to cover all the knowledge that could possibly be had. Step by step, however, some fields set out on their own paths, never to look back: physics, chemistry, and others. The children of AI did the same. At the beginning, they were acknowledged as true aspects of intelligence, artificial or otherwise. Later, each came to cultivate its own techniques and algorithms, departing farther and farther away from the mainstream.

At the moment, they tend largely to depend on *deep learning*, a breakthrough approach to machine learning. So overwhelming is deep learning's domination that it is often identified (or confused) with AI itself. In this author's view, this is a misunderstanding. While it is true that intelligence is unthinkable without the ability to learn, there is more to AI than just learning.

As for *Prolog* and *Lisp*, modern times still see them as fascinating tools—though no longer indispensable. General-purpose languages of the twenty-first century, with their rich libraries of built-in functions, appear to be good enough. This is why this book provides only such information about *Prolog* as is necessary for explaining the principles of automated reasoning. Readers who want to learn more are referred to specialized textbooks.[3]

---

[3] Among the many available on the market, this author's favorite is Ivan Bratko's *Prolog Programming for Artificial Intelligence* (2001).

Control Questions

If you have problems answering any of the following questions, return to the corresponding place in the preceding text.

- Which year saw the publication of the first paper dealing with chess programming? What do you know about the first scientific meeting that discussed the chances of AI? Where and when was it organized?
- What scientific disciplines historically belonged to AI, and were therefore included in classical AI textbooks? What programming languages targeted the needs of AI?

1.2 Problem Solving

Core AI is what is left if we eliminate computer vision, machine learning, natural language processing, and AI programming languages. For the needs of this book, core AI includes two major topics: first, algorithms and techniques for problem solving and second, algorithms and techniques for automated reasoning. Let us begin with the former, relegating the latter to the next section.

1.2.1 Typical Problems

Figure 1.1 shows a few toy domains popular in early AI textbooks: sliding-tile puzzles, Hanoi tower, all sorts of mazes, teasers, and simple games such as tic-tac-toe. One typical goal is to identify an ideal sequence of actions leading to a predefined state. What the puzzles have in common is that the solution is discovered by trial and error, by experimental procedure known as *search*.

Attempts to develop problem-solving algorithms are in line with the original intentions of AI's founding fathers. In traditional computation, every detail of the program's function must be encoded by the programmer. By contrast, AI studies techniques that allow computers to deal with problems for which algorithms are not known, or if they *are* known, their complexity and computational costs render them all but useless.

1.2.2 Classical Approaches to Search

At each moment, a puzzle finds itself in a specific *state*. One state can be converted to another by executing an *action*. Thus in the sliding-tiles puzzle, an action can slide a tile to the neighboring empty square. In the Hanoi tower, an action may transfer the topmost disk from one vertical bar to another. Each state may allow two or more (sometimes many) different actions. The set of all possible states constitutes the puzzle's *search space*.

Chapter 2 explains the baseline algorithms of what is known as *blind search*. Its goal is to systematically explore the *search space* so that "nothing is missed." We label these algorithms as blind because they do not seek to optimize the search: They do not pay attention to its efficiency or computational costs—these would rather be the focus of the *heuristic search* addressed by Chapter 3. Most typically, efficiency is improved by mechanisms to evaluate the quality of each state. Preference is then given to actions that result in higher-quality states. In the case of game programming, the agent's control over the system's state is limited due to the opponent's interference. This calls for specific search algorithms that fall in the category of *adversary search*, a topic of Chapter 4.

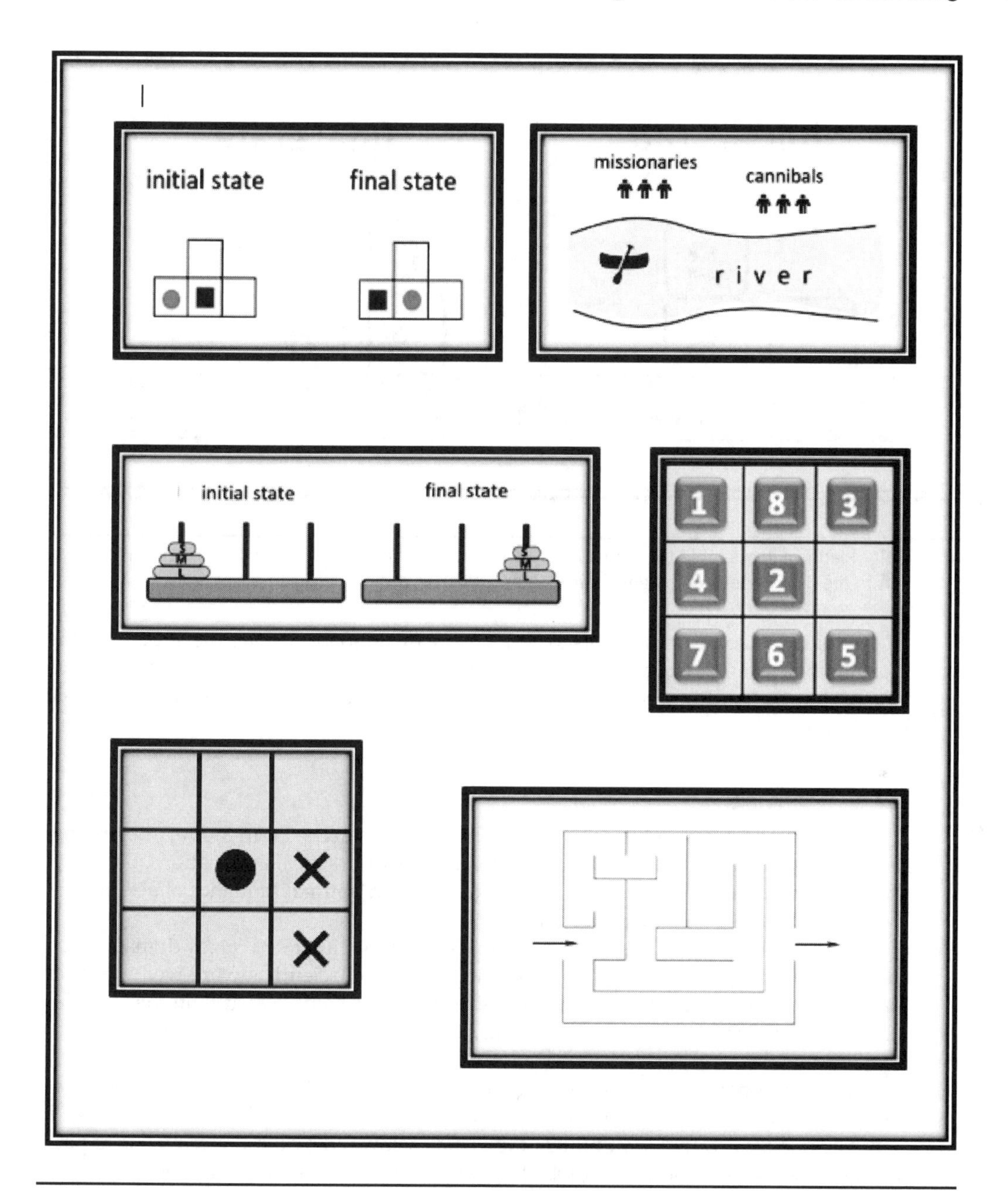

Figure 1.1 AI scientists like to illustrate the behavior of problem-solving techniques using simple puzzles. Modern science may shrug off some of them as too simple, but they are still useful for instruction purposes.

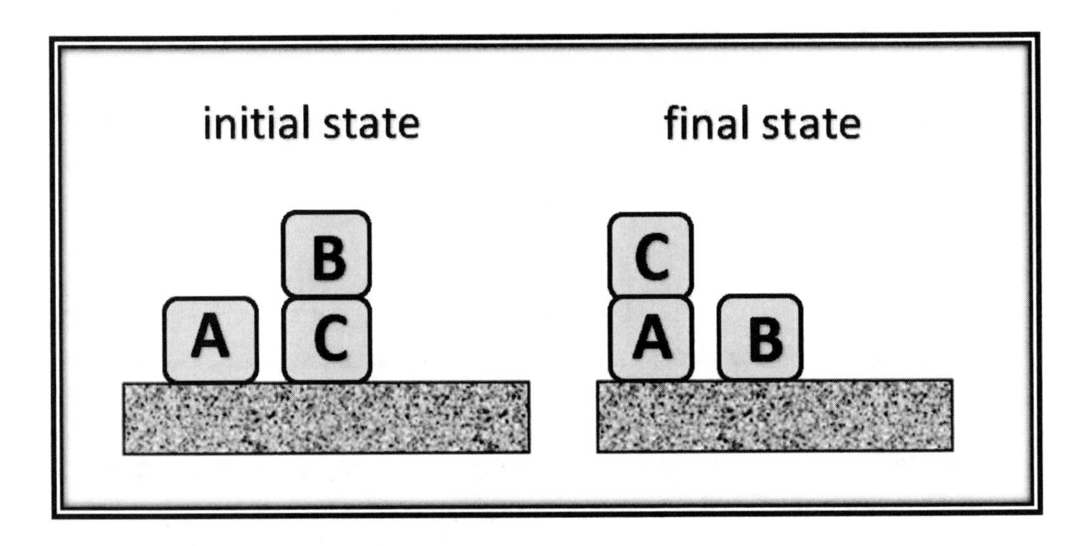

Figure 1.2 Planning in a simple toy domain: Find a sequence of actions that will convert the initial state on the left to the final state on the right by moving one block at a time.

In the 1980s, some alternatives to search were developed. Among these, perhaps the most famous came to be known as *simulated annealing*. This approach is somewhat advanced, but the reader needs to know about its existence. Section 3.4 offers some basic information.

1.2.3 Planning

AI pioneers had higher ambitions than solving simple puzzles. From the very beginning, they understood that AI should target a broad range of practical applications. One category of such applications, *planning*, attempts to find an optimal sequence of actions that would satisfy a certain non-trivial goal.

The behaviors of early planning programs used to be illustrated on toy domains such as the blocks world from Figure 1.2. At the same time, scientists were well aware of the fact that practical applications would be much more complicated than that. Gradually, they converged on a set of model tasks, each representing a whole group of typical applications.

One such model, *job-shop scheduling*, seeks to optimize the distribution of *jobs* over a set of *machines* in a way reminiscent of the needs of assembly lines. Another model, *travelling salesman*, wants to minimize the distance covered by a traveler who is to pass through a set of cities. Both tasks, job-shop scheduling and the traveling salesman, have for decades been analyzed by mathematicians and computer scientists alike, and are known to be more than challenging. Typical goals, applications, and solutions of AI planning are discussed in Chapter 5.

1.2.4 Genetic Algorithm

Search algorithms dominated the field for more than a generation. Gradually, though, many engineers became aware of this paradigm's limits. In the face of truly difficult tasks, search tended to be prohibitively expensive and sometimes failed to find the solution altogether. This is why the late 1980s experienced an outburst of interest in alternatives hoped to do better. One of them was the simulated annealing mentioned earlier in this section. Much more influential, however, was the *genetic algorithm*.

Inspired by the processes underlying Darwinian evolution, the technique proved successful in such diverse fields as identification of extremes of mathematical functions, problem solving, and optimization of engineering design. Enthusiasts soon demonstrated that many problems that used to be addressed by search techniques could just as well be solved by the genetic algorithm, but with superior efficiency. Moreover, the GA, as it came to be known, often succeeded where search techniques failed. The message spread, feeding the new technology's popularity. This book explores it in Chapter 6.

1.2.5 Swarm Intelligence

This is not the end of the story, though. After the eras of search and the GA, another was ushered in: the age of multi-agent systems seeking to implement what came to be known as *swarm intelligence*.

Just as with the GA, inspiration came from biology. The neural system of an insect is too primitive to permit anything that would resemble decision making—and yet these creatures are capable of incredible feats of organization, food gathering, and construction. Biologists know the secret: Each individual executes a very simple job addressing the needs of the whole community. For instance, owing to their tendency to leave pheromone trails after discovering a food cache (and to follow these trails in search for food), ants are hailed as the heroes of foraging and food gathering.

Similar observations have been used by scientists who developed a whole family of *swarm intelligence* algorithms that seek to emulate not only the behavior of ants, but also of birds, honey bees, and quite a few others. The new problem-solving techniques soon attracted the attention of practically minded engineers, and even of journalists and the public in general. This book introduces swarm intelligence in Chapter 8.

1.2.6 Emergent Properties and Artificial Life

The success of swarm intelligence is explained by the philosophical concept of *emergent properties*. The same idea plays a critical role also in the field of *artificial life* that is the subject of Chapter 7. True, its practical benefits are less conspicuous than those of, say, planning. Still, it is good to know about it because the paradigm will help the reader appreciate the secret behind the powers of swarm intelligence.

Control Questions

If you have problems answering any of the following questions, return to the corresponding place in the preceding text.

- Suggest examples of problem-solving tasks addressed by AI. What do they have in common? What models of more advanced applications do scientists commonly use?
- What is the goal of classical search techniques? What alternatives have been invented?

1.3 Automated Reasoning

Problem solving is not enough. An intelligent computer program should be able to answer sophisticated questions and support its answers by convincing explanations. It should perhaps even be capable of participating in arguments and discussions.

This should not be about trivialities. Asked about President Lincoln's date of birth, the machine can simply find the information in a database, and no intelligence is needed. What we have in mind, rather, are queries related to advanced technological problems or, say, to medical diagnosis. The answer then cannot be just looked up in a dictionary; it has to be *inferred* from some kind of background knowledge. To accomplish this feat is the task for *automated reasoning*, the second major goal of AI.

The following examples will give us an idea of the scope of applications.

1.3.1 Zebra

In the 1960s, many newspapers entertained their readers with a novel type of puzzle that came to be known as *zebra*. The first exemplar, published in the *Life International* magazine in 1962, offered a list of facts, such as "there are five houses," "the Englishman lives in the red house," "coffee is drunk in the green house," and "the man who smokes Chesterfields lives in the house next to the man with the fox." Based on many hints of this kind, the readers were to answer two questions: Who drinks water? Who owns zebra?

The latter gave the puzzle its name.

1.3.2 Can a Computer Solve Zebras?

On the surface, the problem looks like just another version of the puzzles from Figure 1.1. In reality, however, things are not so simple. A person seeking to solve the puzzle will subconsciously carry out many "operations" over his or her *background knowledge*. For instance, we know that fox is an animal, green is a color, coffee is a beverage, and Chesterfield is a cigarette. We know that cigarettes are smoked and beverages drunk. From the way the facts have been formulated, it is clear that the houses are arranged in a row, and this gives a concrete meaning to the term, "next house."

This is what *we* know—but the computer is completely ignorant of all of this. If we want to write a program to solve a zebra, we must first find a way of conveying to the machine all the requisite background knowledge, and tell the AI program how to use it.

1.3.3 Family Relations

Textbooks of AI like to illustrate the principles of automated reasoning by simple problems drawn from the field of family relations. The primary advantage is that such problems are simple, and their solutions unambiguous. Besides, the underlying concepts are broadly known. Background knowledge typically consists of facts and definitions of higher-level concepts. The facts may state that "Bill is John's father," and "Jane is Eve's mother." Some concepts may be defined by rules such as "*if* x is father of y and z is wife of x *then* z is y's mother." Other such rules may define an uncle, a sister, a grandmother, and so on.

From these facts and rules, the computer should be capable of inferring new pieces of information. For instance, the user may want the program to verify that Jane is Bob's grandmother from his father's side, or that the number of Bill's aunts exceeds the number of his uncles. To answer queries of this kind, the AI program combines problem-solving techniques (e.g., search) with logic and the ability to draw conclusions from available knowledge.

1.3.4 Knowledge Representation

Seeking to make the computers draw inferences like those in the previous paragraphs, the programmer first has to decide how to represent knowledge. The most popular approach relies on *if-then* rules: *if* a set of circumstances have been satisfied, *then* something else follows ("*if* x is a prime number and is greater than 2, *then* x is odd"). This way of representing knowledge is described in Chapter 9.

Apart from *if-then* rules, other representation paradigms have been developed, the best-known among them being *frames* and *semantic networks*. Throughout the 1980s and the 1990s, these two were almost as common as *if-then* rules. Later, their popularity faded, which is why this textbook only briefly characterizes them in Chapter 12. Rules receive much more space.

1.3.5 Automated Reasoning

The main advantage of knowledge in the form of rules is that logicians have developed for them some easy-to-implement inference mechanisms. Two such mechanisms dominate the field: the time-tested *modus ponens*, and the relatively new *resolution principle*. The latter forms the basis of the programming language *Prolog*. A typical program written in *Prolog* consists of facts and rules. Based on these, the system seeks to answer user's queries by a systematic application of the resolution principle and search. *Prolog* does so automatically, as if behind the scene. The programmer can thus focus on the formulation of the facts, rules, and queries.

The information provided by Chapters 10 and 11 will help the reader write programs capable of *Prolog*-like reasoning. Chapter 12 briefly mentions how to handle automated reasoning in the context of frames and semantic networks.

1.3.6 Less Clear-Cut Concepts

In the world of family relations, elementary terms such as `uncle` or `ancestor` are easy to define and straightforward to use. Outside the realm of simple concepts, things get complicated.

Suppose the user poses the following query: "Is Bill's father richer than Bill's uncle?" In the absence of explicit information about each individual's worth, the answer can be inferred from indirect sources. For instance, if the father is a bus driver and the uncle is a cardiologist, we expect the latter to make more money. The same conclusion can be inferred from their addresses, especially if the uncle lives in a fancy upscale neighborhood. Even their age can offer a guideline: A man in his primes is usually better-off than a fresh college graduate; he is likely to have had many salary raises and may have paid off his mortgage.

1.3.7 Imperfect Knowledge

Yet there is no guarantee. A cardiologist may have lost a bundle in a wild speculation on the stock exchange while the bus driver has not. A man in his primes may have squandered his fortune in women and alcohol while his younger relative has always been thrifty. Someone who has just filed for bankruptcy may still occupy his pricey mansion, in which case his enviable address will be misleading. There is no doubt that many rules we are subconsciously using are less than perfect. Does it mean they are useless?

Far from it. We people are very good at drawing conclusions that rely on rules that are imperfect at best. We do not have much choice, knowing as we do that perfection is hard to come by. Outside mathematics, rock-solid rules are rare and we have to make do with what we have at our disposal: rough-and-ready conventions and tentative rules of thumb. True, these may every now and then lead us astray. More often than not, however, they serve us well.

1.3.8 Uncertainty Processing

In AI, too, perfection is all but impossible, for reasons outlined in Chapter 13. Bowing to reality, many scientists set out to explore the possibilities of drawing conclusions from knowledge that is uncertain, incomplete, and sometimes downright wrong. Decades of concentrated efforts resulted in major breakthroughs.

The earliest techniques, while successful experimentally, were criticised for appearing rather *ad hoc*, lacking as they did in solid mathematical foundations. This was the case of the *certainty factors* described in Section 13.3. Later attempts focused on approaches that rested on solid mathematical foundations, mainly the time-tested theory of probability. This book devotes the entire Chapter 14 to the possibilities this theory offers to automated reasoning. The same chapter introduces the more recent ideas from the *Dempster-Shafer* theory of evidence combination. Above all, however, the public imagination was captivated by the principles of the *fuzzy-set theory* that is the topic of Chapter 15.

1.3.9 Expert Systems

Developments in the field of search techniques, alongside with advances in logic and uncertainty processing, soon bore fruit. New technology appeared on the market. The astonished world learned about the existence of *expert systems*, software packages that sought to emulate human specialists.

The principle was simple. First, create a knowledge base, typically consisting of thousands of rules and facts, some rock solid, others less reliable. Then, in response to the user's query, combine the principles of automated reasoning with uncertainty processing, and infer from the knowledge base an adequate answer. Add to it functions capable of navigating the user through a quasi-intelligent dialog, and of explaining the machine's reasons behind the responses, and you obtain a program exhibiting brand new behavior.

The oldest of such programs, *Mycin*, was introduced in 1969. While fascinating, it still left a lot to be desired in terms of functionality. Weaknesses, however, were promptly removed, and the improved versions from the 1970s were convincing enough to inspire new enthusiasts—and convert old skeptics. Other research groups jumped on the bandwagon, working hard on alternatives that would be more user-friendly, reliable, flexible, and capable of addressing a broader range of problems. The efforts payed dividends.

Some widely publicised demonstrations led many to believe that the key to AI had thus been uncovered.

The sober twenty-first century is less easily impressed. No wonder. Early expectations ran just too high. Failure to deliver on exaggerated promises led to disappointments. Nowadays, stand-alone expert systems are a rarity—but this does not mean that the work has been squandered! Some ideas found their way into modern-day software to an extent that indicates that the once-abandoned technology has now been vindicated. This software, however, is no longer packaged under the once-so-fashionable name that has long since lost its ability to sell.

Some lessons are discussed in Chapter 16.

Control Questions

If you have problems answering any of the following questions, return to the corresponding place in the preceding text.

- Suggest examples of tasks that cannot be addressed by search techniques alone because they cannot be solved without background knowledge.
- What is the dominant method of representing knowledge in modern AI? What principles are used in automated reasoning using this knowledge?
- Why do we need mechanisms capable of dealing with uncertainty?
- What is an *expert system*?

1.4 Structure and Method

The book can informally be divided into four parts. The first part focuses on problem solving. The classical approaches based on search techniques are discussed in Chapters 2 through 5 that explain the principles of blind and heuristic search, simulated annealing, adversary search, and some typified model applications.

The second part complements these topics by modern (biologically inspired) alternatives: the GA in Chapter 6, artificial life in Chapter 7, and swarm intelligence in Chapter 8. These novel techniques are not just fashionable; they have been shown significantly to outperform classical search.

The third part consists of chapters that address knowledge representation and automated reasoning, including discussions of the limitations of pure logic. Proper understanding of these limitations motivates the introduction of techniques for uncertainty processing: probabilistic reasoning in Chapter 14 and fuzzy-set theory in Chapter 15. Chapter 16 summarizes some major experiences from the once famous application of knowledge representation and reasoning: expert systems.

The fourth part deals with issues that do not belong to core AI, but remind us of its broader context. Specifically, Chapter 17 briefly discusses those aspects of computer intelligence that have long since gone their own way: computer vision, natural language processing, machine learning, and a very brief mention of agent technology. After this, Chapter 18 offers some AI-related observations made by famous philosophers. True, these are unlikely to impact an engineer's work, but it is always good to see the larger picture.

Each chapter is divided into sections short enough to be absorbed at one sitting. Each section is followed by three to four control questions to help the students decide whether they have grasped the issues. Each chapter concludes with a *Practice Makes Perfect* section meant to reinforce the main ideas by exercises and by suggestions for independent thought. To put the chapters into broader context is the task for the *Concluding Remarks* sections.

The *Bibliography* at the book's end lists the titles mentioned in the *Concluding Remarks* of the individual chapters. The author of this book does not claim originality, he did not invent any of the techniques (although he did gain experience with most of them over the three decades of his research and teaching). However, each AI paradigm has a history that has to be appreciated. This is why it was necessary to mention at least the names of some of this discipline's greatest contributors.

CHAPTER 2
Blind Search

Early approaches to problem solving relied on mechanisms that explore the space of possible solutions. The simplest of these techniques, those of *blind search*, investigate this space in a systematic manner that guarantees that nothing has been overlooked. While doing so, the techniques do not attempt to optimize the procedure, nor do they try to emulate human ways of problem solving.

The chapter begins with the introduction of the blind-search principles, using to this end a simple puzzle. Once these principles are clear, three algorithms are described: depth-first search (DFS), breadth-first search (BFS), and iterative deepening. Special attention is devoted to aspects that affect the algorithms' efficiency and computational costs.

2.1 Motivation and Terminology

Many problems encountered in realistic applications of artificial intelligence (AI) can be cast as *search* through the space of potential solutions. A simple puzzle will help us establish relevant terminology.

2.1.1 Simple Puzzle

Consider the problem illustrated by Figure 2.1. The playing board consists of four fields. On these, two pieces are placed: a *square* and a *circle*. At each moment, a piece can be moved to the unoccupied square to the left, right, up, or down, but the piece must not leave the four-field board. In the terminology of AI, we say that only then the moves are *legal* (or *permitted*).

The left part of the picture shows the *initial state* in which the square is located to the right of the circle. Our ambition is to find a sequence of moves that convert the initial state to a *final state* (sometimes called *terminal state*), defined here as any configuration where the two pieces have been swapped so that the square is now located to the circle's left.

This definition of the final state is satisfied by two different states, one of which is shown in the right part of Figure 2.1. Some puzzles know just one final state, perhaps specified by a concrete picture such as this one. Others allow for two or more final states, any of which satisfies a condition or a list of conditions (in our case, the condition is, "the square is to the circle's left").

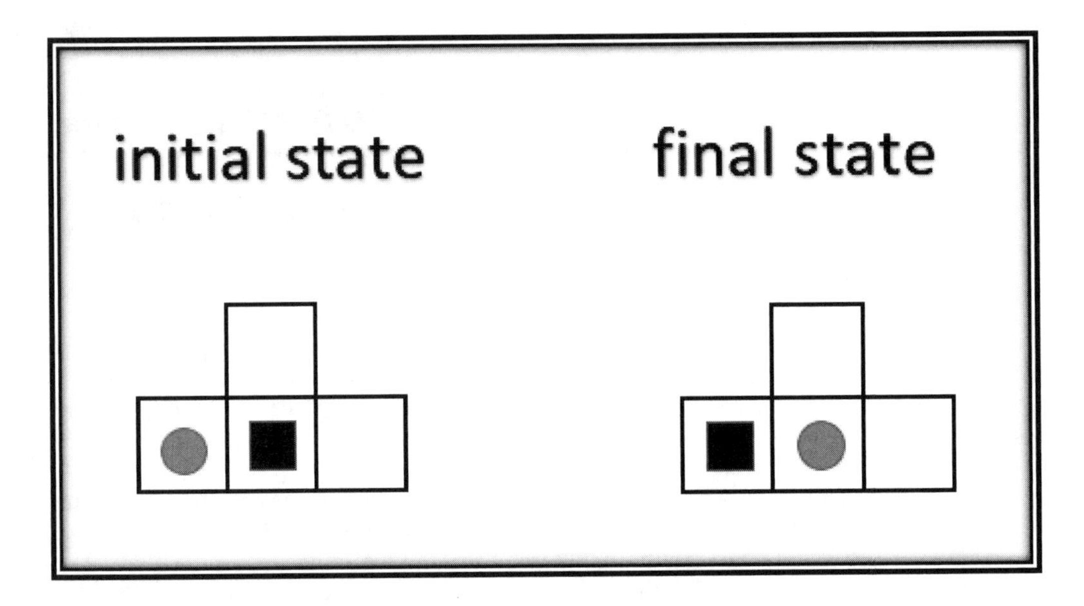

FIGURE 2.1 The puzzle consists of four fields and two pieces: a *square* and a *circle*. The goal is to swap the two pieces. Permitted actions: move a piece by one square to an empty field to the left, right, up, or down, never leaving the four-field board.

2.1.2 Search Tree

The initial state from Figure 2.1 permits only two legal actions: "move the square up" and "move the square to the right." Each of these actions results in a new state which, in turn, allows other actions, and so on. The step-by-step changes are captured by the *search tree* in Figure 2.2, a graphical illustration of the transitions.

At the top is the initial state. Underneath, the two arrows point to states that will result from the two legal actions permitted in the initial state. In each of these new states, two different actions are possible. However, a closer look reveals that the execution of "move the square down" in the state on the top of the left branch would only bring us back to the initial state. Something similar is the case also in the right branch where only one of the two legal actions results in a new state, whereas the other would only bring us back to the initial state.

We can see that not every action results in a new state. Actions that lead to states that have been seen (or *visited*) earlier are not helpful, which is why they are not shown in the picture. Note that the tree from Figure 2.2 is incomplete in the sense that none of the branches has yet reached a state that satisfies the definition of the final state.

2.1.3 Search Operators

In the context of AI, the actions are carried out by *search operators*. Among the search operators that are legal, in the given state, some can be ignored because their application results in states that have been visited before. In the following text, we will use the two terms, action and search operator, interchangeably.

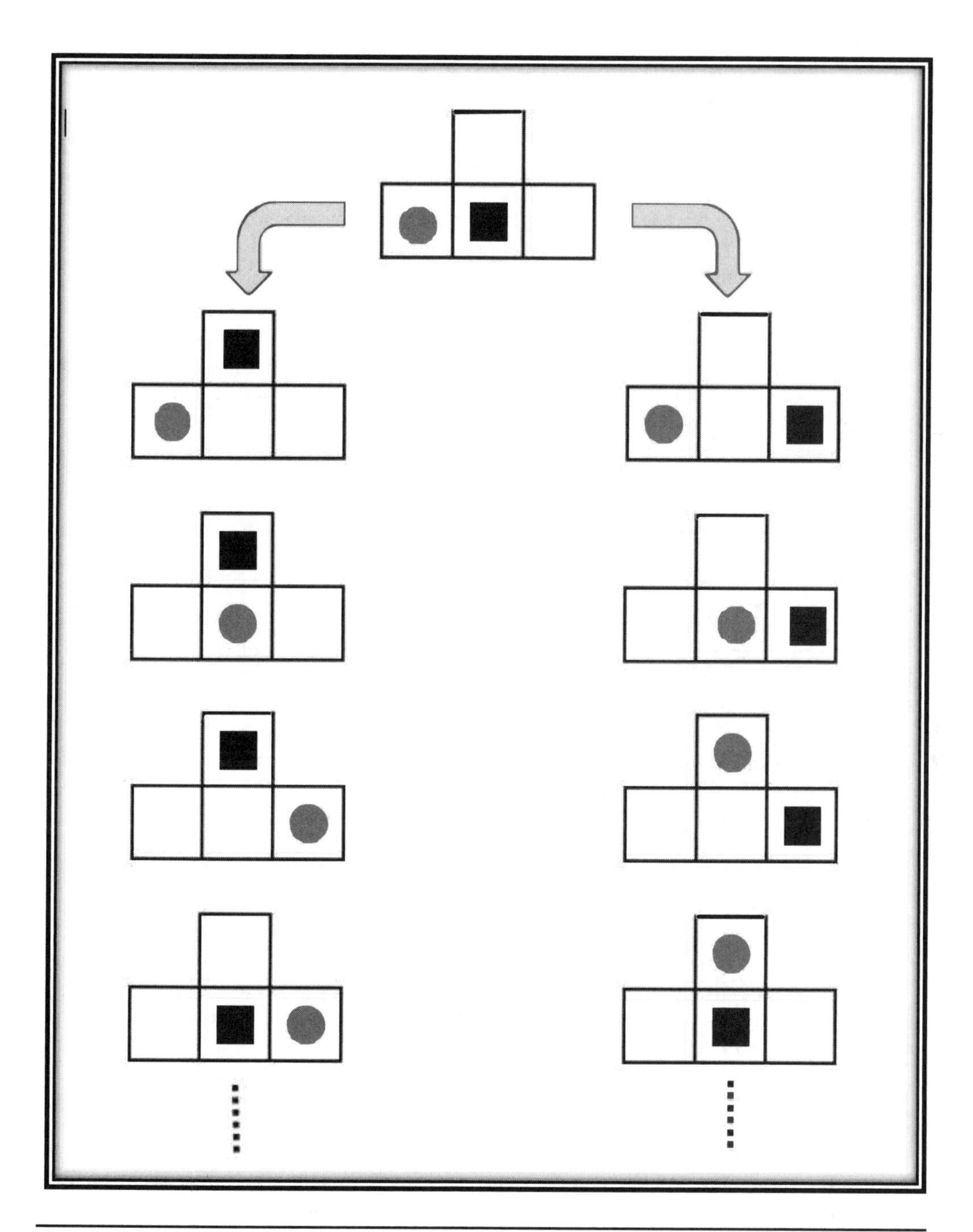

FIGURE 2.2 In the initial state, two different moves are possible. Each of the next states allows only one legal action if we disable the possibility of revisiting a previously seen state.

Classical AI techniques involve the following five aspects:
1. Initial state(s)
2. Final state(s)
3. Interim states
4. Search operators to convert one state to another
5. Rules to chose a concrete search operator in a given state

TABLE 2.1 Formal Definition of Search Techniques

In most of the states in Figure 2.2, only one legal search operator is useful (not leading to a previously seen state). This was due to this toy domain's extreme simplicity. In more realistic domains, the number of useful actions in each state is higher (sometimes *much* higher) but the point remains the same: The search should not revisit a previously seen state.

2.1.4 Blind Search in Artificial Intelligence

Let us summarize what we have learned thus far. The search process begins at some initial state, and the goal is to reach a predefined final state (a.k.a. terminal state). States that are neither initial nor final are called *interim states*. The search process is represented by a sequence of search operators, each converting the current state to another state. For the reader's convenience, Table 2.1 provides the list of these basic terms.

The concrete search operator to be applied to a given state can be picked at random, but engineers prefer algorithms that make sure the search space is explored systematically. Two fundamental algorithms satisfying this requirement are introduced in the next section: the depth-first search and the breadth-first search. Their behaviors are often illustrated by the following simple puzzles.

2.1.5 Sliding Tiles

Figure 2.3 shows a simple version of the popular sliding-tiles puzzle played on a 3-by-3 board.[1] Of the nine squares, eight are covered by tiles labeled with integers from **1** to **8**, and one square is empty. The picture shows a randomly generated initial state, and a predefined final state. Generally speaking, we have to consider also the possibility that the game does not have a solution, in which case the final state can never be reached from the given initial state.

Any action in this puzzle slides a tile from its current location to its neighboring square—if the square is empty. The initial state shown in the left of Figure 2.3 permits three legal actions: move **3** down, move **2** to the right, and move **5** up. The number of legal operators depends here on the concrete state. If the empty square is in the board's center, four legal operators exist. A state with the empty square in the corner permits only

[1] More common is the 4-by-4 board. For the needs of a textbook, however, the larger board allows impractically many interim states.

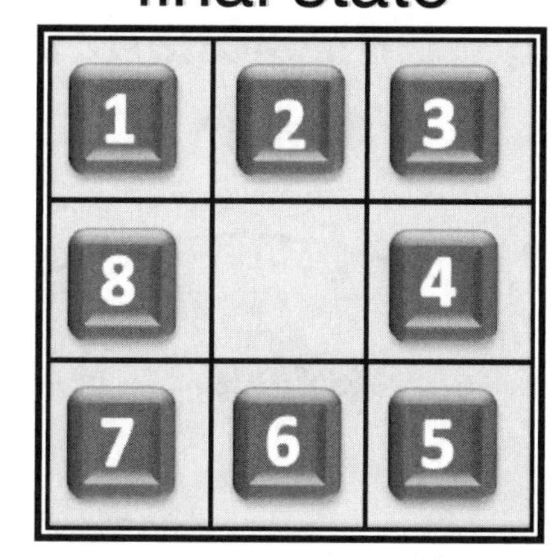

FIGURE 2.3 The task is to convert the initial state to the terminal state by a sequence of actions, each of which slides a tile to the neighboring empty square.

two legal search operators; all other situations permit three legal operators. Of course, the search should only consider search operators that do not result in a previously visited state.

2.1.6 Missionaries and Cannibals

Figure 2.4 depicts a puzzle with which parents liked to tease their kids in the pre-internet era. Three missionaries and three cannibals reach a river they need to cross. The boat at their disposal can take only two persons. Besides, an important constraint has to be kept in mind: the missionaries are safe only when not outnumbered by the cannibals. For instance, when two missionaries leave the north bank, the last one remaining could be overcome by the three cannibals. This applies to either bank.

The goal is to find a sequence of actions that transport all six men safely to the south under the above-mentioned rules. The problem is not difficult, but it requires a minor "trick" that a kid does not immediately see: one has to figure out which person on the south bank is to row the boat back to the north.

2.1.7 Programmer's Perspective

The first thing to decide is what data structure will best represent the states. This representation should be flexible enough to allow easy modifications.

As for the program, the engineer needs a Boolean function that accepts a state description and returns *true* if the state is final and *false* if it is not. Another function will accept the description of a state and creates the state's children by the application of search operators that are legal in this state. Also needed is a function that eliminates previously seen children states. Finally, there has to be a mechanism to control the order in which to process those children that have not been eliminated.

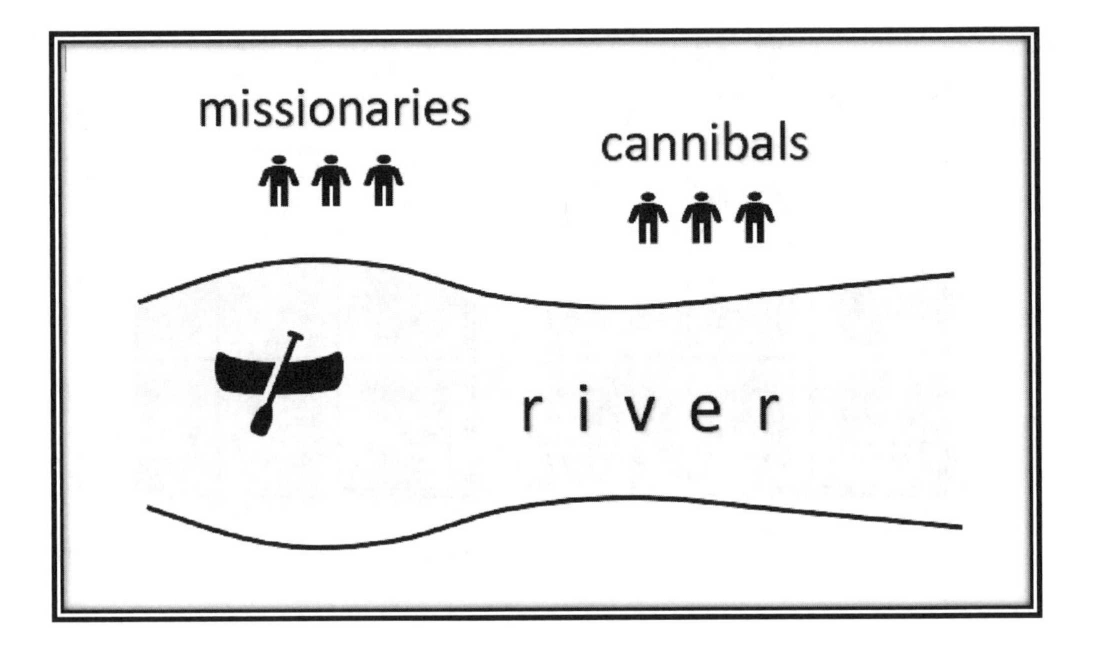

FIGURE 2.4 The goal is to move the six persons across the river in a boat that carries only two. The cannibals must never outnumber the missionaries on either bank.

Control Questions

If you are unable to answer the following questions, return to the appropriate place in the preceding text.

- Explain the meaning of the basic terms of this paradigm: initial, interim, and final states, search operators, search tree, previously seen states.
- Summarize the basic functions that have to be written by a programmer who wants to implement a search technique.

2.2 Depth-First and Breadth-First Search

Let us now turn our attention to the two basic blind-search algorithms. They are so simple as to be almost impractical, but they rely on a useful pattern of thought. Once we absorb it, we will find it easy to grasp the more advanced techniques in the coming chapters.

2.2.1 Example Search Tree

Consider the problem represented by the search tree from Figure 2.5. Let us suppose the tree is complete in the sense that any search operator applied to any of the *leaf* states, **E, I, J, C, G,** and **H**, would only result in a previously seen state. The tree is intentionally small to make it easy to illustrate the basic search algorithms. In realistic domains, search trees will of course be much larger, easily comprising millions or billions of interim states.

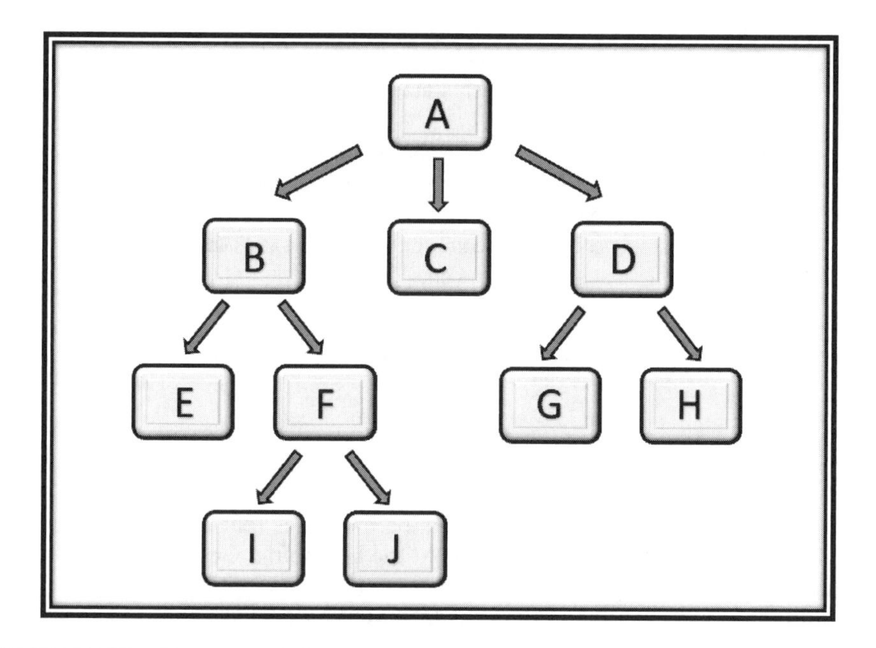

FIGURE 2.5 Example of a search tree of a very simple search problem.

2.2.2 Depth-First Search: Principle

Let us start with the initial state in Figure 2.5. Suppose that a randomly selected search operator has been applied, and that it has converted the initial state to state **B**. What next? If **B** is not final, depth-first search simply applies a legal search operator to this new state, and then to the next, progressing deeper and deeper, ignoring the states in the search tree's other branches. At a certain moment, the process will reach one of the two critical situations: First, the final state has been reached; second, the application of any legal search operator to the current state would only result in a previously visited state. Either way, no further expansion of the tree is possible.

The order of the children (from left to right) in the search tree from Figure 2.5 is random. At any situation where two or more legal search operators could be used, at some state, the concrete choice was here made by a random-number generator, and this is the tree that we thus obtained. In this event, the DFS would investigate the states in the following order: **A, B, E, F, I, J, C, D, G, H.** The reader can see that the search always moves from a parent to a child, backtracking only if the end of a branch (a leaf) has been reached.

2.2.3 Depth-First Search Algorithm

The pseudo-code in Table 2.2 summarizes the algorithm just described. Two lists are employed: L and L_{seen}.[2] The first, L, contains the states that we expect to be checked for

[2] Recall that the difference between a set and a list is that the elements in a list are ordered.

Input: List L, containing the initial state
Empty list L_{seen} of previously seen states

1. Let s be the first state in L. If s is a final state, stop with success.
2. Apply to s all legal search operators, thus obtaining the state's children.
3. Ignore those children that are already in L_{seen}.
4. Put the remaining children, randomly ordered, **at the beginning** of L.
5. Remove s from L and put it at the end of L_{seen}.
6. If $L = \emptyset$, stop with failure, otherwise, return to step 1.

TABLE 2.2 Pseudo-Code of Depth-First Search

being final. The second list, L_{seen}, contains the states that already have been explored and have been shown not to be final. At the beginning, L contains only the initial state whereas L_{seen} is empty.

In each round, DFS picks the first element in L, denotes it by s, and submits it as input to the function that decides whether s is a final state. Note the indefinite article: line 1 of the pseudo-code says "a" final state and not "the" final state. This is because some problems may have more than just one final state; the simplest version of the DFS stops when at least one of them has been found. Of course, some applications may require that *all* final states be found.

If s is not a final state, the program places it in the list L_{seen} of previously seen states, and then applies to s all legal search operators, thus creating all its children states. Some children can already be found in L_{seen} which means they have been seen before, and thus do not need to be investigated again. These are ignored, while the remaining ones are placed *at the beginning* of L. The reason they are placed at the beginning is that this makes sure these children states are investigated before those states that have been placed in L previously (this is the fundamental principle of DFS).

2.2.4 Numeric Example

Let us illustrate the behavior of DFS using the simple search tree from Figure 2.5. At the beginning, L contains only the single initial state, **A**, and L_{seen} is empty.

Realizing that the first element in L is not the final state, DFS transfers it from L to L_{seen} and replaces it in L with its children states, randomly ordered. Suppose that this results in L containing the following sequence: $L = \{\mathbf{B}, \mathbf{C}, \mathbf{D}\}$. Since the first element **B** is not a final state, DFS transfers it from L to L_{seen} and replaces it at the beginning of L with its children, **E** and **F**, again randomly ordered. The procedure continues as shown in Table 2.3 (here, **H** is assumed to be the final state).

We can see that this results in the same sequence of states that was earlier specified for the DFS: **A, B, E, F, I, J, C, D, G, H**. The reader will have noticed that, with the exception of the final state **H**, this is the contents of L_{seen} at the end of the search procedure.

Consider the search tree from Figure 2.5, and suppose that **H** is a final state.

At the beginning, let $L = \{\mathbf{A}\}$ and $L_{seen} = \emptyset$.

First element in L is **A**. Since this is not a final state, it is transferred to L_{seen} and replaced in L with its three children, randomly ordered:

$L = \{\mathbf{B, C, D}\}$ and $L_{seen} = \{\mathbf{A}\}$.

First element in L is now **B**. Since this is not a final state, it is transferred to L_{seen} and replaced at the beginning of L with its two children, randomly ordered:

$L = \{\mathbf{E, F, C, D}\}$ and $L_{seen} = \{\mathbf{A, B}\}$.

First element in L is now **E**. Since this is not a final state, it is transferred to L_{seen}. No new state is added to L because **E** does not have any legal, previously unseen children:

$L = \{\mathbf{F, C, D}\}$ and $L_{seen} = \{\mathbf{A, B, E}\}$.

In the following steps, the two lists evolve as follows:

$L = \{\mathbf{F, C, D}\}$	$L_{seen} = \{\mathbf{A, B, E}\}$.
$L = \{\mathbf{I, J, C, D}\}$	$L_{seen} = \{\mathbf{A, B, E, F}\}$.
$L = \{\mathbf{J, C, D}\}$	$L_{seen} = \{\mathbf{A, B, E, F, I}\}$.
$L = \{\mathbf{C, D}\}$	$L_{seen} = \{\mathbf{A, B, E, F, I, J}\}$.
$L = \{\mathbf{D}\}$	$L_{seen} = \{\mathbf{A, B, E, F, I, J, C}\}$.
$L = \{\mathbf{G, H}\}$	$L_{seen} = \{\mathbf{A, B, E, F, I, J, C, D}\}$.
$L = \{\mathbf{H}\}$	$L_{seen} = \{\mathbf{A, B, E, F, I, J, C, D, G}\}$.

Since **H** is the final state, the search stops here.

Note that the contents of L_{seen} indicates the order in which the individual states have been examined. It also tells us how many states have been investigated throughout the process.

TABLE 2.3 Illustration of Depth-First Search

2.2.5 Breadth-First Search: Principle

The philosophy of this second algorithm for blind search can be summarized by the following requirement: *never proceed deeper, in the search tree, before having investigated all states at a given level.* For instance, state **E** will only be investigated after all the states at the previous level (**B, C,** and **D**) have been tested for being terminal and rejected for having failed the test.

Let us assume that, in the search tree from Figure 2.2, the random generation of each state's children is the one indicated by the ordering from left to right. In this event, the BFS investigates the states in the following order: **A, B, C, D, E, F, G, H, I, J.**

2.2.6 Breadth-First Search Algorithm

Table 2.4 summarizes the algorithm that implements the principle just described. Note that the only difference from the previous algorithm, DFS, is in step 4 which places the given state's children *at the end of L* rather than at the beginning (which is what DFS did). The similarity of the two algorithms is convenient. Once you have written a program for DFS, the code is easy to modify so that it becomes BFS.

Input: List L, containing the initial state
Empty list L_{seen} of previously seen states

1. Let s be the first state in L. If s is a final state, stop with success.
2. Apply to s all legal search operators, thus obtaining its children.
3. Ignore those children that are already in L_{seen}.
4. Put the remaining children, randomly ordered, **at the end** of L.
5. Remove s from L and put it at the end of L_{seen}.
6. If $L = \emptyset$, stop with failure, otherwise, return to step 1.

TABLE 2.4 Pseudo-Code of Breadth-First Search

In the simple version from Table 2.4, the procedure is halted once it has reached one of the final states. In applications where two or more (or all) final states are requested, the program has to be modified accordingly.

2.2.7 Numeric Example

Again, we will illustrate the algorithm's behavior using the simple search tree from Figure 2.5. At the beginning, L contains only the single initial state, **A**, and L_{seen} is empty.

Since **A** is not final, it is moved from L to L_{seen}, and replaced in L with its three children: $L = \{\mathbf{B}, \mathbf{C}, \mathbf{D}\}$. Since **B** is not final, it is transferred from L to L_{seen}. Its two children, **E** and **F**, are placed at the end of L so that $L = \{C, D, E, F\}$. The procedure then continues in like manner. The evolution of the two lists is detailed in Table 2.5.

Consider the search tree from Figure 2.5, and suppose that **H** is a final state.

At the beginning, let $L = \{\mathbf{A}\}$ and $L_{seen} = \emptyset$.

First element in L is **A**. Since this is not a final state, it is transferred to L_{seen} and replaced in L with its three children, randomly ordered:

$L = \{\mathbf{B, C, D}\}$ and $L_{seen} = \{\mathbf{A}\}$.

Since **B** is not a final state, it is transferred to L_{seen} and its two children (randomly ordered) are placed at the end of L. Here is how the two lists then evolve:

$L = \{\mathbf{C, D, E, F}\}$ $\quad L_{seen} = \{\mathbf{A, B}\}$.
$L = \{\mathbf{D, E, F}\}$ $\quad L_{seen} = \{\mathbf{A, B, C}\}$.
$L = \{\mathbf{E, F, G, H}\}$ $\quad L_{seen} = \{\mathbf{A, B, C, D}\}$.
$L = \{\mathbf{F, G, H}\}$ $\quad L_{seen} = \{\mathbf{A, B, C, D, E}\}$.
$L = \{\mathbf{G, H, I, J}\}$ $\quad L_{seen} = \{\mathbf{A, B, C, D, E, F}\}$.
$L = \{\mathbf{H, I, J}\}$ $\quad L_{seen} = \{\mathbf{A, B, C, D, E, F, G}\}$.

Since **H** is the final state, the search stops here.

TABLE 2.5 Illustration of Breadth-First Search

Control Questions

If you are unable to answer the following questions, return to the appropriate place in the preceding text.

- Both algorithms introduced in this section rely on a pair of lists. What is the motivation behind the two lists? What do they contain?
- Summarize the basic steps of DFS. What is the only difference (from a programmer's perspective) between DFS and BFS?

2.3 Practical Considerations

Toy domains conveniently illustrate basic principles, but their simplicity is treacherous. When dealing with a realistic problem, the engineer may run into a whole range of hurdles and complications. It is good to be prepared.

2.3.1 Generic Model of Search

It is one thing to learn the principles of search algorithms; it is another to convert them, mentally, into generic models applicable to a wide range of problems. When casting a concrete engineering task in a form that can be addressed by blind search, some creativity is needed. Fortunately, the requisite skills tend to improve with practice. After repeated attempts to formulate diverse engineering tasks in terms of states and search operators, the student becomes proficient. It is also useful to be mindful of a few circumstances neglected by simple classroom exercises. Let us take a look.

2.3.2 Exact Form of the Final State May Not Be Known

In the domains from the previous section, the termination of the search process could be determined by comparing the current state with the precisely defined final state. Reality is rarely so simple. To appreciate the difficulties, consider the problem of a *magic square*, a grid of nine squares arranged in a 3-by-3 matrix. The goal is to fill each square with a different integer from 1 to 9 in a way that makes the sums in all rows, all columns, and the main diagonals to be the same.

You see? We do not have an idea what the solution looks like; we are only told what has to be satisfied if we are to call a state final. For all we know, the solution may not even exist. There can be several solutions, even many solutions—or none at all. Checking a state for being final is not accomplished by plain comparison, but rather by systematic testing whether it satisfies a set of predefined properties. In the case of the magic square, this is still fairly easy; in other applications, such evaluation can be expensive. This is worth remembering.

2.3.3 Unknown Form of the Final State: Examples

The final state does not have to be defined by a concrete pattern such as the one in Figure 2.3. Rather, the solution can be specified verbally. For instance, in the magic-square puzzle from the previous paragraph, this can be, "find a state where at least one column contains only even numbers or at least one row contains only odd numbers."

Again, we do not know how many such solutions can be found, or whether the solution exists at all.

As another example, suppose we want to identify the best arrangement of operations on an assembly line, perhaps a sequence that guarantees the fastest manufacture. All sorts of practical constraints have to be considered, not all of them immediately obvious. Can the discovered sequence be realistically implemented in the given factory? Will it affect the quality of the final product? What impact will it have on labor force? Will the process be cheap? The reader now understands that the generic model of blind search has to be employed creatively.

2.3.4 Testing a State for Being Final Can Be Expensive

In the examples from the previous paragraphs, the decision whether the current state is final could not be made by mere comparison of two states; additional tests and evaluations were needed.

In realistic applications, these tests and evaluations can prove to be computationally expensive. Sometimes, the state can be labeled as final only after time-consuming experimentation. In this event, list L_{seen} helps prevent not only infinite cycles, but also wasteful repetitions of costly state evaluations.

2.3.5 Goal 1. What Does the Solution Look Like?

So far, we have simplified our considerations by assuming that our only ambition is to reach a terminal state. Sometimes, this is enough. When a 10-years-old shows us the solved sliding-tiles puzzle, we are unlikely to want to verify the exact sequence of moves that have led to it. In the magic square, the primary requirement is to establish that the solution exists. If it does, we want to know what the solution looks like.

In many technological applications, our ambitions do not go any further. Sometimes, we may want to find *all* solutions, not just one; but again, the procedure that has led to finding them may be unimportant.

2.3.6 Goal 2. What Path Leads to the Solution?

Other applications go further. For instance, certain puzzles expect us to find a path through a maze. This means that it is not enough to declare that the path has been found; rather, we are supposed to show what exactly it looks like. We may even be asked to find the *shortest* sequence of actions. By way of another illustration, recall the missionaries and cannibals: Seeking to minimize time and physical exertion, we need a solution with the smallest number of river crossings.

In an industrial application, the engineer may want to minimize the number of operations needed to assemble a certain product or to minimize the attendant economical costs.

2.3.7 Stopping Criteria

We have made another simplifying assumption; to wit, we have pretended that the search sooner or later *does* reach a state to be labeled as final. In many realistic domains, however, this is not the case. Sometimes, we do not even know that the solution exists.

And when it does exist, the blind-search procedure may be so expensive that no computer will ever complete it in realistic time. It is thus necessary that the programmers implement, in their programs, some criteria to instruct the machine when to stop.

Here are some termination criteria to consider. First, the search stops when a terminal state has been reached. Second, the search stops if the length of the L_{seen} list has exceeded a user-specified maximum, say, 10^6 states. Third, the search stops if the program has failed to reach the final state within a certain allotted solution time, say, ten hours.

2.3.8 Examining L_{seen} Can Be Expensive

Each time the children of a given state have been created, blind-search programs are instructed to ignore those states that have previously been visited and rejected as non-terminal. These previously seen non-terminal states are found in list L_{seen} which is thus scanned each time a new child has been created. The scanning is affordable if the size of L_{seen} is manageable. In practical applications, however, the size grows to millions of states, or even more. Scanning the list is costly if carried out sequentially, one state at at time, starting at the list's beginning.

2.3.9 Searching Through an Ordered List

Checking L_{seen} for containing a specific state can be more efficient if its contents are ordered. Efficient scan can then be accomplished by such techniques as *interval halving*.

Suppose we want to know whether the list in Figure 2.6 contains number 16. A mechanical search "from left to right" would have to check seven numbers before reaching 16. Interval halving is faster. The idea is always to identify the middle point of the range, and then ask what the number at this middle point is. In the specific case in Figure 2.6, the middle point is 14. Since this is smaller than 16, we will have to look to the right. The middle point in the right half of the sequence lies between 17 and 21. The smaller, 17, is greater than 16, which means that our object should be in the sub-interval to the left—and indeed it is.

We have convinced ourselves that interval halving can reduce the number of tests carried out before finding the given state in an ordered list. Of course, this requires that the list *can* be ordered.

2.3.10 *Hash* Functions

To order the elements in L_{seen}, we need a mechanism to convert state descriptions to integers. This is accomplished by *hash* functions that are commonly available

1 4 6 10 12 14 16 17 21 23 29

↑

Figure 2.6 An ordered list is easier to search through than an unordered one. One possibility is to use the technique of *interval halving*.

in general-purpose programming languages. A hash function converts a text, or a higher-level data structure, to an integer that is (almost) guaranteed to be different for each text.

In our context, this means adding to the search algorithms' pseudo-codes one additional line: The state that has just been rejected as non-terminal is converted by the hash function to an integer that then determines the location at which the state should be inserted in L_{seen}.

Control Questions

If you are unable to answer the following questions, return to the appropriate place in the preceding text.

- Is the final state always clearly defined? Does each of the puzzles in this chapter have one and only one final state? How can the final state be characterized? Elaborate on these questions.
- Apart from reaching the final state, what other requirements can the search algorithm be expected to fulfill?
- What is a *hash* function? Why and how are hash functions employed in a search?

2.4 Aspects of Search Performance

To be able to decide which of the two basic algorithms to prefer in a concrete application, the engineer needs to understand what determines a search algorithm's performance.

2.4.1 Measuring the Costs

Computational costs have two aspects: memory requirements and the time it takes to complete the calculations. Both are affected not only by the algorithm's inherent nature, but also by the programmer's prowess. For instance, the time it takes to find a specific search state in L_{seen} can be reduced by hash functions and interval halving, but these techniques are external to the search algorithm. Seeing that just measuring computation time can be misleading, the engineer will ask for criteria that do not depend on the program's implementation and on the details of the code.

In the context of search algorithms, memory requirements are reflected by the maximum number of states in the two lists. As for computation time, this depends on the total number of states visited before the terminal state has been reached. Since all rejected states have been transferred to L_{seen}, the length of this list gives us a fairly objective picture.

2.4.2 Branching Factor

In the trivial puzzle from Figure 2.1, the choice of actions is poor: two possibilities in the initial state and only one in each of the follow-up states. In the sliding tiles, the number of options is higher; and in yet other domains, each state may permit a rich choice of search operators.

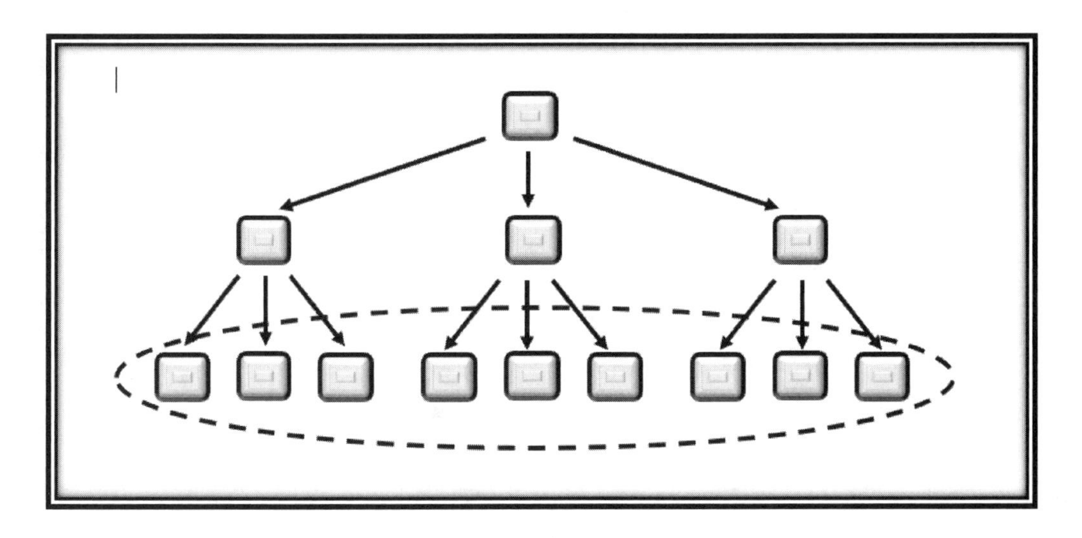

Figure 2.7 In BFS, the maximum length of list L is b^d where b is branching factor and d is depth. This means that memory costs grow exponentially with search depth.

The breadth of this choice is specified by the puzzle's *branching factor*. This factor is defined by the average number of the children of the individual states. Usually, branching tends to diminish in the course of the search process because many of the newly created children states have already been visited before, and as such are ignored.

2.4.3 Depth of Search

Another quantitative aspect relevant to performance analysis is the *depth* of a state in the search tree. This depth is defined by the number of actions taken before the state could be reached from the initial state. Sometimes, it makes sense to speak of the depth of the entire search tree.

2.4.4 Memory Costs of BFS

In this approach, any state at depth d can be reached only after the evaluation of all states at level $(d-1)$.[3] The fact that any rejection of a state at level $(d-1)$ results in the addition of all its children *at the end* of list L means that all d-level states are present in L before the first state at this level can be investigated.

Suppose, for simplicity, that each state has the same number, b, of children. A cursory look at Figure 2.7 tells us that there are b^d states at depth level d, assuming that the depth of the initial state is $d = 0$. This means that the memory costs of BFS grow exponentially with the depth.

[3] Evaluation means here submitting the state description to a function that returns *true* if the state is terminal and *false* if it is not.

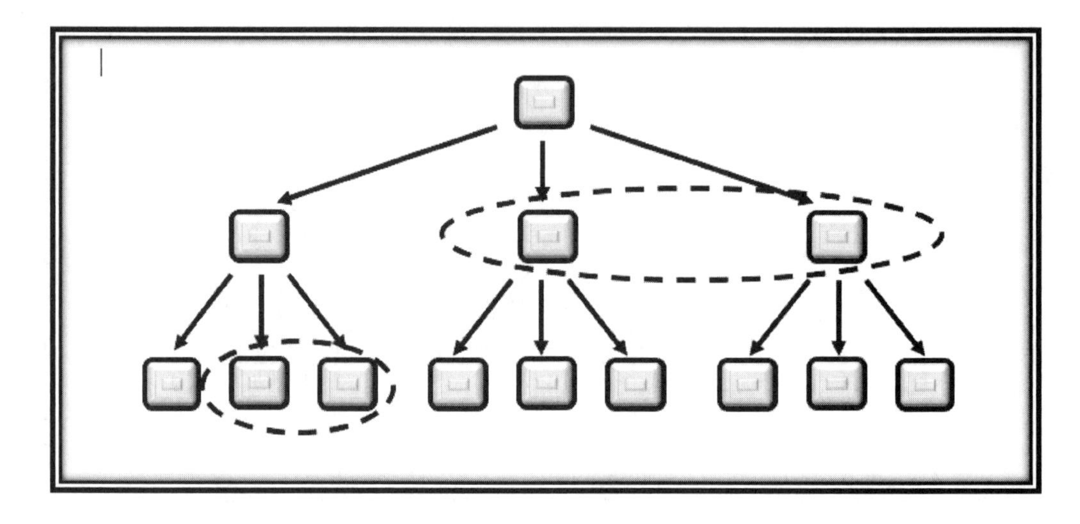

FIGURE 2.8 In DFS, the maximum length of list L is upper-bounded by $d(b-1)$ where b is branching factor and d is depth. This means that memory costs grow linearly with search depth.

2.4.5 Memory Costs of DFS

Again, let us suppose that each state has the same number of branches, b. In DFS, each state that has been rejected as non-terminal is replaced with its b children to be inserted *at the beginning* of L, which is necessary if we want to make sure these children will be investigated before the higher-level states.

As illustrated by Figure 2.8, each time the search proceeds one level deeper, b new states are added to list L. Of these, one is evaluated. After its rejection, its b children are placed into L in front of the state's $(b-1)$ siblings. Since each level adds $(b-1)$ new states, the number of states at level d can be approximated by $d(b-1)$. We see that memory costs of DFS grow linearly with the search depth.

2.4.6 Computational Costs of the Two Algorithms

Suppose we decide to quantify these costs in terms of the number of nodes evaluated before a final state has been reached, and that the depth of this final state is d. Simple analysis would reveal that in both BFS and DFS, the time needed to reach the final state grows exponentially in d, but BFS is $(b+1)/b$-times more expensive (b being branching factor). This difference can be neglected in the case of high-valued b. In the extreme, where $b = 1$, BFS is twice as expensive as DFS. Such domains, however, are rare. We conclude that the difference in these costs is not serious.

2.4.7 Which of the Two Is Cheaper?

Each of the two basic techniques has its strengths and weaknesses whose nature the engineer needs to understand before making a choice between them. Previous paragraphs convinced us that the computation time is in both approaches comparable, but that BFS is saddled with much higher memory costs than DFS.

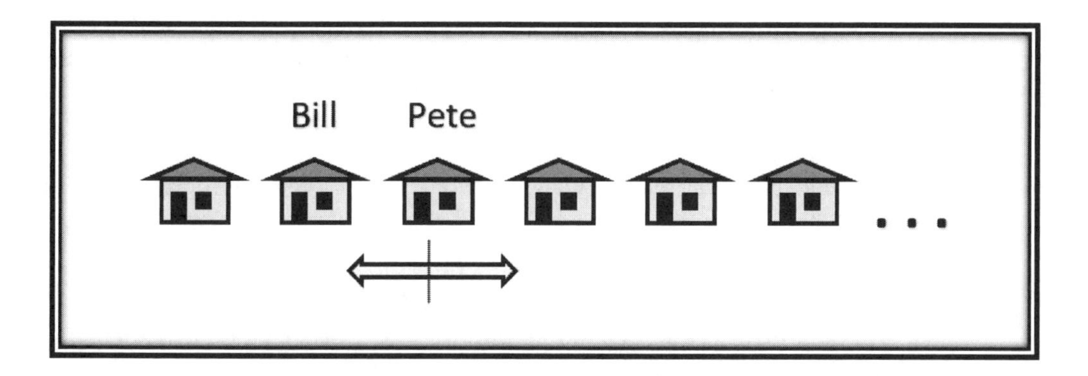

FIGURE 2.9 Pete has decided to find Bill's house using depth-first search. If the very first random action is to go east, rather than west, he will circumnavigate the globe before reaching Bill's house.

This said, we must not forget that this is true only under the assumption that the search depth is in both cases the same or at least comparable. Only then will it matter that memory costs grow linearly in DFS, and exponentially in BFS, which appears to be a major difference. However, this comparable-depth assumption is rarely satisfied, and the math can just be misleading.

2.4.8 Searching for Bill's House

Figure 2.9 illustrates a situation where the exponential memory costs of BFS are much lower than the linear costs of DFS. The initial state is Pete's house, and the terminal state is Bill's house. Only two search operators are allowed: move east and move west.

Suppose the very first randomly chosen action is move east. The house to the east from Pete's house is not the final state; the search thus has to continue. In this new location, the same two search operators exist—but one of them would bring us back to Pete's house, which is already in L_{seen}. This forces us always to continue east; there is no other possibility. The same is the case in each subsequent state.

If the very first random choice of the search operator is unlucky, DFS will not reach Bill's house before having circumnavigated the globe. Here is the lesson. Whereas the memory costs grow only linearly, the depth of the search is in the end so enormous that the incurred costs exceed the exponentially growing costs of BFS which reaches Bill's house at $d = 1$.

True, this is an extreme example, perhaps even an exaggerated one. Yet it serves as a reminder that caution has to be used even when a certain point seems supported by solid mathematical analysis. Concrete circumstances may have unexpected consequences.

2.4.9 Domains with Multiple Final States

In domains with multiple final states, BFS is guaranteed to find the one at the smallest depth—with the shortest path from the initial state. By contrast, DFS will just reach *some* final state. Which of the final states this is depends largely on the whims of the random-number generator that determines the order in which the children states are placed in L.

If our task is to find at least one final state, DFS may be cheaper. However, if the task is to find the shortest solution, then BFS usually does better (unless the branching factor is so high as to drive memory costs beyond reasonable bounds).

Control Questions

If you are unable to answer the following questions, return to the appropriate place in the preceding text.

- Why do we need to be careful about criteria to gauge performance of search algorithms? What criteria were suggested in this section?
- Compare BFS and DFS in terms of memory costs and in terms of computational costs. In what sense can theoretical results be misleading?
- Under what circumstances will the engineer give preference to BFS, and under what circumstances will DFS be the more appropriate choice?

2.5 Iterative Deepening (and Broadening)

Having grasped the performance aspects of the basic search algorithms, scientists sought to develop a technique that would eliminate the weaknesses of DFS and BFS while maintaining their strengths. Perhaps the most successful among these attempts is *iterative deepening*, ID.

2.5.1 ID Algorithm

The pseudo-code in Table 2.6 formalizes *iterative deepening's* basic principle. The idea is to run DFS that is not allowed to go beyond a certain maximum depth. This limited-depth DFS is then re-run again and again, each time with the maximum depth incremented by 1.

In the earliest stage, limiting the DFS to $d = 1$ is essentially the same as running the BFS for this single level. The next step then re-runs DFS, allowing it to reach $d = 2$. Then comes another re-run, this time with maximum depth $d = 3$, then $d = 4$, and so on until the final state has been reached, or until some other termination criterion has been satisfied.

Inputs: List L containing the initial state.
Empty list L_{seen} of previously seen states.
User-specified maximum search depth, *MAX*.

1. Let $c = 1$.
2. Run depth-first search whose maximum must not exceed c.
3. If the search has reached the final state, stop with success.
4. Increase $c = c + 1$. If $c > MAX$, stop with failure.
5. Return to step 2.

Table 2.6 Pseudo-Code of Iterative Deepening

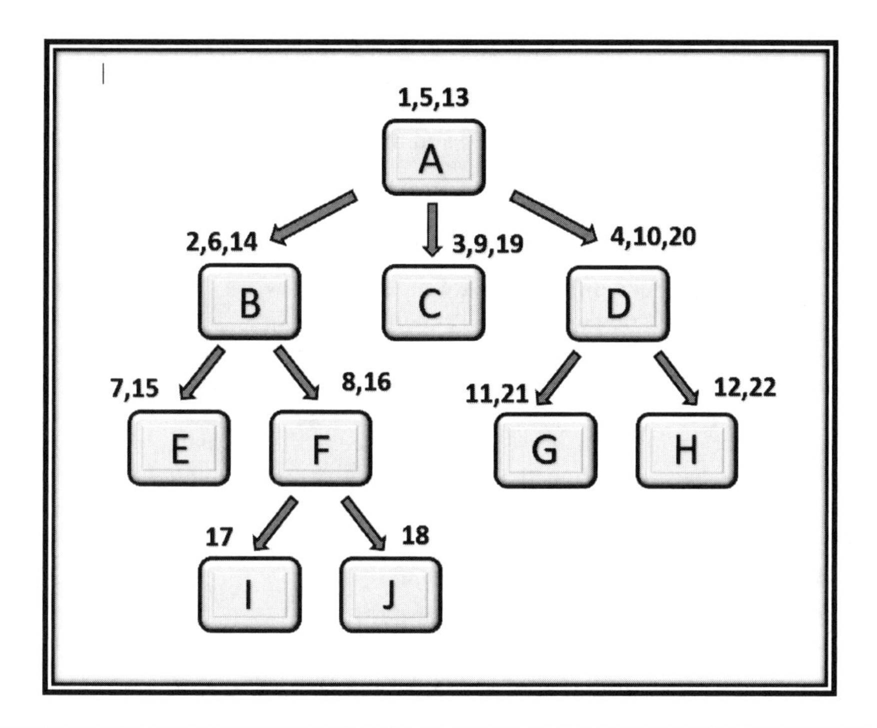

Figure 2.10 The order in which the states are visited by iterative deepening.

2.5.2 Why the Technique Works?

The technique retains DFS's main advantage, the linear growth of memory costs. At the same time, the limited maximum depth eliminates DFS's main shortcoming: the danger of reaching absurd depths while missing a solution early on. Recall the example where Pete found his neighbor's house only after circumnavigating the globe. This absurdity is in iterative deepening prevented. The reader may want to verify this statement by hand-simulating ID's behavior in this particular domain.

We will remember that the maximum depth (*MAX*, in the pseudo-code) prevents the search from going too deep.

2.5.3 Numeric Example

For the sake of illustrating ID's behavior, Figure 2.10 shows the search tree we have seen before. The integers at the individual states indicate the order in which these states are visited by iterative deepening (note that most of the states are visited repeatedly).

In the first stage, ID runs DFS to the cut-off point $c = 1$, which means depth 1.[4] With this limitation, DFS only visits states **1** through **4** because it is not permitted to go any deeper. After this, the cut-off point is increased to $c = 2$, and DFS is re-run with the permission to reach depth 2, which means that it will go through states that are

[4] Recall that the depth of the initial state, **A**, is $d = 0$.

in the picture labeled with integers **5** through **12**. Next, the cut-off point is increased to $c = 3$, and DFS goes through states that are in the picture labeled with integers **13** through **22**.

In summary, the states are visited and revisited in the following order: **A, B, C, D, A, B, E, F, C, D, G, H, A, B, E, F, I**, . . ., etc.

2.5.4 What Contributes to Search Costs?

When comparing our competing search algorithms, we suggested that an objective criterion will count the number of visited states. This, however, was meant only for the sake of these comparisons. When evaluating the costs in a practical application, we need to take into consideration also the costs incurred by the evaluation of an average state.

These costs are largely determined by the programmer's skills, but they essentially involve three components. First is the time needed to evaluate the state for being terminal. The second is the time needed to create this state's children. And the third is the time associated with checking which children already exist in L_{seen}, and as such should be ignored. Any of these three components can potentially be either cheap or expensive.

2.5.5 Is Iterative Deepening Wasteful?

The need repeatedly to traverse the search tree may appear uneconomical. Yet these additional costs are less extravagant than a superficial consideration might suggest. Suppose the branching factor is $b = 10$. Levels at depths 0, 1, and 2 then contain $1+10+100 = 111$ states, whereas the next, third level, contains 1,000 states. This means that in a domain with a high branching factor, the number of states at the next depth level is much higher than the sum total of all previous states. Revisiting those earlier 111 states adds to the overall costs only slightly more than 10%, a tolerable price for the advantages that iterative deepening brings.

Besides, revisiting, say, state **A** can be cheaper than processing it for the first time. At the moment when ID returns to **A**, this state is already in L_{seen}, which means it has already been proved non-terminal (and no new evaluation is thus necessary). Also the costs of creating the state's children can possibly be neglected because a well-written program may have stored information about them after the first visit. What can be expensive, however, is the need to scan a very long L_{seen}, even when a hash function is employed.

We conclude that the overhead of iterative deepening *does* add certain computational costs. Most of the time, however, these costs are acceptable and can be outweighed by the advantages discussed in the next paragraph.

2.5.6 Comparing ID with the Basic Approaches

Mathematicians have established that, at a given depth iterative deepening visits $\frac{b+1}{b-1}$ more states than DFS (for $b \neq 1$). For small values, say, $b = 2$, this fraction is $\frac{b+1}{b-1} = \frac{3}{1} = 3$, which means that ID visits three times as many states as DFS. This may seem significant, but the increase in the number of visited states is negligible for higher branching factors. For instance, if $b = 10$, then we get $\frac{b+1}{b-1} = \frac{11}{9} = 1.2$. By and large, we can consider the memory requirements of the two approaches to be comparable.

When compared to BFS, iterative deepening retains its main advantage: the ability to find the shortest path from the initial state to the final state. This is important in domains where the shortest path is deemed important (in those domains, DFS would be less beneficial).

In summary, we observe that iterative deepening finds the shortest path to the problem's solution, and that it does so with computational time comparable to those of BFS, and with memory costs comparable to those of DFS. It is in *this* sense that iterative deepening is sometimes regarded as an optimal blind-search algorithm.

2.5.7 Word of Caution

In reality, iterative deepening is not guaranteed always to outperform the other two algorithms; it does so only *most of the time*. Depending on the circumstances encountered in a concrete application, experimental evaluation reveals that DFS or BFS are sometimes more efficient than ID.

When choosing the technique believed to be appropriate for the task at hand, we must not forget that a lot depends on our specific goals and intentions (see Section 2.3). For instance, if we want the shortest path, BFS will be best, with ID a bit slower, and DFS rather risky. Similarly, DFS will hardly be our choice if we want to find all terminal states and all paths leading to them.

2.5.8 Alternative: Iterative Broadening

It is only fair to mention the existence of another attempt to marry the advantages of DFS and BFS: *iterative broadening*, IB. Instead of repeatedly running DFS to a depth that gradually increases, IB employs BFS in a way that limits (and gradually increases) the maximum breadth. First, it runs BFS with maximum breadth set to $c = 2$, then it re-runs the search with maximum breadth increased to $c = 3$, then $c = 4$, and so on and so forth (of course, with special attention devoted to the stopping criterion).

Some theoreticians believe this approach to be as powerful as iterative deepening. This said, its advantages are somewhat less intuitive than those of ID. Another disadvantage is that shallow goals (those at small depth in the search tree) are not as readily discovered as in the case of iterative deepening. For all these reasons, ID seems to be more valuable in practical implementations. Iterative broadening is mentioned here only for the sake of completeness, and no further analysis is therefore needed.

Control Questions

If you are unable to answer the following questions, return to the appropriate place in the preceding text.

- Summarize the general algorithm of *iterative deepening*. Discuss in what sense it combines the advantages of DFS and BFS while mitigating their individual shortcomings.
- Discuss the question of computational costs. Why is iterative deepening less wasteful than an analyst may suspect, even though the technique has to revisit many search states?
- Summarize the general principle of *iterative broadening*.

2.6 Practice Makes Perfect

To improve your understanding, take a chance with the following exercises, thought experiments, and computer assignments.

- Consider the *Hanoy-tower* puzzle from Figure 2.11. The task is to convert the initial state into the final state using search operators defined as follows: "transfer a single disk from one vertical bar to another under the constraint that a larger disk can never be placed on a smaller disk." Suggest an appropriate data structure to represent the search states, then hand-simulate DFS and BFS.
- Consider an imaginary puzzle whose search tree has branching factor $b = 3$. How many different states will BFS and DFS have to keep in memory when they reach depth $d = 10$?
- Suppose that a program has been running DFS or BFS, and that it has now reached a final state. Throughout the search process, is the length of L_{seen} always greater than the length of L? Support your answer with convincing arguments. If needed, illustrate these arguments by examples.
- Write a computer program that implements blind search in a way that allows the user to choose between DFS, BFS, and iterative deepening.
- Run the program developed in the previous task on a selected puzzle from this section. Compare the observed computational costs of the three algorithms along two criteria: the number of states visited before the final state has been reached and the maximum memory requirements. The former equals the length of L_{seen} at the moment of the program's termination; the latter equals the maximum length of L throughout the search process. Explain the observed performances based on your understanding of the three search techniques.
- Identify a realistic engineering task that can be cast as a search problem. How will you represent the individual states? How will the final state be defined?

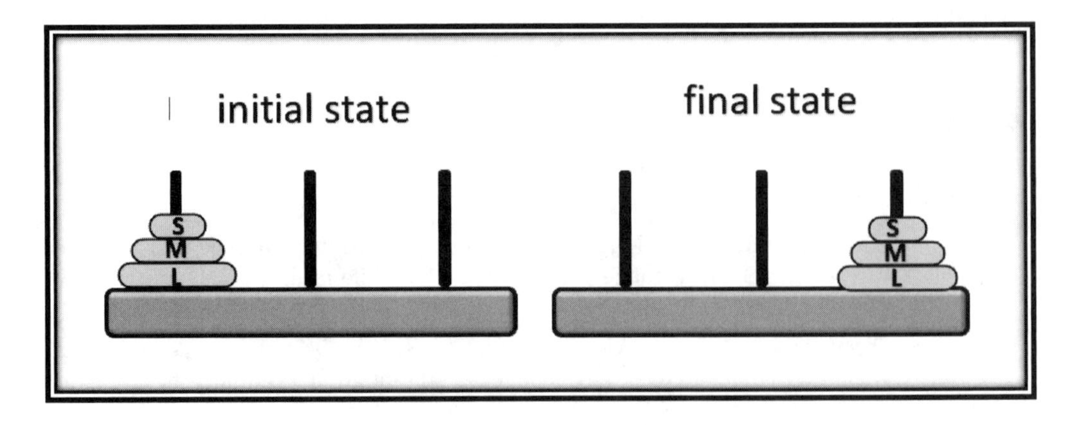

Figure 2.11 The task is to convert the initial state to the final state by moving one disk at a time. Constraint: larger disk must never be placed on a smaller disk.

What will be the initial state? Will this domain require the search to find the shortest path? Will it be necessary to find all final states? Which of the three blind-search techniques is likely to be best-suited here?

- In what kind of domains will iterative deepening be less efficient than DFS and BFS?

2.7 Concluding Remarks

Blind-search algorithms rank among the oldest tools aimed at the mystery of computational intelligence. They are elementary, and they are so simple as almost to appear trivial. Worse than that: Knowing that they have long since been superseded by the more efficient heuristic-search techniques from the next chapter, the reader may be tempted to dismiss them as less-then-useful antiquities, irrelevant to modern needs. Such rash dismissal, however, would be premature.

True, blind-search techniques *are* mechanical, and they do not pretend to be intelligent. Still, at least two reasons recommend their inclusion in this book. First, understanding their principles will help the reader develop thinking patterns that will come handy when we start discussing more sophisticated approaches. Second, blind-search techniques belong to the arsenal of any self-respecting programmer. They can be used to address simple problems where excessive sophistication would do more harm than good. Harley Davidson is faster than a bicycle and may be our tool of choice for long-distance travel. On the other hand, bicycles are cheaper, do not make so much noise, are better for our health, and are easier to carry around. This is the way of thinking we must learn to cultivate when choosing the most appropriate technique for a problem at hand.

Besides, the two basic techniques, DFS and BFS, are often discussed in undergraduate programming courses. For instance, they are used to illustrate the advantages of certain fundamental data structures such as stacks or queues or linked lists.

CHAPTER 3
Heuristic Search and Annealing

Discussions of blind search acquainted us with the fundamental principles of artificial intelligence (AI) search. This said, one has to admit that elementary techniques are useful only in simple domains. When it comes to something more realistic, they suffer from their very blindness. Having created children states, they place them in list L at random, irrespective of their individual advantages. For engineering purposes, this is not enough.

Humans, when seeking to solve a problem, are less mechanical. They tend to rank the "children states" by a heuristic, a rule of thumb or a guideline, that helps them steer the solution process toward more promising avenues. In the context of search, the heuristic takes the form of an *evaluation function* which for each state returns its value—for instance, the state's proximity to the terminal state. Choosing preferentially actions that result in higher-valued children improves the efficiency of the search process.

This, then, is the idea that forms the backbone of a whole family of *heuristic-search* algorithms discussed on the following pages. Apart from these, this chapter briefly mentions also one major alternative, *simulated annealing*.

3.1 Hill Climbing and Best-First Search

Let us begin with two basic algorithms. Both rely on the ability to estimate the intrinsic values of search states.

3.1.1 Evaluation Function

The engineer's way of conveying to the machine an idea of a state's quality is the *evaluation function*. The input is the state's description and the output is a value that helps estimate the state's proximity to the given problem's solution.

3.1.2 Numeric Example: Sliding Tiles

The simplest way of measuring the similarity to the final state counts the number of tiles found on the wrong squares—squares that are different from those prescribed by the final state. Thus in the case depicted in Figure 2.3, some tiles in the initial state are on "correct" squares: **1**, **3**, **5**, **6**, and **7**. The remaining three tiles, **2**, **4**, and **8**, are on the "wrong" squares. The initial state's distance from the final state is therefore $d = 3$.

Admittedly, this is a simple-minded criterion. A better one will also consider the *distances* of the misplaced tiles from the correct locations. In other words, the criterion will ask how many steps each tile would need to reach the correct location if unimpeded by other tiles. In the case of tiles **2, 4, 8**, the individual distances are 1, 2, and 2, respectively. The state's distance from the final state is therefore $d = 1 + 2 + 2 = 5$.

Even this is not perfect. A tile's distance from the correct square is not the same as the difficulty of getting it there. An even better criterion should attempt to quantify this aspect, too. Of course, designing such function is not an easy undertaking; the programmer cannot succeed without certain knowledge, experience, and creativity. It is in this sense that the evaluation function is said to convey to the program some part of the human expert's insight.

3.1.3 Sophisticated Evaluation Functions

Section 2.3 explained that the final state may be defined not just by an exact pattern, but rather by conditions that the final state is to satisfy. Thus in the case of the magic square we may be asked to find a "state where at least one column contains only even numbers or at least one row contains only odd numbers." In this event, the development of a useful evaluation function may require a healthy dose of creativity.

3.1.4 Maximize or Minimize?

There are two essential approaches to the evaluation function. The one from the previous paragraphs relied on the *distance* between the current and the final states. The value is to be minimized: the smaller the distance, the closer the state is to the ultimate solution. The other possibility is to quantify the *similarity* between the two states. This is to be maximized: the greater the similarity, the better.

Whichever possibility the engineer chooses, it is necessary to be consistent. It would be silly to forget whether the evaluation function employed in the developed software is to be maximized or minimized.

3.1.5 Hill Climbing

One shortcoming of depth-first search is that it places the generated children in list L at random order. Evaluation function allows us to do better. One improvement, called *hill climbing*, places the children at the beginning of L in the order determined by the evaluation function so that the most promising states come first. The principle is summarized by the pseudo-code in Table 3.1.

Again, we must not forget that if the evaluation function measures the state's similarity to the final state, the first child should be the one with the highest value. If the function measures distance, then the first child should be the one with the smallest value.

3.1.6 Best-First Search

Hill-climbing forces the search always to proceed from parents to children; only after each child has reached a dead end is the search allowed to return to an earlier state. In some applications, however, the engineer realizes that all children of a rejected state have values inferior to those of an earlier state. In that event, it makes sense to abandon

Input: List L that contains the initial state.
Empty list L_{seen} of previously seen states.
Evaluation function defined by programmer.

1. Let s be the first state in L. If s is a final state, stop with success.
2. Apply to s all legal search operators, thus generating its children.
3. Ignore those children that are already in L_{seen}.
4. Sort the remaining children according to the values returned by the evaluation function, and insert them **at the beginning** of L.
5. Remove s from L and put it at the end of L_{seen}.
6. If $L = \emptyset$, stop with failure. Otherwise, return to step 1.

Table 3.1 Pseudo-Code of Hill-Climbing Search

the current path (which does not offer any immediate improvement), and return to that earlier state whose value is more promising.

This is the principle of the *best-first search* algorithm that is summarized by the pseudo-code in Table 3.2. The reader will note the difference: Each time a new state is to be checked for being terminal, best-first search selects from the entire list L the state with the best value, regardless of where in the search tree this state is located. By contrast, hill climbing only picks the best child of the just-rejected state.

3.1.7 Two Ways to Implement Best-First Search

A simple implementation always appends the newly created children at some predefined place in L (the beginning or the end) and attaches to each child the value it has received from the evaluation function. In principle, the list does not have to be ordered. When the next state is to be tested for being terminal, the program chooses it by the *find-the-best* function. This is the approach indicated by the pseudo-code from Table 3.2.

An alternative implementation takes a different approach. Each time new children have been created, and evaluated, the program inserts them in the appropriate place of

Input: List L contains the initial state.
Empty list L_{seen} of previously seen states.
Evaluation function defined by programmer.

1. Let s be the state in L with the best value returned by the evaluation function. If s is a final state, stop with success.
2. Apply to s all legal search operators, thus obtaining its children.
3. Ignore those children that are already in L_{seen}, and put the survivors in L.
4. Remove s from L and put it at the end of L_{seen}.
5. If $L = \emptyset$, stop with failure, Otherwise, return to step 1.

Table 3.2 Pseudo-Code of Best-First Search

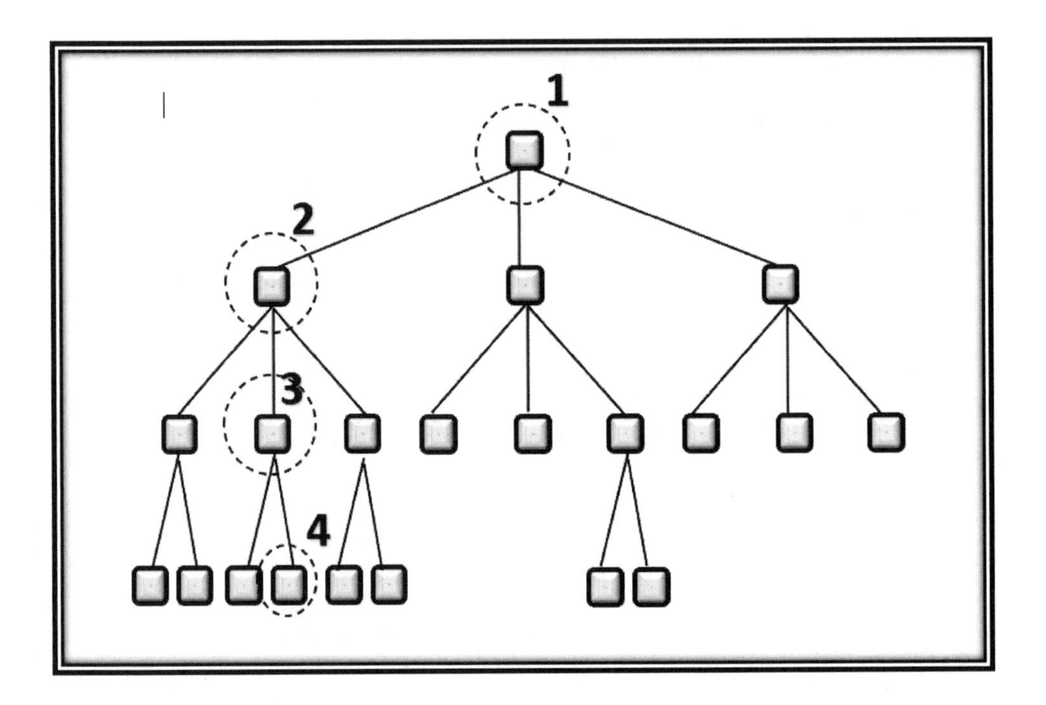

FIGURE 3.1 The circled states have been checked for being final. The integers indicate the order in which *hill climbing* has visited them. Note that this algorithm always proceeds from parent to child.

L so that the list is always ordered from the best state to the worst.[1] When the next state is to be selected, the program picks the first element of *L*.

The concrete solution will depend on the circumstances of the concrete domain, as well as on the engineer's personal preferences and common sense.

3.1.8 Comparing the Two Techniques

Figures 3.1 and 3.2 illustrate the main difference between hill climbing and best-first search. In hill climbing, the next state to be investigated is always chosen from among the current state's children. Best-first search is more flexible: It proceeds to the best state in the entire list *L*. This winner does not have to be the current-state's child.

Does the higher flexibility guarantee faster search? It does, most of the time. Yet there is no guarantee. In some domains, the children are always better than the parents, in which case hill climbing does the right thing while best-first search's flexibility only slows the process down by the extra overhead incurred by always having to identify the best next state.

3.1.9 Human Approach to Search

A schoolboy presented with the sliding-tiles puzzle will move one tile at a time, converting the current state into another, and then to yet another still—which means always

[1] For this, the linked-list data structure is often employed.

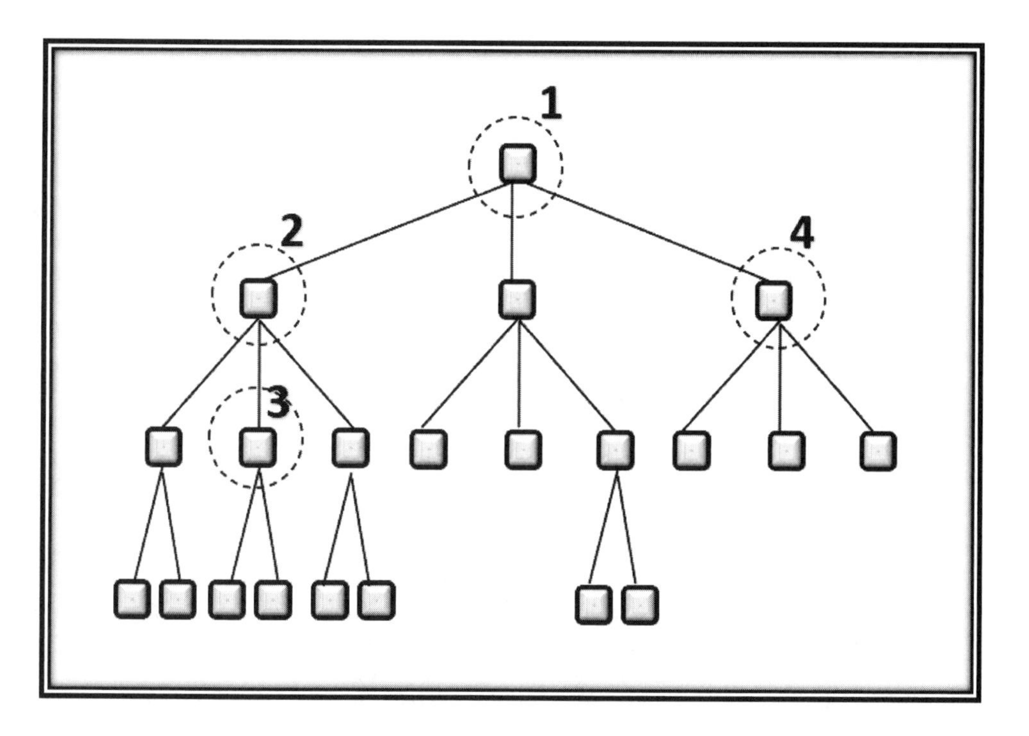

Figure 3.2 The circled states have been checked for being final. The integers indicate the order in which *best-first* search has processed them. Note that this algorithm does *not* have to proceed from parent to child.

proceeding from parent to child. In other words, he adheres to the hill-climbing principle. At the same time, the schoolboy will often revisit the same state, perhaps without even noticing it because he lacks the luxury of list L_{seen}. With a pinch of salt, we may say that human problem-solving follows some kind of imperfect hill climbing.

Control Questions

If you are unable to answer the following questions, return to the appropriate place in the preceding text.

- What does the term *heuristic* mean? In what way does the evaluation function serve as a heuristic?
- Explain the principles of hill climbing and best-first search. Comment also on issues related to their practical implementation.
- Compare the behaviors of hill climbing and best-first search. Under what circumstances will one of them outperform the other?

3.2 Practical Aspects of Evaluation Functions

Every programmer knows that grasping the principles of an algorithm is not enough. To succeed, one has to be prepared to deal with many situations that classroom examples avoid because they have to be simple for educational purposes. Ignoring these complications, however, may cause a seemingly well-thought-out project to fail dismally. Let us therefore take a look at what the engineer may run into.

3.2.1 Temporal Worsening of State Value

Let us return to the sliding-tiles puzzle. Suppose that the initial and final states are those from Figure 3.3. Suppose, further, that the evaluation function is defined as follows. Create a counter, initialized to 0. For each tile whose location is different from what the final state requires, add to the counter the number of steps that would be needed if the tile were to reach the correct location on an empty board. In the case from Figure 3.3, the distance is $d = 1 + 2 + 2 = 5$ because the misplaced tiles, **2**, **4**, and **8** would have to be moved by 1, 2, and 2 squares, respectively. The reader will recall that this formula was proposed in an earlier section.

Let us now take a look at how this evaluation function may affect the solution process. In the initial state, three actions are possible. One of them slides tile **3** down. Since this tile is already on the correct square, the action will only increase by 1 its distance from the correct location, with the corresponding increase in the new state's value as returned by the evaluation function. The same will happen if we move **5** up. The last possibility is to move **2** to the right. This tile does *not* find itself on the correct location; nevertheless, sliding it will increase the state's distance, from 1 to 2.

We have made an important observation: While searching for the solution, the program may be forced to accept interim states whose values appear to be a step back—just as a tourist may have to traverse an intervening valley before scaling a mountain peak.

3.2.2 Many States Can Have the Same Value

The same example illustrates another frequently encountered circumstance: Each of the three search operators increases the distance from the final state by the same amount. In

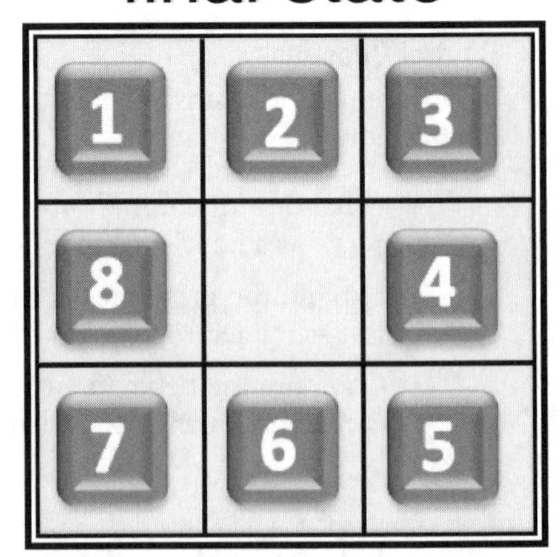

Figure 3.3 Moving any single tile in the initial state will increase by 1 the state's distance from the final state.

a situation like this, the hill-climbing algorithm is unable to decide (except by random choice) which of the three search operators to apply.

Admittedly, the evaluation function we have used is rather crude. The search program is less likely to face this last difficulty if it relies on a more subtle formula which might give to each of the state's children a different value. On the other hand, we must not forget that evaluation functions are nothing but approximations of the engineer's intuition, and as such may be unreliable; they may even lead the search astray.

3.2.3 Look-Ahead Evaluation Strategy

Previous paragraphs convinced us that the agent sometimes has to choose between equally valued states. We now also understand that the search process may have to overcome local peaks (or valleys) without knowing what comes later. These observations inspire what is known as the *look-ahead evaluation strategy*.

The idea is that the agent should not base the choice on the evaluation of the state's immediate children, but rather on the evaluation of more distant offspring: grandchildren, great-grandchildren, and so on. Thus in the search tree from Figure 3.4, the value of state **b** is 2.5 but the look-ahead strategy of depth 2 will operate with 3.5, which is the value of the maximum-valued grandchild, **m**.

How deep the agent should look is a user-set parameter. Of course, greater depth tends to mitigate the seriousness of the aforementioned complications (e.g., local

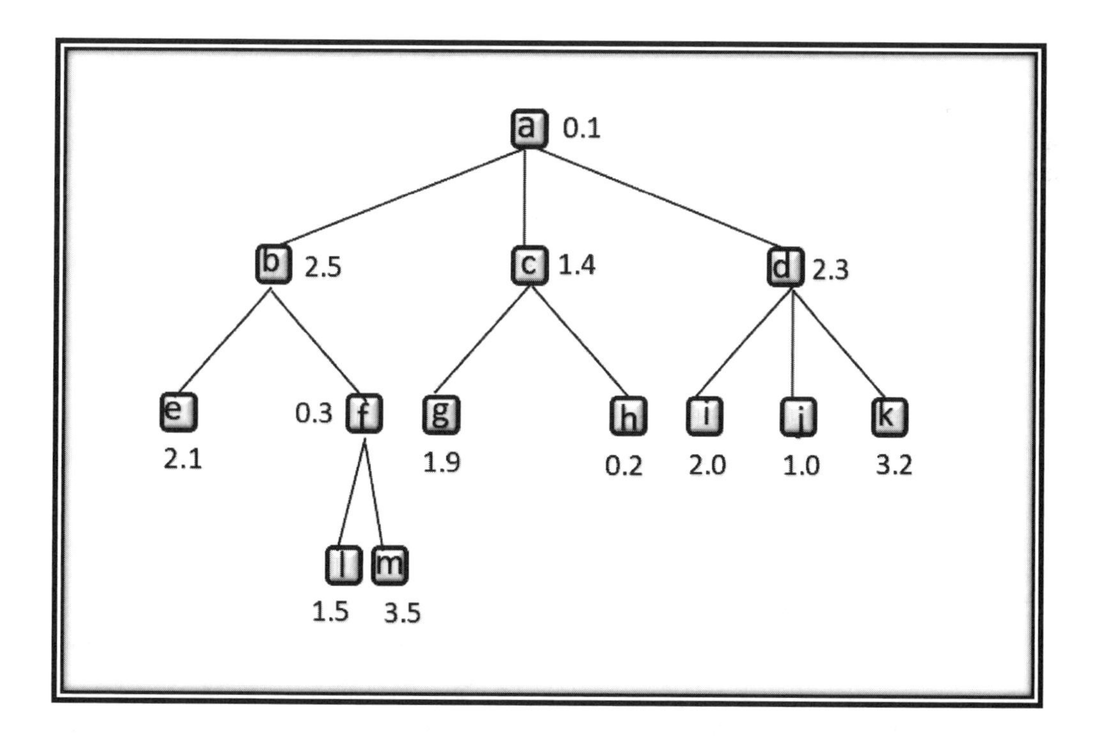

FIGURE 3.4 An example of search tree. Each node represents a state. For each state, the value returned by the evaluation function is provided.

Input: List L containing the initial state.
Empty list L_{seen} of previously seen states.
Evaluation function defined by programmer.
User-specified value of N.

1. Order L by the evaluation function. If $L = \emptyset$, stop with failure.
2. Let $N' = \min\{N, \texttt{length}(L)\}$. If at least one of the first N' elements in L is a final state, stop with success.
3. Find all children of these N' states. Ignore those that are already in L_{seen}; insert the remaining ones in L.
4. Move the first N' states from L to L_{seen}.
5. Return to step 1.

Table 3.3 Pseudo-Code of Beam Search, a "Parallel Version" of Best-First Search

extremes of the evaluation function). On the other hand, the costs of having to evaluate a great many "descendants" may be high.

3.2.4 Beam Search

Yet another way of dealing with the aforementioned difficulties is to run two or more searches in parallel. This principle, known as *beam search*, is summarized by the pseudo-code in Table 3.3. The idea is always to pick not just a single best state (e.g., the best child, in the case of hill climbing) but N best states. If none of them is final, these N states are summarily transferred from L to L_{seen} and replaced in L by all their children, appropriately ordered. The same principle can of course be employed also in the context of best-first search.

The reader may find the pseudo-code's step 2 a bit non-intuitive. Here is the motivation behind the "strange" term, $N' = \min\{N, \texttt{length}(L)\}$. Sometimes, the user-specified value of N exceeds the momentary length of list L. For instance, this will happen if the user specifies $N = 3$ but L contains only two states, so that $\texttt{length}(L) = 2$. In this event, the agent can only evaluate $\min\{3, 2\} = 2$ states.

3.2.5 Role of *N* in Beam Search

What value of N do we recommend? The answer, as always, depends on certain trade-offs. Higher-valued N helps the agent overcome local extremes of the evaluation function and also helps reduce the danger of overlooking shallow solutions (those with short path from the initial state). However, these benefits may be outbalanced by high memory costs.

Studying this algorithm, the reader will realize that beam search with very high values of N may degenerate to breadth-first search—which is no longer heuristic but blind.

3.2.6 Numeric Example

Consider the simple problem represented by the search tree shown in Figure 3.4. Each node in the tree represents one state. Next to each state is a value returned for it by the

Best-first search

L	L_{seen}
$\binom{a}{0.1}$	$\emptyset$
$\binom{b}{2,5}\binom{d}{2.3}\binom{c}{1.4}$	$\binom{a}{0.1}$
$\binom{d}{2.3}\binom{e}{2.1}\binom{c}{1.4}\binom{f}{0.3}$	$\binom{a}{0.1}\binom{b}{2,5}$
$\binom{k}{3.2}\binom{e}{2.1}\binom{i}{2.0}\binom{c}{1.4}\binom{j}{1.0}\binom{f}{0.3}$	$\binom{a}{0.1}\binom{b}{2,5}\binom{d}{2.3}$
$\binom{e}{2.1}\binom{i}{2.0}\binom{c}{1.4}\binom{j}{1.0}\binom{f}{0.3}$	$\binom{a}{0.1}\binom{b}{2,5}\binom{d}{2.3}\binom{k}{3.2}$

Hill climbing

L	L_{seen}
$\binom{a}{0.1}$	$\emptyset$
$\binom{b}{2,5}\binom{d}{2.3}\binom{c}{1.4}$	$\binom{a}{0.1}$
$\binom{e}{2.1}\binom{f}{0.3}\binom{d}{2.3}\binom{c}{1.4}$	$\binom{a}{0.1}\binom{b}{2,5}$
$\binom{f}{0.3}\binom{d}{2.3}\binom{c}{1.4}$	$\binom{a}{0.1}\binom{b}{2,5}\binom{e}{2.1}$
$\binom{m}{3.5}\binom{l}{1.5}\binom{d}{2.3}\binom{c}{1.4}$	$\binom{a}{0.1}\binom{b}{2,5}\binom{e}{2.1}\binom{f}{0.3}$
$\binom{l}{1.5}\binom{d}{2.3}\binom{c}{1.4}$	$\binom{a}{0.1}\binom{b}{2,5}\binom{e}{2.1}\binom{f}{0.3}\binom{m}{3.5}$

Beam-search version of best-first search, N=2

L	L_{seen}
$\binom{a}{0.1}$	$\emptyset$
$\binom{b}{2,5}\binom{d}{2.3}\binom{c}{1.4}$	$\binom{a}{0.1}$
$\binom{k}{3.2}\binom{e}{2.1}\binom{i}{2.0}\binom{g}{1.9}\binom{c}{1.4}\binom{j}{1.0}\binom{f}{0.3}$	$\binom{a}{0.1}\binom{b}{2,5}\binom{d}{2.3}$
$\binom{i}{2.0}\binom{g}{1.9}\binom{c}{1.4}\binom{j}{1.0}\binom{f}{0.3}$	$\binom{a}{0.1}\binom{b}{2,5}\binom{d}{2.3}\binom{k}{3.2}\binom{e}{2.1}$

TABLE 3.4 First Few Steps of the Heuristic-Search Techniques Applied to the Search Tree From Figure 3.4. The State Values Are to be Maximized

evaluation function. For instance, the value of state **b** is 2.5. Then, Table 3.4 details the behavior of the three heuristic-search techniques we have seen so far. For each, the step-by-step evolution of the two lists, L and L_{seen}, is provided.

It is a good exercise to hand-simulate these behaviors, just with a pencil and paper, in the way indicated in the table. The reader is also advised to take a closer look and figure out in what ways the behaviors of these algorithms differ and why.

3.2.7 Expensive Evaluations

In realistic applications, to assess any state's distance from the final state may not be easy. Very often, it is impossible to rely on just some mathematical formula. Even if such formula *can* be designed, the engineer may opt for the look-ahead strategy; to obtain the value of a single state may then necessitate that the program repeat the calculations for the hundreds of the offspring states in later generations. Finally, in some domains the only way to evaluate a state is by systematic experiments, which can be even more expensive.

Impractical costs of these state evaluations may become yet another critical factor to be considered by the engineers who want to identify the most appropriate search algorithm for the task at hand.

Control Questions

If you are unable to answer the following questions, return to the appropriate place in the preceding text.

- Discuss the main difficulties of evaluation functions: local peaks, equal values of many states, subjectivity, high costs.
- What is meant under the term, *look-ahead strategy*? What are its advantages and disadvantages?
- Explain the principle of *beam search*, and the practical impact that its breadth (i.e., parameter N) has on its behavior.

3.3 A-Star and IDA-Star

People lack patience. Whenever the solution process seems to lead nowhere even after a long series of steps, the average person loses interest in the current avenue and abandons it in favor of some other approach. This lack of consistency is not as bad as a moralist may suspect. Similar behavior has proved useful in one of the most powerful search algorithms.

3.3.1 Motivation

Consider the maze shown in Figure 3.5. Having entered it on the left (see the arrow), the agent is to reach the exit on the right, and do so with the minimum number of steps.

It is easy to see that the shortest path is the one where the agent, right after the entry, turns down. However, this is not a first step recommended by heuristic search whose evaluation function (to be minimized) measures the agent's distance from the goal. Such search will first go right, not down, because moving right decreases the distance and moving down increases it. Unfortunately, the move to the right will condemn the agent to a long path full of inevitable twists and turns, an inefficient try to say the least. We realize that hill climbing and best-first search will not do a good job here.

A human problem solver behaves differently. Having pursued an unpromising path for some time, we lose heart, and sooner or later backtrack to some earlier location where some other turn could have been chosen. This may seem reminiscent of best-first search, but the similarity is only on the surface. Best-first search backtracks if some earlier state is closer to the exit than any of the current state's children. Humans, however, are motivated by the costs that have been accrued by the search so far.

3.3.2 Cost Function

The main innovation of the approach introduced in this section is the intention to minimize not only the distance of a state, s, from the goal, but also to minimize the costs

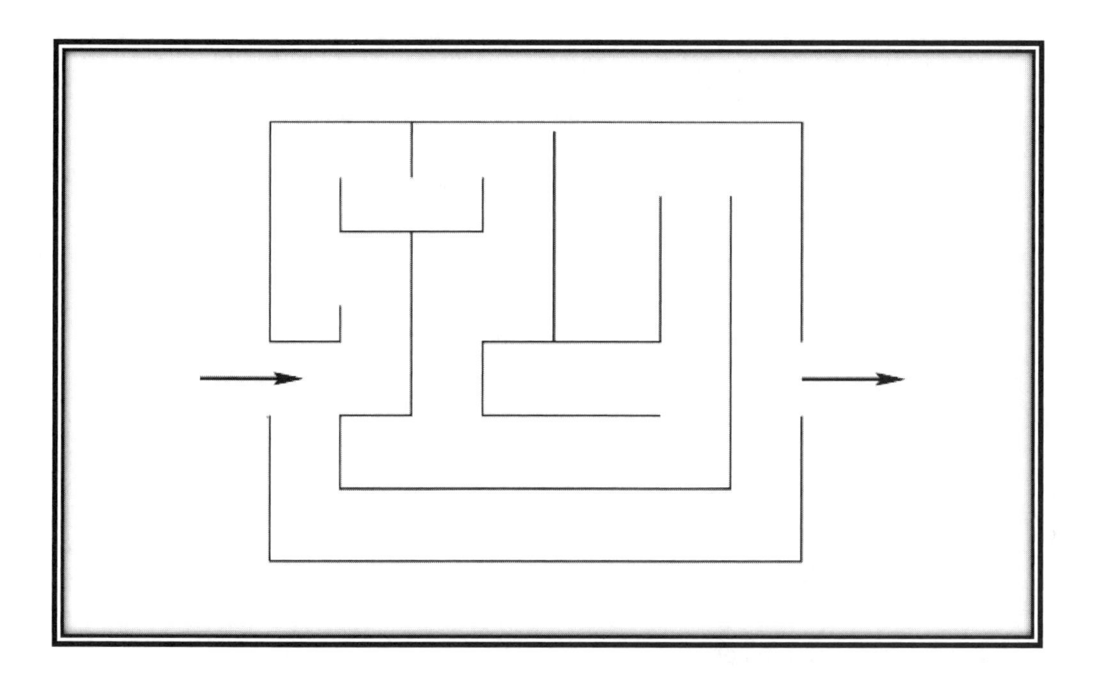

FIGURE 3.5 Mere distance from the final state is not enough. What matters are also the costs that have been accrued before any interim state has been reached.

accrued before s has been reached from the initial state. In a simple application, such as the maze mentioned in the previous paragraph, these costs can be identified with the number of actions (search operators) that have been executed along the way from the start to s. Later, we will discuss more sophisticated ways of assessing these costs.

3.3.3 A* Algorithm

The technique known as A* (read, "A star") relies on an evaluation function as well as on a cost function. If the evaluation function returns $g(s)$, and the cost function returns $h(s)$, then the search agent wants to minimize the sum, $g(s)+h(s)$, if the two carry equal weight. Note that in A*, the values returned by the evaluation function are to be minimized (in the previous search techniques, the engineer could choose between minimization and maximization).

The pseudo-code in Table 3.5 summarizes the principle just described.

In some applications, the two components (search costs and the state's distance from the terminal state) cannot be assumed to be equally important. In that event, the programmer may prefer the following, slightly more general, version of the formula: $c_g g(s) + c_h h(s)$. Here, the values of coefficients c_g and c_h reflect the relative importance of the two components.

3.3.4 Numeric Example

Figure 3.6 shows the same search tree that we already know from Figure 3.4. This time, however, the values of the individual states are obtained as $g(s)+h(s)$; this means adding to the evaluation function the number of steps needed to reach s from the initial state. For

Input: List L with the initial state; empty list L_{seen}.
Evaluation function, $g(s)$, estimating the distance from state s to the goal.
Cost function, $h(s)$, measuring the costs accrued before reaching state s.

1. Let s be the state in L that has the lowest value of $g(s) + h(s)$.
2. If s is a final state, stop with success.
3. Apply to s all legal search operators, thus obtaining its children.
4. Ignore those children that are in L_{seen}. Place the remaining ones in L.
5. Remove s from L and put it in L_{seen}.
6. If $L = \emptyset$, stop with failure, Otherwise, return to step 1.

Table 3.5 Pseudo-Code of A* Algorithm: The Best-First Search Version

instance, in Figure 3.4 the value of state **g** was $g(\mathbf{g}) = 0.9$. Observing that **g** is reached from the tree's root in two steps, $h(\mathbf{g}) = 2$, we establish the value of this state as $0.9 + 2 = 3.9$. This is the value shown in Figure 3.6. It is easy to verify that this is how the values of all the other states have been obtained.

3.3.5 Two Versions of A*

Note that the algorithm from Table 3.5 is only a minor improvement of best-first search. This is actually the most common version of A* in engineering practice. On the other

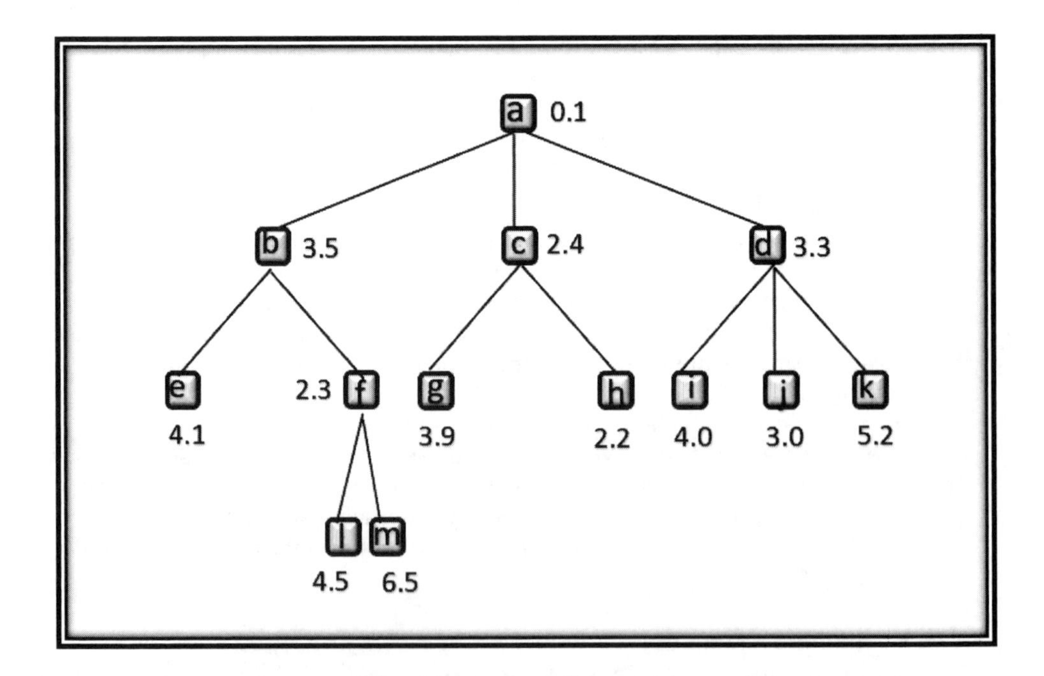

Figure 3.6 This is almost the same search tree as the one from Figure 3.4. The only difference is in the state values: those provided by the evaluation function have been increased by each state's depth.

hand, the reader will find it easy to apply the same modification (adding search costs to the value returned by the evaluation function) to hill climbing. This version, too, is sometimes employed.

3.3.6 More Sophisticated Cost Functions

For simplicity, this section has so far assumed that the search costs are adequately approximated by the number of actions executed along the way from the initial state to the current state—in other words, by the number of edges traversed in the search tree. In the real world, things may not be so simple.

In some applications, each action is associated with different costs. In this event, it makes sense to label each edge in the search tree with a number that quantifies the costs of the search operator represented by this edge. The costs associated with state **s** are then obtained by adding up the costs of the edges traversed from the root to the node representing **s**.

3.3.7 "Leaps and Bounds"

Sometimes, the engineer finds it impossible to design the evaluation function because any similarity between a given state and the final state is only speculative, and quantifying it would be misleading. In a situation of this kind, it may be safer to ignore the evaluation function altogether, $g(s) = 0$, and to seek to minimize only the costs, $h(s)$. For historical reasons, this approach is sometimes referred to as the technique of *leaps and bounds.*

3.3.8 IDA*

The ultimate heuristic-search technique is the iterative-deepening version of A*. The motivation is the same as in the case of blind search (see Chapter 2).

The idea is to run the hill-climbing version of A* down to a certain depth. If the solution has not been reached, the search is repeated with the maximum depth incremented by 1. While the hill-climbing version of A* seems more natural here, one can just as well apply iterative deepening to best-first search.

Control Questions

If you are unable to answer the following questions, return to the appropriate place in the preceding text.

- Under what circumstances are search costs an important criterion that should not be ignored?
- Explain the basic principle of A*, and discuss also its minor variations such as a flexible cost function, and the way to determine the relative importance of search costs versus the evaluation function.
- Discuss the motivation and principles behind the technique of *leaps and bounds.* How will you implement IDA*?

3.4 Simulated Annealing

The attentive reader has noticed that the heuristic-search algorithms discussed so far were nothing but variations on the theme of classical blind search. There is an alternative, though: a technique that has been inspired by certain physical processes known from metallurgy.

3.4.1 Growing Defect-Free Crystals

Suppose you decide to manufacture perfect, or at least almost-perfect, silicon crystals. For this, you need the atoms of silicon to settle in their lowest-energy arrangement—the one that is represented by the crystal's well-known lattice. This arrangement can be obtained by what is known as *annealing*.

The process consists of two stages. During the first, the material is heated up to some high initial temperature. In the second, this heated material is subjected to slow cooling. Quantum physics explains what is going on during this cooling stage. At each moment, the set of atoms finds itself in a certain arrangement that we may call a state, s. This state is subject to random modifications, each re-arranging the atoms, thus giving rise to a new state, s_n. Sometimes, the new state is stable; sometimes it is not, and the arrangement may revert itself back to the previous state, s.

3.4.2 Formal View

Whether the new state holds, or whether the system reverts back to the previous state, depends on the overall energy, E. We know that nature prefers lower-energy states. This means that, if $E(s_n) < E(s)$, the system remains in the new state s_n until the next random change. In the opposite case, where the new state's energy is higher than that of the previous state, $E(s_n) > E(s)$, things are more interesting. Most of the time, s_n turns out to be unstable and the system reverts to the previous lower-energy state, s. With a certain probability, however, the system retains the higher-energy state until the next random change.

The probability of the system's remaining in the higher-energy state depends on temperature. The higher the temperature, the higher the chances that the system will stay in the higher-energy state. In the initial stages, when the temperature is high, the system is thus highly unstable. As the temperature goes down, the system less frequently stays in the higher-energy state, which means that most of the system's random modifications result in lower-energy states. Physicists have formulated the following principle: If the temperature is being reduced very slowly, the atoms almost always arrange themselves in perfect crystals.

In summary, a heating period followed by carefully controlled cooling period results in big and almost defect-free crystals.

3.4.3 AI Perspective

The process just described reminds us of something we experienced in our discussions of heuristic search. There, the task was to reach a well-defined final state (here, the lowest-energy state of perfect crystals), but to reach this state, the problem solver had to pass through many sub-optimal states. This was illustrated on the example of the sliding-tiles puzzle, and this is what motivated the introduction of such strategies as best-first search, beam search, and look-ahead state evaluation. The power of these techniques lay in their ability to carry the solution process over the sub-optimal states.

The ambition to manufacture perfect silicon crystals does not appear to have much to do with the goals of AI search. Still, the efficiency of the underlying processes is intriguing. No wonder that computer scientists have sought to cast the principles in the form of an algorithm—and they succeeded!

3.4.4 Simple View of Simulated Annealing

The complete version of this technique is for an introductory text somewhat advanced. However, even a greatly simplified formulation of simulated annealing will help the engineer deal with challenging problems.

Let us modify hill climbing in the following manner: When choosing the next state, the agent *most of the time* picks the search operator that leads to the best-valued child. Occasionally, however, the agent gives preference to a sub-optimal child state. Suppose the evaluation function measures the state's distance from the final state, which means the process should minimize it (just like nature seeks to minimize a state's energy). Denote the current state's value by v and the selected child's value by v_n. Let k be a constant, and let T be *temperature*, a parameter whose specific role will be detailed later. The probability with which the action leading to this child is taken is calculated by the following formula:

$$P = \begin{cases} e^{-\frac{v_n - v}{kT}}, & \text{if } v_n \geq v \\ 1, & \text{otherwise} \end{cases} \tag{3.1}$$

3.4.5 Impact of State Values

Equation 3.1 specifies that the new state should be retained with 100% probability if its value is lower than that of the current state, $v_n < v$. If the new state's value is higher than that of the current state, $v_n > v$, the new state may still be accepted. The probability of this happening depends on the difference between the values, $v_n - v$: The more serious the value's worsening, the lower the probability that the new state will be retained. Of course, the engineer also has to instruct the computer program what to do in the case of equality, $v_n = v$. This, however, is from the perspective of this section irrelevant.

3.4.6 Impact of Temperature

The chances that a worse state will be accepted are controlled by the system's *temperature*, T. It is high at the beginning, and gradually reduces in time. A closer look at Equation 3.1 reveals that high temperature means high odds that a worse state will be accepted. As the temperature goes down, perhaps even approaching zero, the worse state is rarely accepted. Note also the constant k in the exponent's denominator. Its role is to control the formula's sensitivity to changes in T.[2]

The success or failure of simulated annealing depends on how quickly the temperature is reduced. Metallurgists know that the reduction should be slow. This said, how

[2] In the original physical formula, k is the Boltzmann constant.

slow is slow enough? Besides, what should be the initial temperature at which the process begins?

3.4.7 Cooling

Theoreticians have derived a formula that guarantees success, but the temperature reductions are then so minuscule that the process seems to take forever. A compromise is needed. The practically minded engineer will accept sub-optimal behavior (i.e., the process may find only an "almost-best" solution) if the search can be completed within a reasonable time frame. While some very sophisticated formulas based on rigorous analysis have been proposed, experience shows that good results are often achieved even if the following one is used (T_k denotes temperature at the k-th time step, and we use $\alpha \in [0.90, 0.99]$):

$$T_{k+1} = \alpha T_k \tag{3.2}$$

In baseline applications, temperature is reduced after each state change. Under these circumstances, engineers prefer higher values of α (approaching 1). Other domains may require that the temperature be reduced only after a predetermined number of steps, say, after each ten state changes. In this event, lower values of α are employed, perhaps close to 0.9.

3.4.8 Initial Temperature

Initial temperature, T_0, has to be sufficiently high so as to allow the system to make many "bad decisions" early on. This is essentially the only recommendation the author can offer here. There does not seem to be any rock-solid general prescription as to what the initial temperature should be. In real-world applications, the choice can be guided by preliminary experimentation.

Control Questions

If you are unable to answer the following questions, return to the appropriate place in the preceding text.

- Explain the principle of the *annealing* process that metallurgy uses to grow defect-free silicon crystals.
- Describe the computational approach that implements *simulated annealing*. How will you generate the random state changes? When will a new state be accepted and when will it be rejected?
- What can you say about temperature initialization, and how will you go about temperature reduction?

3.5 Role of Background Knowledge

The search techniques described in the previous sections have solved interesting puzzles, and they have proved useful in many non-trivial tasks. But we always have to be wary. Early success tends to thwart criticism, and it has a way of clouding our ability to see things in perspective. When the ideas underlying heuristic search were first presented, some enthusiasts went so far as to claim that the approach holds the keys to all computational intelligence.

Today we know better. The faith in the omnipotence of search techniques has been shaken, and experts now agree that a lot remains to be done. The following examples are meant to show in what ways the idea of search is limited when confronted by true intelligence.

3.5.1 Magic Square Addressed by AI Search

Figure 3.7 shows the popular magic-square puzzle. The task is to fill the nine fields with nine integers from 1 to 9 so that each field contains a different number, and the sums in all columns, rows, and the main diagonals are all equal. This is how the final state is defined. The initial state is the empty 3-by-3 square on the picture's left; the search operator places one integer in one of the empty squares. Having studied the previous sections, the student will find it easy to write a program to deal with the task.

Will such program be efficient? Let us take a look. Suppose the search begins with placing a random integer in the upper-left corner. There are nine possibilities to choose from. For each of them, the next field (say, the one in the middle of the upper row) will accept one of the remaining eight integers. This makes $9 \times 8 = 72$ combinations. Continuing in this way, we realize that blind search may have to confront up to 9! different ways of filling the magic square! This is prohibitive. True, we may reduce the number significantly if we find a way to deal with various symmetries (e.g., turning the board by 90 degrees, or inverting it horizontally). Moreover, a well-designed evaluation function will help us employ the more-efficient heuristic search, which may further expedite the process. Still, the costs remain high; in domains of this kind, search seems hopelessly mechanical.

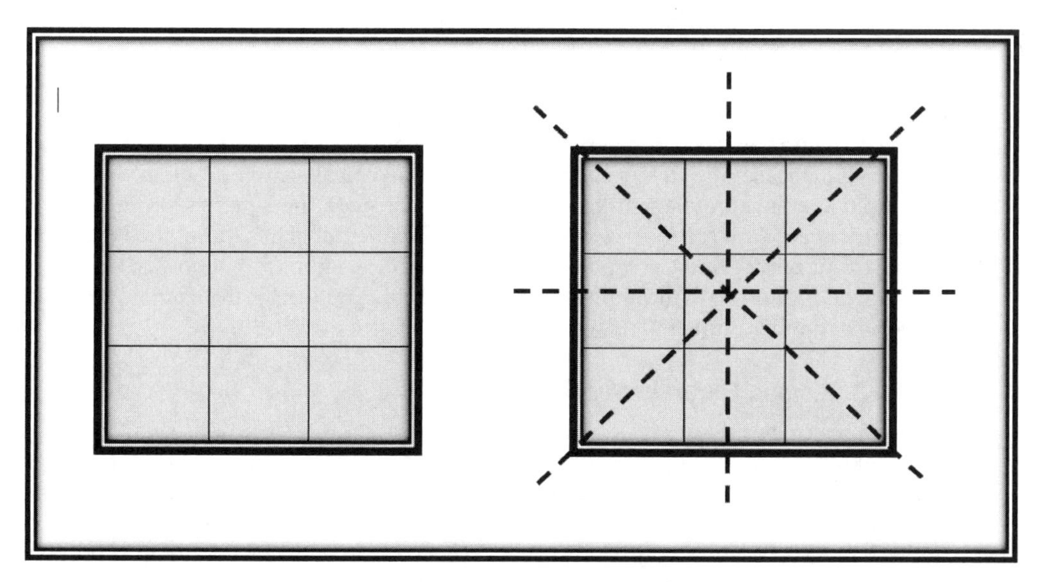

FIGURE 3.7 Magic square: The task is to fill the nine fields with integers from 1 to 9 in such a way that the sums in all columns, in all rows, and along the main diagonals are all the same.

3.5.2 Magic Square Addressed by a Mathematician

Human problem solvers approach the problem differently. A mathematician will first observe that the sum of integers from 1 to 9 is $\Sigma_{i=1}^{9} i = 45$. If each of the three columns is to have an equal sum, then this sum must be $45/3 = 15$. This clearly reduces the number of ways in which the square can be filled.

The next step in the mathematician's reasoning is indicated on the right of Figure 3.7. Suppose we add the fields in the middle row, in the middle column, and in both main diagonals. This comprises all the nine fields, but the one in the square's center (let us denote its value by x) is included four times. If we subtract three of this field's four occurrences ($3x$), then each field is represented only once—and we know, from the previous paragraph, that the sum is 45.

If we do not subtract the three superfluous occurrences of the central field, we have four triplets (the column, the row, and two diagonals). Since the sum in each is 15, the total has to be $4 \times 15 = 60$. Let us again denote the value of the central field by x. Subtraction of its three "superfluous" occurrences results in $60 - 3x = 45$. From here, we obtain $x = 5$. We conclude that **5** is the integer to be placed in the central field.

The rest is easy. At the upper left corner, any of the remaining eight integers can be placed. Each of them determines the value for the bottom-right corner. For instance, if the upper left corner contains **9**, then the bottom-right has to be $15-9-5 = \mathbf{1}$. This allows only six possibilities for how to fill the field in the middle of the upper row. Suppose we choose **2**. Then the upper-right corner receives $15 - 9 - 2 = \mathbf{4}$; and the rest of the magic square will be filled similarly, always observing the rule that rows, columns, and main diagonals sum to 15.

3.5.3 Lesson: The Benefits of Background Knowledge

The analysis from the previous paragraph reduced the number of possibilities to $8 \times 6 = 48$. This is much less than the original 9!, so much so that we do not even need a computer; the problem can now be solved with a pencil and paper in a few minutes' time. In the light of this experience, we are beginning to suspect that classical approaches to search are indeed too mechanical to be efficient. Something better is needed.

The reduction in the search-space size was achieved thanks to *background knowledge*. On the surface, the magic square may seem a typical search problem. However, we noticed that a more elegant approach would consider the square's geometry, and an even more efficient solution would rely on the mathematician's thoughts. We are beginning to appreciate that the problem-solving computer should somehow be endowed with knowledge and with the ability to manipulate this knowledge in a meaningful way. Brute-force number crunching is wasteful.

3.5.4 Branching Factor in Sudoku

Another aspect that deserves our attention is the branching factor. A casual glance at the Sudoku problem in Figure 3.8 informs us that the top row has used only three out of the nine integers. A thoughtless beginner may be tempted to assume that, say, for the square to the right of **3**, any of the remaining six integers can be considered. This, of course, is not a reasonable way of approaching it. To begin with, **8** is disqualified because it already appears in the same column (in the third row). Likewise, **6** and **9** do not come into question because they already appear in the given 3-by-3 square. This leaves us with only three possibilities, rather than six.

5	3			7				
6			1	9	5			
	9	8					6	
8				6				3
4			8		3			1
7				2				6
	6					2	8	
			4	1	9			5
				8			7	9

Figure 3.8 Branching factor of Sudoku is smaller than it seems.

This is not the end of the story, though. Seeing that there are three possibilities to choose from, without any indication as to which to prefer, the human player will look for a square where the number of options is smaller. Very often, a square can be found that permits only one possibility, which amounts to the reduction of the branching factor $b = 1$, a significant reduction indeed! Again, we observe that the impressive reduction was made possible by our understanding of Sudoku's nature, and not by the computer's brute force.

3.5.5 Zebra

In other puzzles, the need for knowledge representation and knowledge manipulation is even more conspicuous. Consider the *zebra* puzzle from Chapter 1. We were provided with a set of facts about houses, people, animals, and so on, and were asked to answer a question about something that could only be inferred from these facts. Trying to do so by classical search is futile unless we find a way to convey to the machine some of the knowledge available to any reasonable person. This may take the form of *if-then* rules; their successful manipulation by a computer program may then deserve to be called *automated reasoning*.

Zebra has provided us with yet another piece of evidence to convince us that search alone is not enough. In many meaningful domains, search has to be assisted by some kind of knowledge. The questions of how to encode this knowledge, and how to take advantage of it in a problem-solving context, will be addressed later in this book, starting with Chapter 9. Before we reach that point, however, a lot still remains to be said about search itself.

Control Questions

If you are unable to answer the following questions, return to the appropriate place in the preceding text.

- Discuss the difference between the mathematician's approach to the magic-square puzzle and that of a search technique.
- How will a skilled programmer reduce the average branching factor in Sudoku? Can you write a program that implements this reduction?
- Suggest a way to solve the *zebra* puzzle by classical search. How will the facts be represented? What will constitute the search states?
- What is *background knowledge* and in what way can it assist problem solving?

3.6 Continuous Domains

Early AI focused predominantly on domains where each state allowed a finite number of actions, and the states were described by discrete variables. In many realistic applications, though, both the descriptions and actions can only be characterized by numbers from continuous domains. Let us take a look at what this means for search.

3.6.1 Example of a Continuous Domain

Figure 3.9 shows a function known as *Mexican Hat*. Any point, $[x, y]$, in the horizontal plane defines a state for which the vertical coordinate gives the evaluation function's value. We see that the states come from a continuous domain.

Suppose we start at some randomly generated initial state on the horizontal plane. The goal is to find the state where the function reaches its maximum.[3] Suppose we want to use hill climbing: At each state, the agent wants to modify the values of x and y in a way that promises the sharpest increase in the function, following its steepest gradient. The actions, therefore, are continuous valued, too.

The domain does not have to be two-dimensional; the states can be described by any number of variables. Some of them can be discrete, but the domain will be deemed continuous if at least one is continuous.

3.6.2 Discretization

The search techniques we have seen so far expect the agent's actions to be discrete: At each state, the action (search operator) is selected from a finite set of alternatives. By contrast, the number of actions in a continuous domain is infinite because there is an infinite number of possible changes to the state-describing variables.

[3] In this specific case, the state corresponding to the function's maximum is $[0, 0]$, the origin of the system of coordinates.

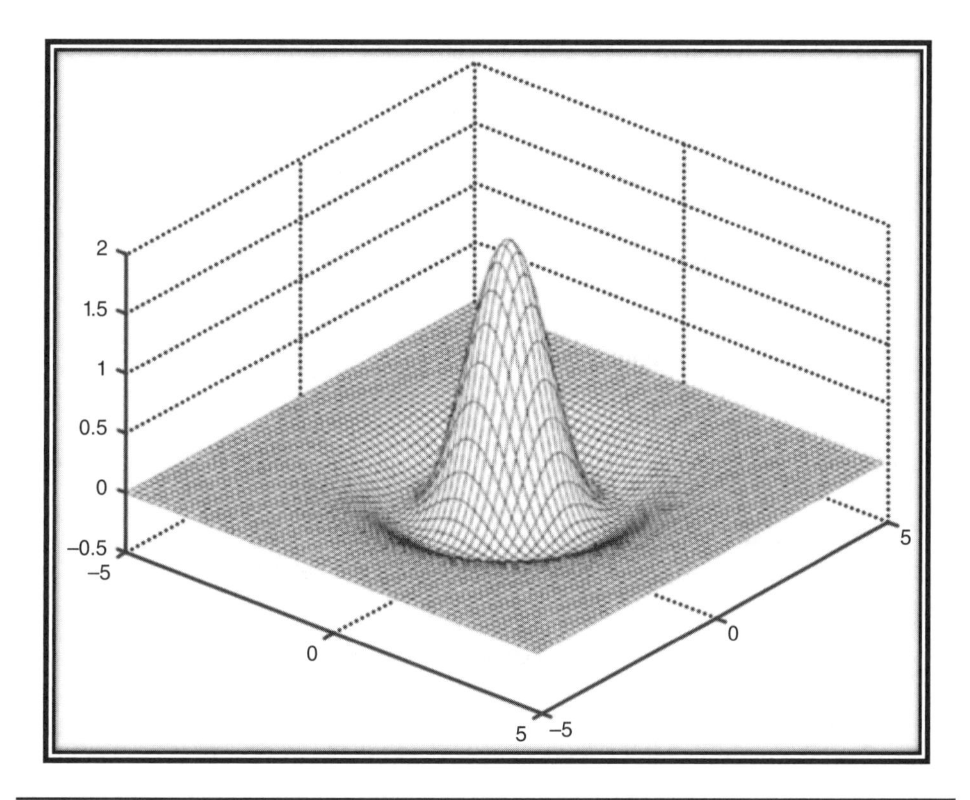

FIGURE 3.9 A continuous domain. Each pair [x, y] in the horizontal plain represents a state. The evaluation function is mapped on the vertical axis.

The simplest way of dealing with this situation is *discretization*. The idea is to divide each variable's domain into a finite number of intervals. Thus in the case from Figure 3.9, we can cover the horizontal plain with a grid of, say, 10-by-10 squares. At each moment, the agent finds itself in a state defined by one of these squares. An action can then be defined by the command, "move by N squares along the x-axis and by M squares along the y-axis."

Discretization allows us to employ heuristic search as we know it. The price for this convenience is an inevitable loss in accuracy: All states within a given square are deemed equal. Quite often, such limitation is acceptable because the engineer may not insist on finding the best solution, being content with "doing reasonably well." This is the difference between the worldview of a mathematician, who seeks perfection, and an engineer, who seeks practical solution.

3.6.3 Gradient Ascent and Neural Networks

Apart from discretization, one can use methods that are mathematically clean and do not lose information. If the formula defining the evaluation function is known, one possibility is to find the function's steepest gradient by calculus. Local extremes are identified by setting to 0 the function's first derivative; what remains is to decide whether the extreme represents a maximum or a minimum. The approach is known as *gradient ascent*—or *descent*, if we want the function's minimum.

Without going into details, let us mention in passing that gradient-ascent techniques are typical of *neural networks*. This technology has recently become popular thanks to the triumphs of *deep learning* that uses them. However, discussions of neural networks and deep learning belong rather to the field of *machine learning*, and not to a textbook of core AI. It is enough if the reader knows that continuous domains *can* be addressed by hill climbing, and that the relevant techniques involve some calculus.

3.6.4 Swarm Intelligence

There is an alternative, though. Chapter 8 will introduce a whole family of powerful techniques that are collectively known as *swarm intelligence*. They are powerful, do not require calculus, and they are no less popular and fashionable than deep learning.

Control Questions

If you are unable to answer the following questions, return to the appropriate place in the preceding text.

- What is the main difference between domains with discrete actions, and those with continuous-valued actions?
- How does discretization help us use classical AI techniques? What is the main shortcoming of discretization?
- This section briefly mentioned two approaches that handle continuous domains without the need to discretize. Do you remember their names?

3.7 Practice Makes Perfect

To improve your understanding, take a chance with the following exercises, thought experiments, and computer assignments.

- For each of the puzzles from the previous chapter, design at least two different evaluation functions. For instance, will the number of missionaries on the southern bank be a good indicator? If not, why? The point is to get used to the idea that many alternatives usually exist, some trivial, others sophisticated; but only rarely is any one of them completely satisfactory in the way it guides the search process.
- Suppose that a puzzle is represented by the search tree from Figure 3.10. Show the order in which the states are visited by hill climbing, best-first search, A*, and beam search (the last with $N = 2$).
- Readers that have implemented some blind-search technique will find it easy to modify it so as to turn it into hill climbing and into best-first search. The most difficult part is to formulate the evaluation function. Try to do so by way of a simple programming exercise.

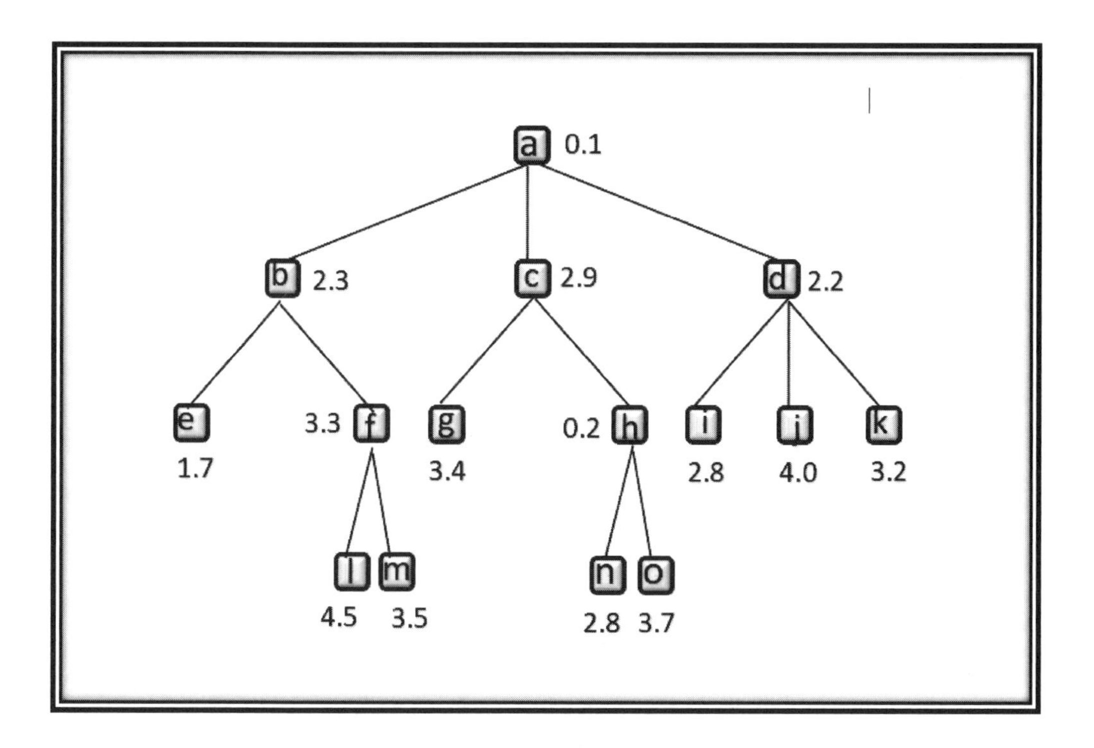

Figure 3.10 An example search tree for practicing. The numbers represent the individual states' estimated similarities to the final state (to be maximized).

- Having implemented a program for hill climbing and best-first search, run experiments to compare their behaviors on various test beds. Write a two-page essay summarizing your observations.
- Another useful study will compare the efficiency of heuristic search guided by competing evaluation functions. The point is to observe that a well-designed evaluation function can significantly speedup the solution process.
- Convert the best-first search program developed in the previous exercises into the A* technique. The only addition this requires is a function that sums the costs accrued before the given state has been reached. In the simplest implementation, this means to monitor the state's depth—the number of actions along the path from the initial state to the current state. In a more sophisticated implementation, the programmer will also consider the possibly different costs of the individual actions along the path.
- How will you apply simulated annealing to the toy domains from this chapter? How will you go about the initial temperature?
- Write a computer program implementing simulated annealing. Run an experimental study that compares its behavior with that of the more traditional heuristic-search techniques. Explore the practical impact of constant k in Equation 3.1.

3.8 Concluding Remarks

History of heuristic search begins with the introduction of the hill-climbing algorithm by Minsky (1961). The addition of the cost function (in A*), is traditionally attributed to Hart, Nilsson, and Raphael (1968). The ideas underlying simulated annealing were proposed by several authors, but Kirpatrick, Gelatt, and Vechi (1983) are usually credited with having achieved the main breakthrough by successfully applying the technique to the traveling-salesman problem.[4] They also seem to be the ones who first used the term, simulated annealing.

For some time, techniques of blind search and heuristic search were believed to offer a credible answer to the mystery of computer intelligence. The optimism was reinforced by reports of successful applications that went well beyond toy domains and elementary puzzles. Among these, perhaps the most impressive were automated planning systems, a topic that this book covers in Chapter 5. Additional excitement was caused by the introduction of simulated annealing: If the principle was good enough to grow defect-free crystals, why should it not be just as instrumental when applied to AI problems?

Gradually, though, scientists and engineers came to be aware of certain limitations such as those mentioned in Section 3.5. To begin with, evaluation function does not really convey to the machine adequate equivalent of human understanding; other ways to reduce the branching factor and to guide the solution process are needed. This was illustrated by the Sudoku example, and even more so by the magic square where we saw that a mathematician follows lines of reasoning that have nothing to do with the essence of search. The last example, zebra, only amplified our suspicion that an intelligent program should be capable of reasoning with explicitly encoded *knowledge*.

Later in this book, several chapters will be devoted to the questions of how to represent knowledge and how to employ it in automated reasoning. Special attention will be paid to the ways of dealing with imprecision, uncertainty, and ignorance, all of these being inseparable from human way of thinking.

For now, however, we are not yet done with the problem-solving agenda. To begin with, a lot can be said about algorithms that help a computer play games. As we will see in Chapter 4, these techniques are fairly different from those we have seen so far. Then, to convince us that classical search, in spite of all its limitations, can successfully address realistic tasks, Chapter 5 will discuss some simple ways of using it in the field of automated *planning*.

After this, major alternatives to classical search are introduced. Chapter 6 explains the nature and use of the popular *genetic algorithm*, Chapter 7 introduces the idea of emerging properties in *artificial life*, and Chapter 8 builds on these issues when explaining a few techniques from what is known as *swarm intelligence*.

---

[4] For more about the traveling-salesman problem, see Section 5.5.

CHAPTER 4
Adversary Search

In some applications, the agent's control of the problem at hand is limited. Thus in chess, the current chessboard position depends not only on the agent's latest move, but also on the opponent's response; and of course, the opponent has no intention to assist the agent in reaching its goals. This is something that the classical search techniques from the previous chapters did not consider. Other algorithms are needed.

The chapter focuses on techniques addressing a problem known as *adversary search*. After the presentation of the *mini-max* algorithm, the text proceeds to its more advanced heuristic versions, and to various methods (such as alpha-beta pruning) to increase its efficiency.

4.1 Typical Problems

The basic task addressed by adversary search is exemplified by two-person games where the players take turns, each of them pursuing the opposite goal. Using the artificial intelligence (AI) terminology, we can say that one player seeks the maximum of the evaluation function, and the other its minimum. For simplicity, we will assume that both players have complete information about the game's state.[1]

4.1.1 Example of a Simple Game

A good example of a game satisfying these requirements is the *ti-tac-toe* whose simple 3-by-3 version is shown in Figure 4.1. At the beginning, the board is empty; this is the game's initial state. The final state is one that contains a sequence of three instances of the same symbol in a row, column, or diagonal. One player wants to create the sequence with circles, the other with crosses. Starting from the initial state, the players take turns, seeking to achieve their goals while interfering with the intentions of the opponent.

In the position depicted in Figure 4.1, the player with crosses wins by placing a cross on the square in the upper-right corner. The opponent may prevent this by placing there a circle.

4.1.2 Other Games

Figure 4.1 is only a simple example meant to facilitate the explanations of the basic terms. In reality, tic-tac-toe is played on larger boards, at least 5-by-5. The reader already has an idea how the size of the board affects the game's branching factor.

---

[1] Note that this requirement eliminates many card games where the players' knowledge is limited. For such domains, adversary search in the form presented here is inappropriate.

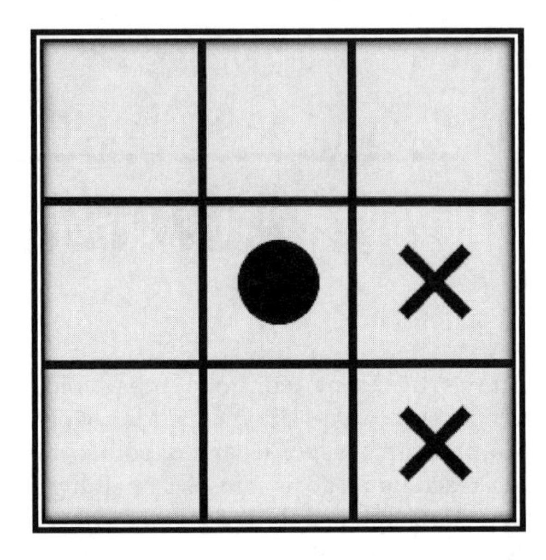

FIGURE 4.1 The game of tic-tac-toe. Two players take turns, one with circles, one with crosses. Whoever first achieves a 3-symbol sequence, wins.

Performance of adversary-search algorithms is usually tested on more ambitious games, such as chess, checkers, Go, backgammon, and some others. One game, Othello, is said to have been invented specifically for the needs of testing AI ideas.

4.1.3 More General View

Games represent a broad scope of applications that can be characterized as follows. An agent finds itself in a certain environment, and seeks to modify the environment in a way that maximizes some pre-defined benefits captured by the evaluation function. The environment reacts in ways beyond the agent's control. These reactions do not have to be intelligent. However, adversary-search techniques follow the worse-case analysis where the opponent is assumed always to choose the strongest move that hurts the agent's interests as much as possible.

4.1.4 Differences from Classical Search

In some games, such as tic-tac-toe, it is impossible to revisit the same state. This essentially eliminates the need for list L_{seen} of previously seen states. Usually, no backtracking is possible, either.

Another difference from the approaches from the previous chapters is that the agent's control of the game's state is limited, and always has to count with the opponent's interference.

Control Questions

If you are unable to answer the following questions, return to the appropriate place in the preceding text.

- What is the basic principle of adversary search? In what sense does it address certain problems where the search techniques from the previous chapters cannot be used?
- Suggest examples of games addressed by adversary search, and suggest examples of games where adversary search may not be appropriate.

4.2 Baseline Mini-Max

Now that we have defined our goals, we are ready to examine the fundamental technique at our disposal: the *mini-max* approach. Here we will introduce essential terminology, and then proceed to the algorithm's baseline formulation. More advanced versions will be the topic of the next section.

4.2.1 Maximizer and Minimizer

Whatever the concrete application, we will conceive it as a game played by an *agent* against an *adversary* (also known as *opponent*). At each moment, the game finds itself in a certain state. The value of the state from the agent's perspective is provided by an evaluation function. The agent wants to maximize this value, and the opponent wants to minimize it—this is why the agent is called *maximizer* and the opponent *minimizer*.

4.2.2 Game Tree

Let us leave tic-tac-toe, for a while, and instead consider a much simpler game where the number of states is so small that all possible developments are captured by the *game tree* from Figure 4.2.

The concept of the game tree is analogous to the search tree from the previous chapters. Each node represents one of the game's states. As already mentioned, the agent is maximizer, the opponent is minimizer. The squares in the picture represent states where it is the maximizer's turn; the circles represent states where it is the minimizer's turn. The game's initial state is at the tree's root. For each state, the game tree shows all possible actions (search operators) that lead to new states. Common terminology distinguishes between parent states and child states.

At the bottom of the tree are *leaf* nodes, states that do not have children. These are the terminal states where the game's outcome has been decided. Note that the leaf nodes are labeled either with **1** or **−1**. By convention, the value of a state in which the agent maximizer has won is **1**, and the value of a state in which the agent has lost is **−1**. Some games, such as tic-tac-toe played on the 3-by-3 board, allow for a third possibility: a draw, defined as a state in which none of the players can ever win. The value of a drawn state is **0**. For the time being, we will consider only these three values.

4.2.3 Parents Inherit from Children

Suppose that the entire game tree has been developed, from the initial state at the root, all the way down to the leaf nodes at the bottom. Suppose that an evaluation function has labeled all leaf nodes as states where the agent has won, lost, or drawn.

One such game tree is shown in Figure 4.2. Take a look at the node denoted by **i**. This state has two children: **l**, evaluated as **1**, and **m**, evaluated as **−1**. The fact that **i**

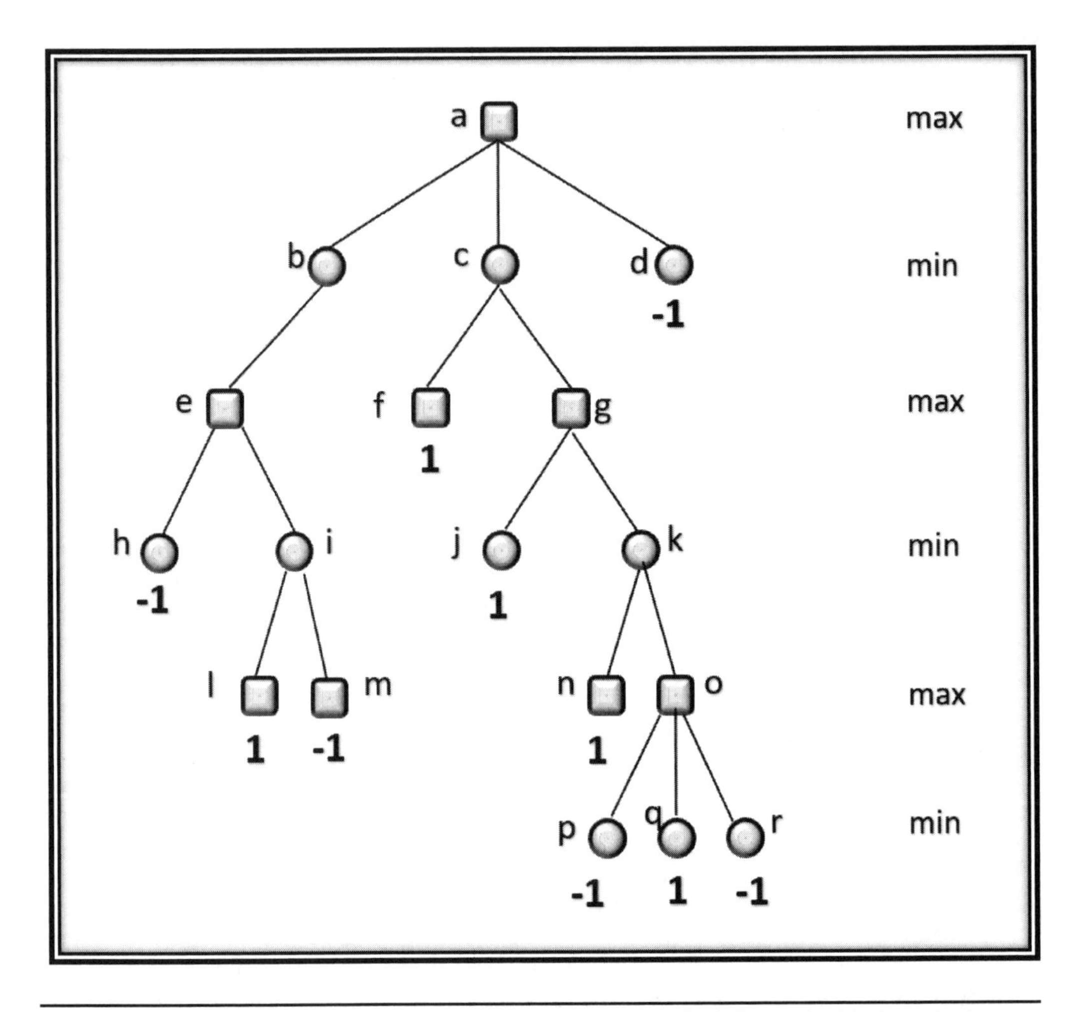

FIGURE 4.2 An example game tree. At the root is the initial state. Maximizer and minimizer take turns until a final state has been reached. In the final state the maximizer is winning (**1**) or losing (**−1**).

is represented by a circle indicates that it is the opponent's turn. Being minimizer, the opponent is bound to choose the action that leads to the smallest-valued child. In this specific case, this smallest value is **−1**, and it thus makes sense to evaluate **i** as having value **−1**, too, because this will be the result once the opponent has executed its chosen action.

The node denoted by **o** has three children, two of them evaluated as **−1** and one, **q**, evaluated as **1**. Being maximizer, the agent is bound to chose the action that leads to the highest-valued child, which is **1**, and it thus makes sense to evaluate **o** as having value **1**, too, because this will be the result once the agent has executed its chosen action.

4.2.4 Principle of the *Mini-Max* Approach

The ideas outlined in the previous paragraphs form the basis of an adversary-search algorithm known as *mini-max*. Its pseudo-code is summarized in Table 4.1. The first steps

Input: Initial position.
Function capable of evaluating a state as won, drawn, or lost.

1. Starting with the initial state, create the entire game tree all the way down to terminal nodes.
2. Evaluate the terminal nodes as won (**1**), lost (**-1**), or drawn (**0**),
3. If all the nodes in the game tree have been evaluated, stop.
4. Choose a node that has not been evaluated, while all its children *have* been evaluated.
5. If the node is *maximizer*, give it the maximum value found among its children.
6. If the node is *minimizer*, give it the minimum value found among its children.
7. Return to step 3.

Table 4.1 Pseudo-Code of the Simplest Version of *Mini-Max*

expand the entire game tree and evaluate all its terminal nodes. After this, *mini-max* finds a node whose all children have been labeled. If the node is maximizer, it receives the maximum value found among its children; if the node is minimizer, it receives the minimum value found among the children. The same is repeated with all other nodes whose children's values are known.

Once all the nodes in the lower layers have received their values, the same principle is applied to the parents of *these* nodes, and the process then continues all the way up to the initial state at the root. Once the root has been reached, we know the game's value: the outcome that results from the fact that both the agent and the opponent have always chosen their best actions.

4.2.5 Numeric Example

In the specific case shown in Figure 4.3, *mini-max* has evaluated the entire game tree from Figure 4.2, having moved backward from the terminal nodes all the way up to the root. We can see that the initial state has three children, **b**, **c**, and **d**, and that among these, the value of **c** is **1** while the values of the other two states are both $-\mathbf{1}$. The interpretation is that, at the initial state, the agent is winning, and its first action will be the one that leads to **c**.

4.2.6 Backed-Up Value

We have seen that *mini-max* gives each state the value obtained as the maximum or minimum of its children. The value thus obtained is called the *backed-up* value, the name being derived from the mechanism that starts with the values of the terminal nodes, and then back-propagates these values up the game tree.

Control Questions

If you are unable to answer the following questions, return to the appropriate place in the preceding text.

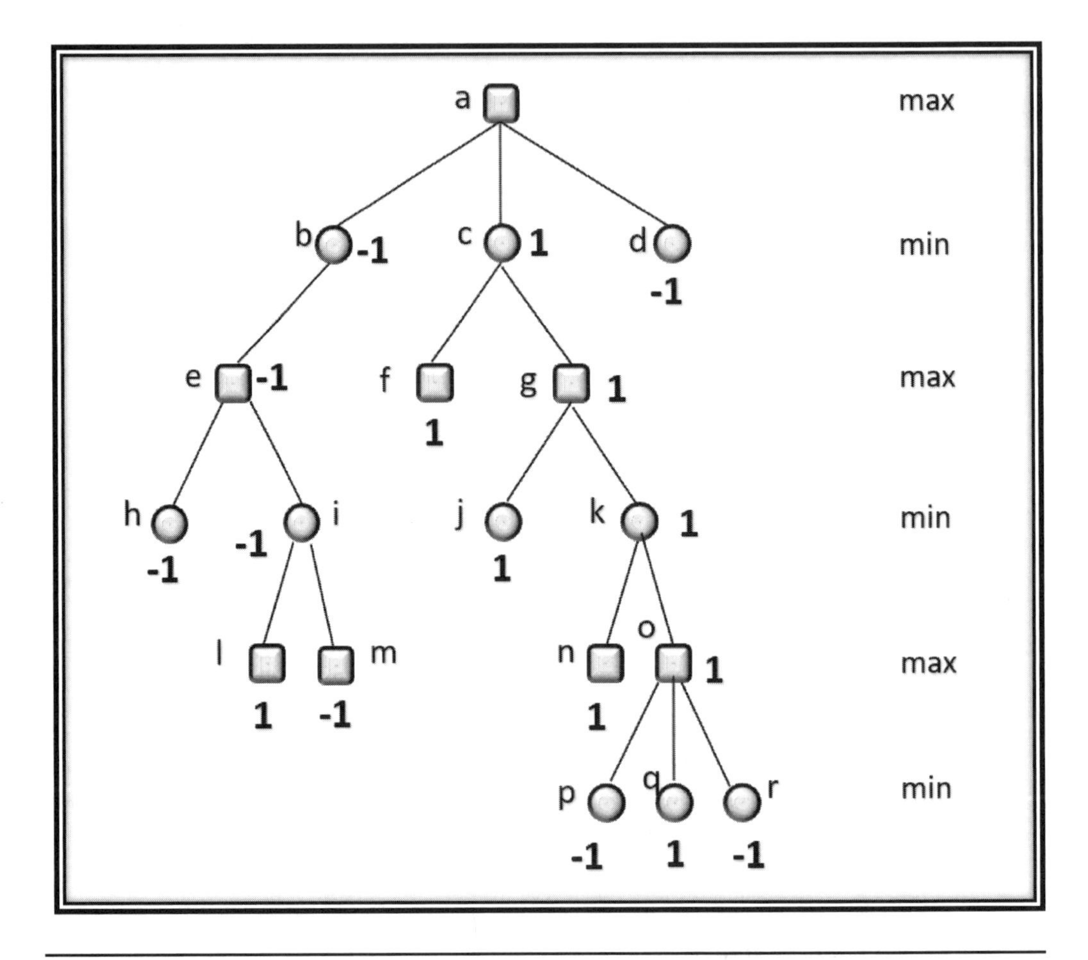

FIGURE 4.3 The same game tree where the values of the terminal states have been backpropagated all the way up to the root node whose value has thus been shown to be 1.

- Why is the agent called *minimizer* and its opponent *maximizer*?
- What is a *game tree*? What does its root represent? How are the leaf nodes evaluated?
- Summarize the principle of the *mini-max*. What is a *backed-up value*?

4.3 Heuristic Mini-Max

The game tree in the previous section was extremely small, which was necessary for the explanation of the algorithm, and for a numeric illustration of its behavior. In any realistic domain, however, the game tree is so huge as to render the baseline version of *mini-max* impractical. This is why heuristic approaches are almost always preferred.

4.3.1 Prohibitive Size of the Game Tree

In a typical chessboard position, the agent can choose from among 30 to 40 moves. The opponent then has the same number of responses. The branching factor is thus very high—and indeed, the number of different chess games allowed by the game's rules has been estimated as around 10^{120}. To call this an astronomical number is an understatement. Decades ago, Carl Sagan estimated that the visible universe contains about 10^{80} atoms. This means that each atom in the universe can be assigned some 10^{40} different chess games.[2]

Chess is complicated; other games are simpler. But knowing that the size of the game tree grows exponentially with its depth, we suspect that any non-trivial domain will require a game tree far too large ever to be stored in a computer memory. As a matter of fact, the tree is almost prohibitively large even in the case of tic-tac-toe if it is to be played on a realistically sized board (much larger than the 3-by-3 from Figure 4.1).

4.3.2 Depth Has to Be Limited

The combinatorial intractability of the baseline *mini-max* compels the programmer to constrain the depth which the exploration of the game tree is to be allowed to reach. Instead of expanding the tree all the way down to its leaf nodes, practical implementations stop at levels that can be reached at reasonable costs. For instance, chess-playing programs rarely analyze more than a few moves from the current position.

4.3.3 Evaluation Function in Adversary Search

Similarly as in the heuristic search from Chapter 3, game-playing programs rely on evaluation functions that accept as input a state description, and return an estimate of the state's value.

This idea was for the first time proposed as early as by Shannon (1950), a truly pioneering attempt to suggest a way of implementing chess in computer programs. In his paper, he devoted considerable space to the design of the evaluation function. The first thing to consider, Shannon said, is the value of the material in the given chessboard position. Each piece is associated with a certain integer: a queen has 9; a rook, 5; a bishop and a knight, 3; and a pawn, 1. The sum of the values of the pieces on the board is then increased or decreased by such positional factors as the castled king, control of the center, free diagonals for bishops and free ranks for rooks, support points for knights, and so on. All these factors find their way into an evaluation formula. In professional programs, this formula can be fairly complicated.

4.3.4 Where Do Evaluation Functions Come From?

In a student term-project, the position-evaluating formula is usually created manually, being based on the student's intuition of what really matters—and how much it matters. The program's playing strength can then be increased by tweaking diverse coefficients and exponents of the formula in response to trial-and-error testing.

---

[2] The number of different chessboard positions is smaller than the number of possible games. 10^{64} is sometimes mentioned as a realistic upper bound.

Authors of commercial software used to consult masters and grandmasters, while also relying on extensive experimentation. In the last decades, impressive improvements have been achieved by the use of machine learning. These techniques, however, belong to a different textbook.

4.3.5 Heuristic *Mini-Max*

When evaluating the position represented by the root of the game tree, heuristic *mini-max* expands the tree only down to a certain depth, evaluates all nodes at this depth, and backs-up the obtained values all the way up to the root. Having thus obtained all the necessary information, the agent will always know which action to take (in the employed part of the game tree): the one that leads to the highest-valued state. Once all of these moves have been made, the final state that has thus been reached becomes the root of a new game tree, and the whole procedure is repeated.

Of course, this is just the basic principle. Practical implementations are usually much more flexible than this. Let us briefly introduce a few ideas employed in commercial software.

4.3.6 What Affects Playing Strength

Section 3.2 introduced the look-ahead strategy for state evaluation in heuristic search. The further the strategy looked, the more reliable the state's evaluation. The same argument can be made in game tree evaluation. Deeper analysis yields more reliable results—at the price of fast-growing costs.

Another important factor is the quality of the evaluation function itself. If a bad state has been incorrectly evaluated as good, the game-playing program is misled. In those attempts where the evaluation function has been created manually, inspired only by the programmer's fallible intuition, its quality usually leaves a lot to be desired. However, the function can later be improved by experimentation and parameter tweaking.

Ultimately, the program's performance depends on a combination of these two factors: We want the maximum depth of evaluation, and we also want the best evaluation function.

4.3.7 Flexible Depth of Evaluation

The depth of the states to be evaluated does not have to be constant. Realistic applications commonly adjust the depth to momentary circumstances, pursuing some of the game tree's branches to much deeper levels than others. Decisions about the concrete depth of analysis rely on various rules of thumb that often work remarkably well. Section 4.5 will take a look at some of the most typical ones.

4.3.8 Evaluations Can Be Expensive

There are two reasons why the evaluation of a game state can be expensive. One of them has already been indicated: Realistic game trees tend to be large, with high branching factors. An attempt to evaluate a state by backing-up the values of states many layers deeper often means that millions of states (or even more) have to be explored. Even if the evaluation of a single state is relatively cheap, the necessity to evaluate so many states may give rise to prohibitive costs.

Second, the evaluation function itself can be expensive. In some domains, the only way to evaluate a state is to run time-consuming experiments. For instance, one can make the computer play a series of chess games (against itself or against other programs), and adjust the evaluation according to the final score. Such costs are high even if the backup analysis does not go very deep.

4.3.9 Success Stories

For the general public, the 1997 defeat of the world champion in chess was a watershed achievement. The triumph was preceded by others, including grandmaster-strength checkers-playing programs. Scientist also valued highly the computers' performance in Othello, a game that is said to have been invented specifically for testing adverse-search techniques.

On the other hand, world-championship performance in backgammon or Go are less relevant here. In these games, the successes have to great degree been achieved thanks to machine learning, especially its sub-fields known as *reinforcement learning* and *deep learning*. While search techniques are being used here, too, their importance for the final success is only secondary.

Control Questions

If you are unable to answer the following questions, return to the appropriate place in the preceding text.

- Why do we almost never develop the entire game tree? Explain the motivation for heuristic evaluation functions. How are these functions created?
- What are the two main factors that affect the playing strengths of programs that rely on heuristic *mini-max*?

4.4 Alpha-Beta Pruning

So far, we have pretended that to backup state values, the entire available part of the game tree has to be analyzed. This, however, is not necessary. Entire branches can be ignored because they do not affect the agent's choice of action anyway. These "superfluous" branches are identified by a mechanism known as *alpha-beta pruning*.

4.4.1 Trivial Case

In Figure 4.3, we observed that, in the root state **a**, the agent is winning because it will choose the action leading to state **c** which has been evaluated as **1**. Note that our opinion about **a**'s value is unaffected by the value of **d**. If that state is worse than **c**, then the maximizer will not go there. If it is as good as **c**, the value of the root node does not change, either.

We see that a state's value is sometimes determined just by a subset of its children. This indicates one way of reducing computational costs: having evaluated **b** and **c**, we can forego the (possibly expensive) evaluation of **d**. However, there is more to it.

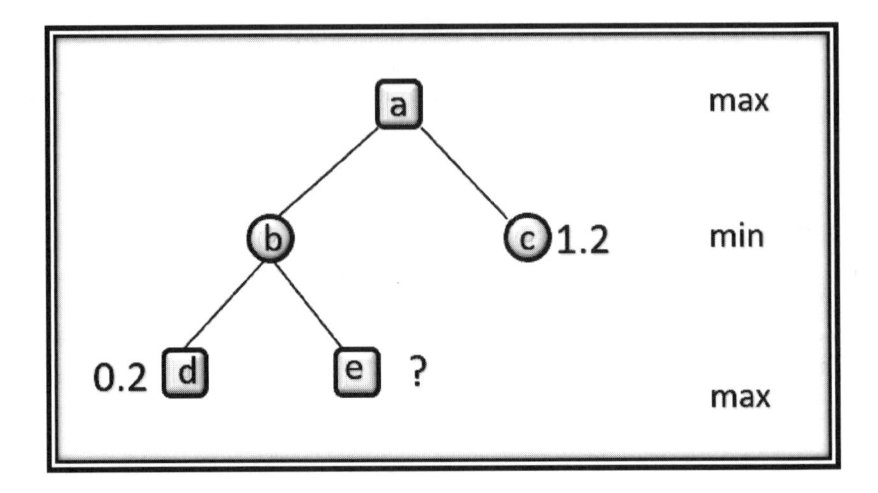

Figure 4.4 In the initial state **a**, the agent will choose the action leading to **c**, regardless of the value of **e**.

4.4.2 Superfluous Evaluation

Take a look at the small game-tree segment in Figure 4.4. Two of the nodes, **c** and **d**, have been evaluated, perhaps by backing-up the values of some huge subtrees underneath each of them (these subtrees are not shown here). If we know these two values, do we also need to run the possibly expensive evaluation of state **e**? Let us take a look.

In state **a**, the maximizer has to decide whether to choose the action leading to **b** or the action leading to **c**. Whereas **c** has been evaluated as 1.2, the value of **b** is not yet known. **b** has two children, **d** and **e**. The value of **d** has been found to be 0.2, which is already worse than the consequence of the action leading from **a** to **c**. In this sense, the value of the other child, **e**, is irrelevant because 0.2 is bad enough. If **e**'s value happens to be higher than that of **d**, the opponent (minimizer) will prefer to go to **d**; if it happens to be lower than that of **d**, then this would only made things worse, which confirm that the maximizer will be better off by choosing at **a** the action leading to **c**.

4.4.3 Another Example

Let us proceed to the slightly more complicated case from Figure 4.5. Here, we want to know whether it is necessary to evaluate state **d** if the values of nodes **c**, **f**, and **g** are known.

To begin, let us obtain the backed-up value of **e**. Since this state is maximizer, its value is 0.2 because $0.2 > -0.5$. This is already worse than the value of **c**, which is 1.2. We conclude that the maximizing agent, finding itself in **a**, will prefer to go to **c**. This is because the value of **e** is bad enough to prevent the maximizer from following the branch beginning with **b**. Evaluation of **d** cannot change this conclusion: The situation at **b** can only get worse than 0.2, another good reason to avoid this state. We conclude that the possibly expensive evaluation of state **d** is unnecessary.

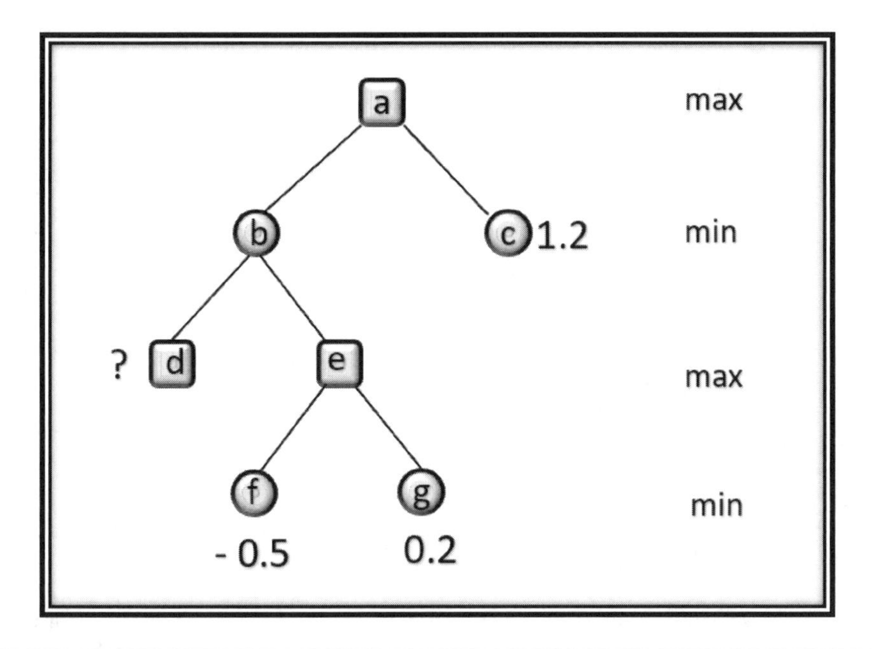

Figure 4.5 In the initial state **a**, the agent will choose the action leading to **c**, regardless of the value of **d**.

4.4.4 Towards the Pruning Algorithm

The last two examples have convinced us that the game-playing program does not have to explore the entire game tree. This observation can lead to considerable computational savings. Practical experience shows that, in many applications, only a small part of the game tree actually needs to be processed, sometimes less than 10%. We see that we would benefit greatly from an algorithm that might carry out this particular analysis automatically and then tell us which branches to ignore.

4.4.5 *Alpha-Beta* Pruning

Our goal is to figure out which states, being unnecessary for the analysis, can be eliminated from the game tree; in other words, how to *prune* the tree. The term *pruning* is commonly used in connection with mechanisms that exclude from further investigation some of the nodes.

How to identify the nodes that can be ignored? In the previous examples, an action by the minimizer led to a state that was worse than what the maximizer would have gotten by choosing another action earlier in the game tree. Thus in Figure 4.4, the value of state **d** was bad enough to suggest that, in the initial state, the agent should go to **c** regardless of the value of state **e** whose investigation could thus be ignored.

Considerations of this kind have inspired an algorithm known as *alpha-beta pruning*. To decide whether to prune a node **n**, the agent takes a look at the path from the root to **n**. This path consists of a series of alternating actions by the minimizer and maximizer.

Input: Initial position.
Evaluation function that returns the values of states.

1. Create a game tree down to a certain depth level and establish the values of the nodes at this level. Use *mini-max* to back propagate the values up the tree.
2. Let **n** be a node where it is the *maximizer*'s turn (e.g., **i** in Figure 4.6).
3. Let **s** be **n**'s sibling, and let its backed-up value be v_s (in Figure 4.6, the sibling is **h**, and its value is $v_h = -0.5$).
4. On the path $[p_0, p_1, \ldots, p_k]$, the *maximizer* is to move in nodes with even indexes, and the *minimizer* in nodes with odd indexes.
5. Any minimizing node that has a sibling with a backed-up value greater than v_s can be pruned.

TABLE 4.2 Pseudo-Code of *Alpha-Beta* Pruning

Alpha-beta pruning considers the siblings of the states on this path. In the case of a minimizer, this sibling's backed-up value may be greater than that of **n**. When this happens, the sibling can be pruned. This is what we observed in Figures 4.4 and 4.5.

The pseudo-code in Table 4.2 summarizes this approach to pruning. For the reader's convenience, the individual steps are illustrated by the tree in Figure 4.6.

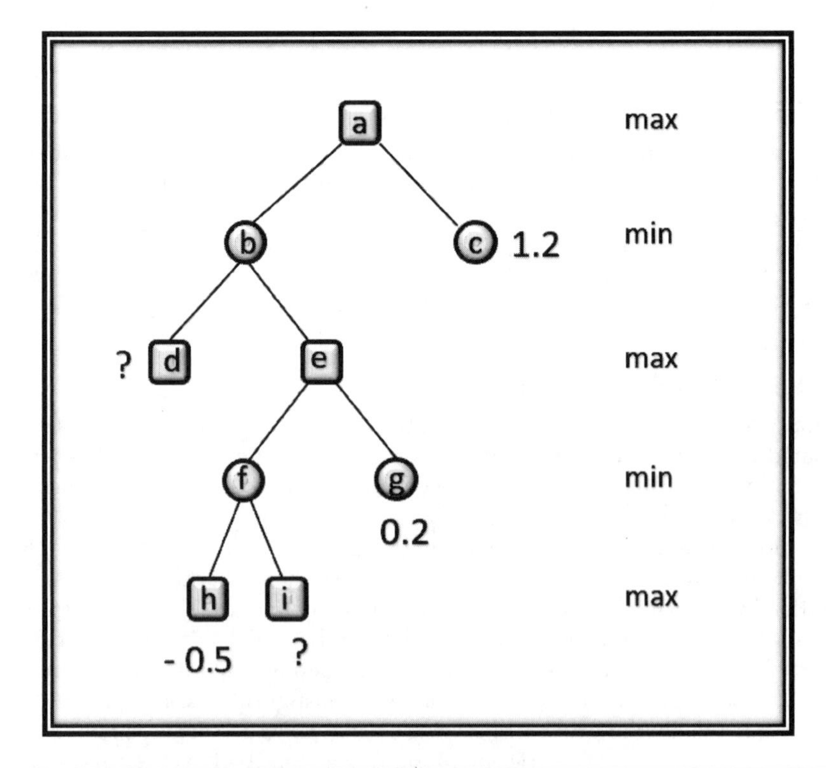

FIGURE 4.6 A game tree to illustrate the algorithm of *alpha-beta pruning*.

4.4.6 Opposite Approach

The algorithm just described focuses on the values observed in the states where it is the *minimizer*'s turn. The reader will find it easy to formulate the analogous criterion to be used in states where it is the *maximizer*'s turn.

Control Questions

If you are unable to answer the following questions, return to the appropriate place in the preceding text.

- We have learned that it is unnecessary to know the values of all states in the game tree. Why is this the case?
- Explain the principle of *alpha-beta* pruning.

4.5 Additional Game-Programming Techniques

Commercial game-playing programs do not rely just on *mini-max*, evaluation functions, and *alpha-beta* pruning. Their strengths are usually further increased by diverse additional techniques. Let us briefly characterize at least some of them.

4.5.1 Heuristics to Control Search Depth

In simple implementations of heuristic *mini-max*, the depth of the exploration is fixed. This, however, is no dogma. More sophisticated programs are less constrained because their authors know that some branches are more promising than others and as such deserve to be explored deeper. This begs the question: How to recognize these more-promising branches and how much deeper to pursue them?

Here is one idea. Let us denote by $e(n)$ the value returned by the evaluation function for node n, let the series of nodes along a specific branch emanating from the game-tree root be $n_1, n_2, \ldots$, and let the values of these states be $e(n_1), e(n_2), \ldots$. If we observe dramatic changes of these values from one state to another, then the branch appears to follow a "tactical variation" whose final outcome is uncertain and as such deserves deeper analysis. If, conversely, the changes observed in this series are barely perceptible, the branch represents a "strategic variation" where the analysis does not have to go too deep.

4.5.2 Peek Beyond the Horizon

Suppose a game tree has been explored using the variable-depth search from the previous paragraph, so that the program has already determined the sequence of actions to be taken. Suppose that these actions result in node n. Before executing these actions, it is usually a good idea to perform a *secondary search* that goes a bit deeper beyond n, something we may call a "peek beyond the horizon." The idea is to prevent unpleasant surprises (under-performing children) lurking beyond n.

4.5.3 Opening Book

In the early stage of a complicated game, it is virtually impossible to form an objective opinion about the quality of this or that move. Put more formally, the qualities of early states are hard to establish by any evaluation function, and the actions have to be chosen by other means. Recall the *magic square* from Section 3.5. The need for expensive search was eliminated by preliminary analysis that revealed that the first action should place **5** on the central square.

Something similar is observed in chess and other games. Experts know that certain opening moves result in good positions, whereas others precipitate disasters. Experienced players take advantage of the knowledge accumulated by generations of theoreticians who, having analyzed thousands of master games (and knowing the results of these games), have put together manuals and textbooks of opening schemes. Grandmasters are known for their prodigious memory that helps them memorize thousands of opening variations. They follow them almost automatically, without second thought. The knowledge of opening variations improves their performance considerably.

Any computer's memory dwarfs that of the most talented grandmasters. The machine is capable of remembering not thousands, but millions of variations stored in an *opening book*, a look-up table that not only helps avoid early-stage blunders, but suggests moves that lead to favorable middle-game positions where the player "knows what to do." On one famous occasion, in 1997, the opening book was targeted at a concrete player, world champion Garry Kasparov. Partly thanks to the well-designed opening book, the computer could for the first time win a match against the strongest human player on the planet.

4.5.4 Look-Up Tables of Endgames

In the last stage of a chess game, the end-game, only a few pieces remain on the chessboard. The game tree that starts from a simple position of this kind is still impractically large for any real-time analysis. In the 1990s, however, computers were already powerful enough to allow chess programmers to run (off-line) exhaustive analyses that reached the point where for each five-piece endgame position, a complete game tree could be developed. In such a complete game tree, each leaf could be labeled with `1`, `0`, or `-1` for won, drawn, or lost (we saw one example in Figure 4.2).

As a result, they had at their disposal very large look-up tables that for any five-piece position specified the best move, the one guaranteeing the fastest path to the player's goal (win or at least draw, if possible). Once any of these positions was reached, the computer no longer needed adversary search; it simply found the position in the database and played the moves that the database recommended. The much faster computers of the twenty-first century made it possible to extend the database to all six-piece positions.

The same principle can be followed in other applications, too.

4.5.5 Human Pattern-Recognition Skills

Heuristic *mini-max* is a simple mechanism that allows us to implement game-playing skills in computers. Some theoreticians have gone so far as to suggest that human chess players calculate variations in much the same way. This, however, is only partially true; the main difference between computers and humans is that the latter can take advantage of their pattern-recognition skills.

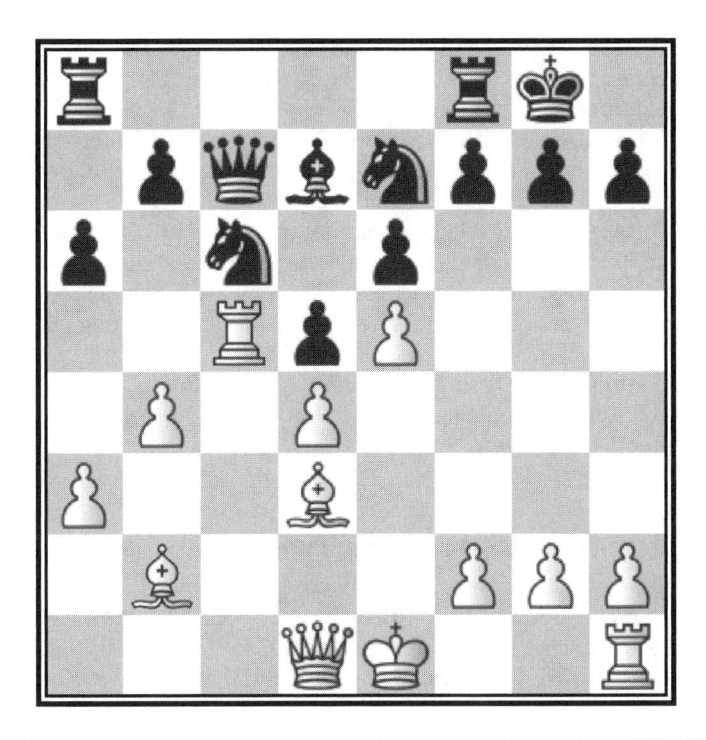

FIGURE 4.7 Human players rely on pattern-recognition skills. In this position, an experienced player immediately sees the possibility of the Bishop sacrifice at h7.

Thus in the chessboard position from Figure 4.7, any experienced tournament player immediately sees the possibility of the Bishop sacrifice on h7.[3] For this reason, the master will start his or her analysis with this idea in mind, and will check whether all variations following this sacrifice lead to checkmate or at least to an otherwise won position. In the process, he or she will largely ignore all other possibilities, planning to return to them only if the current search does not lead anywhere.

By contrast, the computer program only calculates and calculates, checking all variations allowed by chess rules without any special preferences. This is wasteful. In the given position, White can choose from 30 to 40 different moves, each allowing 30 to 40 responses from the Black, and so on, which leads to an incredibly large game tree. This is why, until the 1990s, computers were unable to reach the strength of leading grandmasters. In this particular position, at least 10 moves are needed to land the checkmate, and slower computers simply could not get that deep even after hours, or days, of analyses.

4.5.6 Human Way of "Pruning"

In the nineteenth century, when this particular game was played, Paulsen's ability to discover the Bishop sacrifice over the board was hailed as a mark of genius. Today, you

[3] The position is from the game Paulsen–Schwarz, played in Leipzig, Germany, in 1889.

do not have to be a master. Any average expert will find the move. In the course of his studies, he has encountered many similar sacrifices, and is thus quick to notice the tell-tale signs: locations of the white-squared Bishop and Queen, the Knight's readiness to jump to g5, plus the ease with which the c5-Rook can be transferred to participate in the attack on the opponent's king. Seeing all this, many players will begin their analysis by exploring the chances of this very sacrifice.

Classical AI search develops the game tree, prunes it, evaluates the states down to a certain depth, and then backs up the values. From the human perspective, this does not look smart. Most of the moves investigated by the computer are quite unreasonable. Indeed, why should one even consider any variation beginning with moving the King to the right, Ke1-f1?

Considerations of this kind led many scholars to suspect that the brute force of blind number-crunching may give rise to strong game-playing programs, but these can scarcely be called intelligent.

4.5.7 Pattern Recognition in Game Playing

For a professional player, specific arrangements of pawns and pieces suggest concrete plans (say, a pawn attack on the castled king position). Masters seek to execute their own plans while at the same time doing their best to interfere with the likely intentions on the part of their opponents. Besides, in many positions, a tactical variation is indicated—as in the game from Figure 4.7. Professionals have mastered thousands of such patterns, and these instruct their thought processes accordingly.

Attempts to implement this goal-oriented approach in computer programs were reported as early as in the 1970s.[4] As the speed of computers increased by orders of magnitude, however, number-crunching seemed to be gaining the upper hand. It was only in the twenty-first century that programmers found ways of implementing pattern-learning and patter-recognition in such games as Go or chess, using to this end techniques known from *reinforcement learning* and *deep learning*. These, as already mentioned, do not belong to *Foundations of AI*.

Control Questions

If you are unable to answer the following questions, return to the appropriate place in the preceding text.

- Explain the techniques of variable-depth search. Why are they important? What is meant by the term, *secondary search beyond the horizon*?
- What is an opening book and what are the benefits of endgame look-up tables?
- Discuss the limitations of the *mini-max* approach in comparison with human pattern-recognition skills.

[4] Most famous were the efforts of the team led by ex-world champion M. Botvinnik.

4.6 Practice Makes Perfect

To improve your understanding, take a chance with the following exercises, thought experiments, and computer assignments.

- Consider the game tree from Figure 4.8. Back up the values from the leaf nodes all the way up to the root node. Then answer the following questions:
  1. Once we have established the values of **b** and **f**, is it necessary to determine the value of **g**?
  2. Once we have established the values of **j** and **n**, is it necessary to determine the value of **o**?
- Having analyzed the game tree from Figure 4.6 (Section 4.4), we realized that the agent did not have to investigate state **d**. What would the values of some of the nodes have to look like to make us decide that evaluation of **d** *is* necessary?
- Section 4.5 mentioned that professional game-playing programs often rely on *opening books* and on endgame look-up tables. How would you go about creating these two tools for tic-tac-toe played on the 3-by-3 board?
- Write a program to play tic-tac-toe. Find appropriate representation of states and implement the game-playing program. If only the 3-by-3 board is used, exhaustive search is possible: One can expand the entire game tree and back up the leaf-node values all the way up to the initial state. Consider speeding up the process by alpha-beta pruning.

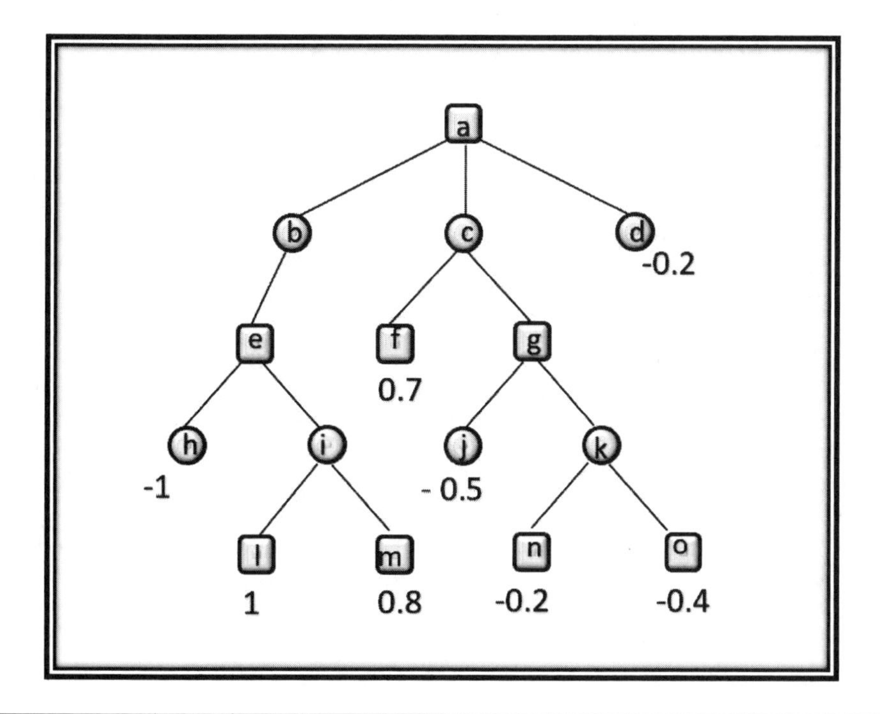

FIGURE 4.8 An example game tree for practicing.

- Proceed to the tic-tac-toe played on a larger board, say, 5-by-5 or 7-by-7. In these, exhaustive search is no longer practical. Suggest an appropriate evaluation function and write a program implementing *heuristic mini-max*.
- The number of different states in tic-tac-toe can be significantly reduced by exploiting the game board's various symmetries. For instance, the game's layout does not change if we rotate the board. Suggest a list of such search-space reducing mechanisms.
- Suggest a way to implement a game-playing program aiming at something more advanced, say checkers or Othello.
- If you are a chess player, write a two-page essay about how the human player's way of thinking differs from the wasteful classical AI approach.

4.7 Concluding Remarks

Claude Shannon (1950) was the first to outline the general principle of what is now known as adversary search. He does not seem actually to have implemented the program, at least not at the time of publishing his paper. But the fact that he was able to formulate these ideas in the days when computer science had scarcely been born is noteworthy. He anticipated the need for evaluation functions, and even offered a few ideas of what such a function might look like in chess.

At about the same time, a criterion was suggested for deciding whether machine intelligence has reached performance comparable to that of the human brain: A computer must beat the world champion in chess.[5] This looked like a severe test indeed, and quite a few thinkers doubted AI would ever pass it.

The victory of IBM's Deep-Deep Blue in a match with Garry Kasparov in 1997 (even though only by a narrow margin) came as a bombshell. The ultimate test had been passed! This said, the event raised serious questions. Particularly disturbing was the realization that the machine had to evaluate, on average, billions of chessboard positions before deciding on a move. The grandmaster's playing strength was not significantly inferior, and yet the number of positions he evaluated per move was lower by orders of magnitude—a few dozen, perhaps a few hundreds, but certainly not more than that. In the light of this sobering observation, claims about the chess-playing program's intelligence remained unconvincing.

The world had learned that the colossal number-crunching power might give rise to feats that *resembled* intelligence, but human way of thinking was of a very different category, relying as it did on insight and pattern-recognition skills. How to compare the intelligence of entities of such disparate nature? In the 1950s beating a world champion seemed almost unachievable. Once the milestone had been reached, people came to suspect that things were not so simple.

On the positive side, the very dissimilarity of the two worlds, human and technological, may hide great promise. If we find a way to marry the machine's massive number-crunching power with human pattern-recognition skills and insight, undreamed-of vistas may open up.

[5] This suggestion seems to have been made by Shannon himself.

CHAPTER 5
Planning

The search techniques from the previous chapters have been deployed in domains that are more practical than simple puzzles. A broad range of applications falls under the rubric of *planning*. Here, the goal is to optimize sequences of actions in engineering processes: in designing an efficient assembly line, optimizing a robot's behavior, or developing useful AI software.

The field is so rich that an entire book would be needed if we wanted to do justice to all its major aspects. Being limited in space, we will narrow our attention to *deterministic* planning where the outcome of each action is unaffected by chance or by an opponent's interference. The core of this chapter focuses on STRIPS, a legendary approach that has been around since the 1970s. Its essence is explained on the simple blocks world, but the approach is easily generalized to much broader settings.

To persuade the reader that planning applications are more than useful, the chapter then goes through a few model applications that range from the popular traveling salesman to the knapsack problem to job-shop scheduling.

5.1 Toy Blocks

The problems facing AI planning are easily illustrated on a domain whose simplicity and clarity makes it popular among teachers at both undergraduate and graduate levels: the blocks world. A typical task is to find a sequence of actions capable of arranging a set of blocks in a predefined way.[1]

5.1.1 Moving Blocks Around

Consider a domain with just three cubes, such as those in Figure 5.1. The task is to convert the initial state on the left to the final state on the right. Only one block can be moved at a time, and we do not care about the distances between the columns.

On the surface, this looks trivial, and the reader no doubt immediately sees the solution: put **B** on the table and then **C** on **A**. The reason this appears easy is that an intelligent being subconsciously carries out an "analysis" that eliminates unpromising moves. For instance, we know it would be silly to begin, in the initial state, by removing **A** from the

[1] This author first got acquainted with this domain when reading the excellent (even if now outdated) textbook by Ginsberg (1993).

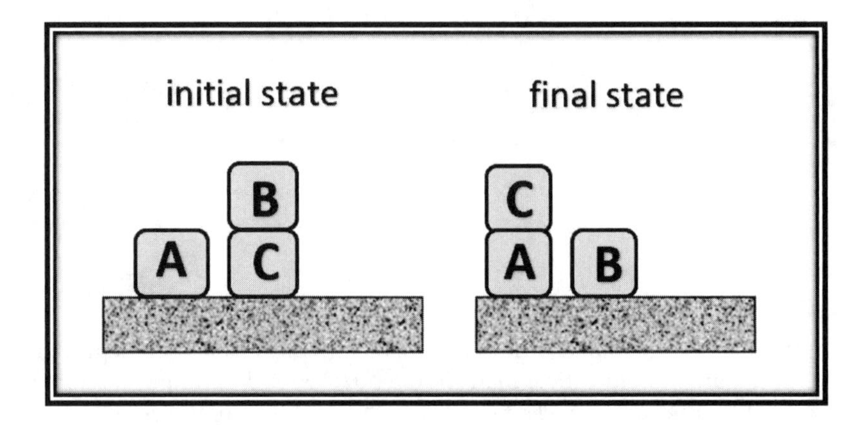

Figure 5.1 Find a sequence of actions which, by moving one block at a time, will convert the initial state into the final state.

table and placing it on top of **B**. We would like the machine to be able to take the same view.

We are dealing here with a classical search problem characterized by an initial state, final state, interim states, and search operators. Let us take a look at one popular way of implementing it in a computer program.

5.1.2 Descriptors

To begin with, we need a set of *descriptors* to characterize each given state. The descriptors should provide all the information that is needed to decide whether an action can be carried out, in this state, and what will be the action's consequences. Recall this domain's chief constraint: At each time, only a single block can be moved. This means that a block is a legitimate candidate for relocation only if nothing is sitting on its top. Our first descriptor will therefore tell the agent whether the top of block `x` is clear:

`clear(x)` (5.1)

For each block, the agent also needs to know where it is currently located, whether on the table or on another block. This motivates the following descriptor which states that `x` is located on `y`:

`loc(x,y)` (5.2)

In this, as in the previous descriptor, the variable `x` can be instantiated[2] to any concrete block: `A`, `B`, or `C`. The second variable, `y`, can be instantiated to any location, which can be a block, `A`, `B`, `C`, or the `table`. Besides, we require that the instantiated value of `y` be different from that of `x`.

[2]By *instantiation* we mean here replacing a variable with a constant.

5.1.3 Examples of State Descriptions

Looking at the initial state in Figure 5.1, we realize that it can be characterized by the following list of descriptors:

$S_I = \{$`loc(A,table),loc(C,table),loc(B,C),clear(A),clear(B)`$\}$

Starting from this initial state, we want to find a sequence of actions that convert S_I into the final state, S_F, that is characterized as follows:

$S_F = \{$`loc(A,table),loc(B,table),loc(C,A),clear(B),clear(C)`$\}$

5.1.4 Comments

The concrete choice of descriptors is the programmer's responsibility. In our specific case, creating them was easy. Two were sufficient: `clear(x)` and `loc(x,y)`. In more realistic settings, a wide range of options may exist, and a lot will depend on the programmer's experience and skills. Some constraints have to be respected, though. The descriptors must make it possible to characterize any possible state unambiguously and in a way that facilitates decisions about the legality of concrete actions.

Note that some descriptors may be redundant. Thus in the simple domain from Figure 5.1, one may decide that any block whose location is not given, will be assumed to sit on the table (rather than on another block). Also the descriptor `clear(table)` is superfluous because it is telling us that something can be put on the table. If the table is large enough, this will always be the case.

Control Questions

If you are unable to answer the following questions, return to the appropriate place in the preceding text.

- What is a descriptor? Give some examples.
- What rules should an engineer adhere to when designing the descriptors for a given application domain?

5.2 Available Actions

Now that we know how to describe the individual states; the next question is how to characterize the search operators that execute the actions available in each state.

5.2.1 Actions in the Toy Domain

In the domain from Figure 5.1, the search operator removes one block from its current location and transfers it to another location. Let us represent this action by the following term.

$$a = \texttt{move(x,y,z)} \tag{5.3}$$

The interpretation is, "remove block x from its current location on y and place it on z." Here, variable x can be instantiated to A, B, or C, whereas y and z can be instantiated to table or to one of the remaining blocks, A, B, C. Each of the values of x, y, and z must be different.

Note that action a is *generic* in the sense that, to be applicable, it first has to be instantiated to a concrete version. The number of versions represented by the generic action equals the number of its instantiations.

5.2.2 List of *Preconditions*

Before executing an action, the agent has to establish whether the action is *legal*, whether it can be executed, in the given state. For instance, if A is on the table, it cannot be removed from the top of B; this eliminates action move(A,B,C) which would be legal only if the state description contained loc(A,B). We realize that for any concrete action to be legal in a certain state, the state has to satisfy certain preconditions that are specific for this action. These preconditions take the form of a set of descriptors that have to be included in the state's description.

This means that when defining an action, the engineer has to define the action's *list of preconditions*. Thus for move(x,y,z), three preconditions have to be satisfied. First, x can be removed from the top of y only if it is present there. Second, the top of x has to be clear because the agent is allowed to move only one block at a time. Third, the top of the destination has to be clear, too, because the agent cannot place x on top of z if something else is already sitting there. With the help of the descriptors from the previous section, the three preconditions are captured by the following list[3]:

$$P(a) = \{\texttt{loc(x,y)},\ \texttt{clear(x)},\ \texttt{clear(z)}\} \tag{5.4}$$

For an action a to be legal in a state characterized by set S of descriptors, the following condition has to be satisfied:

$$P(a) \subseteq S \tag{5.5}$$

5.2.3 *Add* List

Following the execution of an action, the system's state changes. Section 5.1 suggested that any state be characterized by a list of descriptors. The execution of the action results in the modification of this list. Some items are added, others are removed.

To be concrete, action a = move(x,y,z), which removes x from the top of y and places it onto z, adds to the list two descriptors: one of them specifies that x is now located on z, and the other specifies that the top of y is now clear. This is reflected in the following *add* list:

$$A(a) = \{\texttt{loc(x,z)},\ \texttt{clear(y)}\} \tag{5.6}$$

[3]Strictly speaking, this is not a list but a set: In a list, the order of the items is specified, whereas in a set it is not. In the context of *planning*, however, tradition has always relied on the somewhat incorrect term, *list*. The same applies to the *add* list and *delete* list.

a = `move(x,y,z)`:

$P(a)$ = { `loc(x,y)`, `clear(x)`, `clear(z)` }
$A(a)$ = { `loc(x,z)`, `clear(y)` }
$D(a)$ = { `loc(x,y)`, `clear(z)` }

TABLE 5.1 Definition of the Generic Action `move(x,y,z)`

5.2.4 *Delete* List

Apart from adding some descriptors to the state description, the action causes some of the descriptors to be dropped from it. After the execution of a = `move(x,y,z)`, block `x` will no longer be located on `y`, and the top of `z` will no longer be `clear` because `x` has been placed there. This is reflected in the following *delete* list:

$$D(a) = \{\texttt{loc(x,y)},\ \texttt{clear(z)}\} \tag{5.7}$$

5.2.5 Defining `move(x,y,z)`

Let us summarize. Any action is unambiguously defined by the three lists mentioned above: list of preconditions, $P(a)$, the *add* list, $A(a)$, and the *delete* list, $D(a)$. Table 5.1 summarizes these lists for `move(x,y,z)`. Note that all arguments, `x`, `y`, and `z`, are variables. This makes the definition *generic*. In an action that can actually be executed, the variables have to be replaced with constants.

5.2.6 Instantiation of the Generic Action

The definition in Table 5.1 was *generic* in the sense that it took a general form that included variables. To be able to apply `move(x,y,z)` to a concrete state, the program must instantiate the variables `x`, `y`, and `z` to the available constants, `A`, `B`, `C`, or `table`.

The replacement of variables with constants has to be carried out consistently in each of the three lists that define the generic action. For the sake of illustration, here is one such instantiation:

b = `move(A,B,C)`:

$P(b)$ = {`loc(A,B)`, `clear(A)`, `clear(C)`}
$A(b)$ = {`loc(A,C)`, `clear(B)`}
$D(b)$ = {`loc(A,B)`, `clear(C)`}

Note how easy it is to implement instantiation in a computer program. All that is needed is to replace each variable with the constant in the corresponding place in the list of `move`'s arguments. In this specific example, `x` was replaced with `A`, `y` with `B`, and `z` with `C`.

5.2.7 How Many Instantiations Exist?

Theoretically speaking, one generic action, such as `move(x,y,z)`, represents a whole set of instantiations. Thus for each of the three blocks that can take the place of `x` (i.e.,

`A`, `B`, `C`), there are three instantiations of `y` (i.e., the remaining two blocks plus `table`), and two destinations represented by `z` (i.e., the last remaining block plus `table`). This means that the total number of instantiations is $3 \times 3 \times 2 = 18$. This said, we must not forget that not all of these actions are in the given state legal.

5.2.8 Executing an Action

In the course of the solution process, the agent has to decide which of the available actions can be applied to the given state. Our blocks world has only one generic action, `move(x,y,z)`, which, we have seen, has 18 instantiations. The agent eliminates those instantiations whose preconditions the current state does not satisfy. If S is the set of descriptors characterizing the given state, the instantiated action a can be executed only if $P(a) \subseteq S$.

Executing the action means to remove from S all of the descriptors listed in $D(a)$, and to add to it all the descriptors listed in $A(a)$.

5.2.9 Example

Let us consider action $a =$ `move(x,y,z)` for which Table 5.1 specifies the following list of preconditions:

$$P(a) = \{\texttt{loc(x,y), clear(x), clear(z)}\}$$

Suppose the agent faces a state described as follows:

$$S = \{\texttt{loc(A,table),loc(B,table),loc(C,A),clear(B),clear(C)}\}$$

Let us consider two instantiations, $a_1 =$ `move(A,B,C)` and $a_2 =$ `move(C,A,B)`. Here are their preconditions:

$$P(a_1) = \{\texttt{loc(A,B), clear(A), clear(C)}\}$$

$$P(a_2) = \{\texttt{loc(C,A), clear(C), clear(B)}\}$$

Checking these preconditions against the requirement $P(a) \subseteq S$, we easily verify that $P(a_1) \not\subset S$ and $P(a_2) \subset S$. This means that a_2 is legal in S and a_1 is not.

Control Questions

If you are unable to answer the following questions, return to the appropriate place in the preceding text.

- Explain in what sense the three lists, $P(a)$, $A(a)$, and $D(a)$, define action a. What is the role played by each of these lists?
- What is an action's *instantiation*? How does the computer program create an instantiation from an action's generic definition?
- How does a computer program establish that a concrete instantiation is legal in the given state? How are the two lists, $A(a)$ and $D(a)$, used to modify the state's description?

5.3 Planning with STRIPS

Let us take a look at the legendary AI program known under the acronym STRIPS (standing for Stanford Research Iinstitute's Problem Solver). The reason for introducing it here is not just its historical value. Rather, the program illustrates an interesting and useful alternative of implementing search.

5.3.1 Set of Goals

STRIPS characterizes the final state by a set of goals to be satisfied. For instance, in the domain from Figure 5.1, STRIPS would define the final state by the following goals:

$$Z = \{\texttt{loc(C,A)},\ \texttt{loc(A,table)},\ \texttt{loc(B,table)}\}$$

The agent has reached a final state only when all of its goals have been satisfied; that is, when the current-state description contains all of these descriptors. If we denote the list describing a state by S, then the agent recognizes the state as final if $Z \subseteq S$. Note that the list of goals does not contain those descriptors that can be considered to be redundant.

5.3.2 General Philosophy

STRIPS proceeds backwards. It starts with the final state (the set of goals) and then tries to work it way back to the initial state, keeping a tally of the actions taken in the process. Here is the principle.

To begin with, the agent has to consider the possibility that the final-state's requirements are satisfied already in the initial state. If we denote the list of goals by Z, and the description of the initial state by S_I, then $Z \subseteq S_I$ indicates that no further analysis is needed.

In the opposite case, $Z \not\subseteq S_I$, the agent attempts to identify the last action that could have resulted in the final state—an action that might have added the last missing goal(s) from Z. In the specific case of the example from the previous paragraph, this last action might have added one or more of the following goals: `loc(A,C)`, `loc(A,table)`, or `loc(B,table)`.

Once such an action has been found, STRIPS asks what could possibly have been the state in which this last action could be executed: the penultimate state in the search. This penultimate state is characterized by a modified set of goals; a set that does not contain the goal(s) added by the final action. This having been established, the agent (again) asks if these goals are already present in the initial state, S_I. If yes, the procedure is terminated with success. If not, the same procedure is repeated recursively.

5.3.3 Concrete Example

Let us return to the blocks-world example from Section 5.1. For this particular configuration, Figure 5.2 shows the situation just before the final state has been reached. On the right is the final state. To the left of this final state are three possible penultimate states that would allow the execution of such instantiations of `move(x,y,z)` that would add the last missing goal.

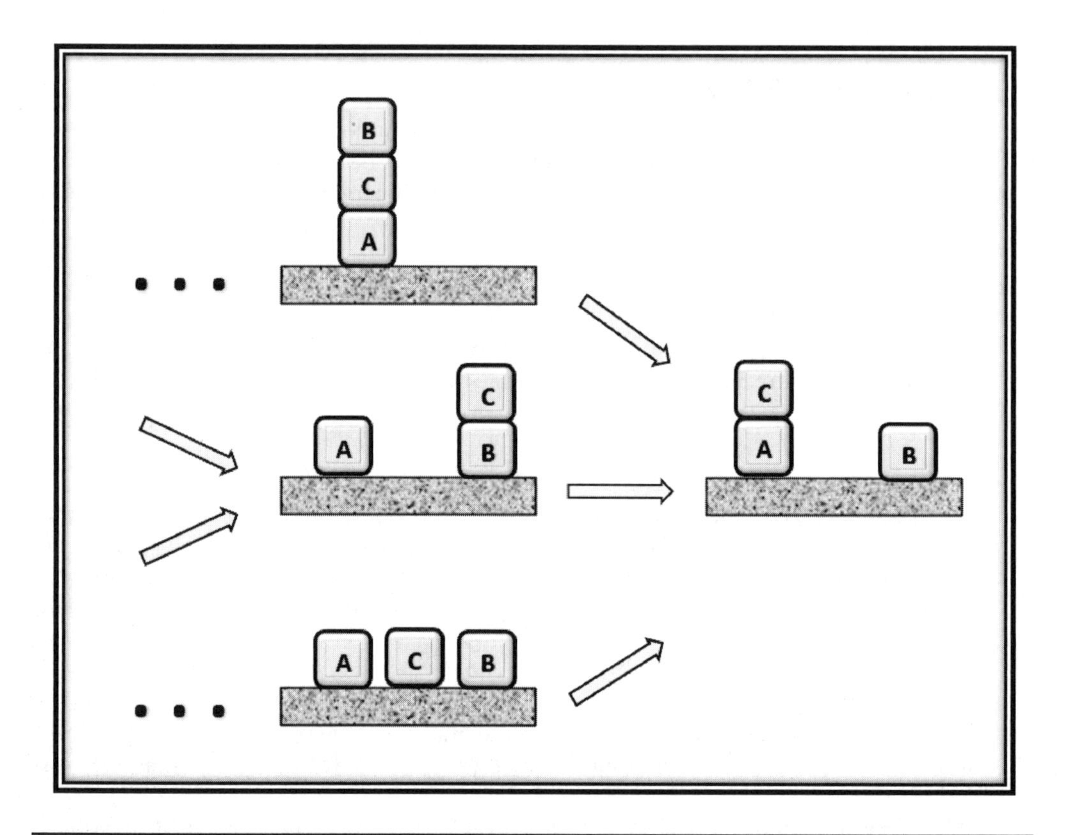

FIGURE 5.2 STRIPS starts from the goals and proceeds backwards, always seeking to identify actions that might have added some goal(s) to the state's description.

Here is one possibility. In the topmost configuration (column of three cubes), `move(B,C,table)` would reach the final state by removing block `B` from the column and placing it on the table, thus satisfying goal `loc(B,table)`. The picture shows two other penultimate states that, similarly, could have allowed appropriate instantiations of the `move(x,y,z)` that would lead to the final state.

None of these penultimate states being identical with the initial state, the agent continues backwards, trying to figure out what actions could have led to the penultimate states, and what the states preceding them could have looked like. The process continues recursively and stops when the initial state has been reached.

5.3.4 How to Identify the Action?

When trying to identify the action that could have completed the final state, STRIPS takes advantage of the lists $A(a)$ and $D(a)$ that were introduced in Section 5.2. Specifically, any action that adds the last missing item to the list of goals has to satisfy the following two conditions:

$$A(a) \cap Z \neq \emptyset \tag{5.8}$$

$$D(a) \cap Z = \emptyset \tag{5.9}$$

The first condition is here to make sure that the selected action adds at least one of the goals listed in Z. The second condition is here to make sure that this action does not remove any of the goals that might have been satisfied previously. STRIPS goes through all possible instantiations of available actions and then chooses those that satisfy both the conditions.

5.3.5 What Does the Penultimate State Look Like?

Suppose the agent has identified an action that satisfies the two conditions from the previous paragraph. This, then, may be the last action in a sequence that results in the final state. But what was the shape of the state to which that final action was applied?

We know that the action could only be executed in a state that satisfies this action's preconditions, $P(a)$. For example, if the final action is `move(B,C,table)`, then the description of the previous state has to contain `loc(B,C)` and `clear(B)` (see Figure 5.2). On the other hand, the description does not have to contain those descriptors that are added by action a—those specified by $A(a)$. Some of these added descriptors may actually be among the goals that define the final state.

Let us characterize the penultimate state by a set of goals, in a way reminiscent of the definition of the final state. Formalizing the previous considerations, we realize that the penultimate state should satisfy goals obtained as follows:

$$Z' = \{ Z \setminus A(a) \} \cup P(a) \tag{5.10}$$

Symbol "\" stands for "set minus." For instance, $\{A, D, F\} \setminus \{A, E\} = \{D, F\}$: the operation removes from the first set all elements found in the second set; in this particular example, the removed element is A.

Formula 5.10 is telling us that the new set of goals, Z', is obtained by removing from Z all descriptors that will be delivered by action a, and then adding those that are required as a's preconditions.

5.3.6 Pseudo-Code of STRIPS

Table 5.2 provides the pseudo-code of a simplified version of STRIPS. The reader recognizes the steps described in the previous paragraphs. At the start, the list, L_A, of actions identified by the algorithm is empty. Each recursion adds, at the beginning of this list, an action that delivers some of the final state's goals. Once the backward search has reached the initial state, the actions are listed in L_A, properly ordered.

The fourth line of the pseudo-code addresses the possibility that the correct sequence of actions cannot be found and the search has to stop with failure.

Control Questions

If you are unable to answer the following questions, return to the appropriate place in the preceding text.

- Explain the general procedure of STRIPS. What is the list of goals? What do we mean by saying that STRIPS is searching backwards?
- What conditions have to be satisfied by an action that is to deliver some of the final state's goals?

Input: List, Z, of goals to be satisfied in the final state.
The description of the initial state, S_I.
Available actions and their instantiations.
Empty list of actions, L_A.

strips(S_I, Z, L_A).

1. If $Z \subseteq S_I$, stop with success, returning L_A.
2. Identify each instantiation, a, that satisfies the following criteria:
   $A(a) \cap Z \neq \emptyset, \quad D(a) \cap Z = \emptyset$
   If no a satisfies these criteria, stop with failure.
3. For each instantiation that satisfies the criteria:
   i. Characterize the previous state by a new list of goals:
   $Z' = \{ Z \setminus A(a) \} \cup P(a)$
   ii. Place a at the beginning of L_A and call, recursively, *strips(S_I, Z', L_A)*.

TABLE 5.2 Pseudo-Code of Simplified STRIPS

- What can we say about the "penultimate" state, the one before the final action has been executed? What descriptors must this state contain?
- When does the STRIPS procedure stop?

5.4 Numeric Example

At first reading, students usually find the STRIPS algorithm a bit confusing. Let us therefore go through a detailed numeric example that clarifies some less obvious details.

5.4.1 Which Actions Should Be Considered?

Let us return to the final state depicted in Figure 5.2. The goals to be met here are summarized as follows[4]:

$$Z = \{\texttt{loc(C,A)}, \texttt{loc(A,table)}, \texttt{loc(B,table)}\} \tag{5.11}$$

Suppose the only available action is `move(x,y,z)`, defined in Table 5.1. At the beginning, the agent creates all instantiations of this action, which also means to specify sets $P(a_i), A(a_i)$, and $D(a_i)$ for each instantiation. The next step then identifies those instantiations that satisfy Conditions 5.8 and 5.9.

The reader will easily verify that in the case of the goals from List 5.11, these conditions are satisfied by the following three instantiations. A quick look at Figure 5.2 will convince the reader that these are indeed the last actions that might have been executed before reaching the final state.

[4] Note that the information about clear tops is here redundant (it is important only in the set of preconditions). This is why such descriptors as `clear(C)` are ignored in Z.

1. a_1 = move(B,C,table)
2. a_2 = move(C,B,A)
3. a_3 = move(C,table,A)

5.4.2 Checking the Lists

For illustration, here are the details of how the first of these actions, a_1 = move(B,C,table), was checked for being eligible. Table 5.1 specifies the *add* list and the *delete* list for the action's generic version. To obtain the instantiation, the program simply replaces variables x, y, and z with constants B, C, and table, respectively, that are found in action a_1. Care has to be taken to make sure the replacements are made in the correct order of arguments. This result in the following pair of lists:

$A(a_1)$ = {loc(B,table), clear(C)}
$D(a_1)$ = {loc(B,C), clear(table)}

Again, clear(table) can be omitted from $D(a_1)$ if the table is large enough always to allow for a block to be placed there. Besides, table is never going to be subject to moving, which means that its clear top will never be investigated in any move command.

We can easily verify that $A(a_1) \cap Z$ = {loc(B,table)}. This is telling us that action a_1, applied to some as yet unknown penultimate state, is capable of delivering loc(B,table), which is one of the items in Z. Satisfaction of the second condition, $D(a_i) \cap Z = \emptyset$, guarantees that a_1 does not remove any of the goals from Z that might have already been present in the penultimate state.

In summary, a_1 has thus been confirmed as a worthy candidate for the last action in the as-yet-unknown sequence capable of converting the initial state to the final state.

By way of a simple exercise, the reader is advised to repeat the same procedure to verify that also a_2 and a_3 satisfy the final action's conditions.

5.4.3 Word of Caution

We humans find it easy to discover these actions. Thanks to our insight and understanding of the problem, we right away "see" the correct action in the picture, our analysis having been carried out subconsciously.

Computers lack the insight. They can only run the programs that instruct them to carry out the requisite algorithms. In the specific case treated here, STRIPS investigates for each action, a_i, whether it satisfies the requirements $A(a_i) \cap Z \neq \emptyset$ and $D(a_i) \cap Z = \emptyset$. If it does, the action can be labeled as a candidate; if it does not, the action is ignored. Unlike humans, the computer does all this mechanically, without any consciousness or intention.

5.4.4 Describe the Previous State

Action a_1 has been identified as one of those that may have resulted in the final state. However, the action could only have been executed in a state that satisfied the action's preconditions, $P(a_1)$. In STRIPS, this previous state (*penultimate* state) is characterized by a list of goals.

Recall that $A(a_1) = \{\texttt{clear(C),loc(B,table)}\}$, $P(a_1) = \{\texttt{clear(B),loc(B,C)}\}$, and $Z = \{\texttt{loc(C,A),loc(A,table),loc(B,table)}\}$. The goals defining the penultimate state are specified by Equation 5.10. Seeing that $Z \setminus A(a_1)$ removes from Z the descriptor `loc(B,table)`, we easily establish the penultimate state's list of goals as follows:

$$
\begin{aligned}
Z' &= \{\, Z \setminus A(a) \,\} \cup P(a) \\
&= \{\texttt{loc(C,A),loc(A,table)}\} \cup \{\texttt{clear(B),loc(B,C)}\} \\
&= \{\texttt{loc(C,A),loc(A,table),loc(B,C),clear(B)}\}
\end{aligned}
$$

Note that Z' characterizes the topmost of the three possible penultimate states in Figure 5.2: the column of the three blocks.

5.4.5 Iterative Procedure

The previous paragraphs illustrated how to create a penultimate state that is defined by the goals listed in Z'. In the next step, the agent checks whether these goals are already satisfied in the initial state. If they *are*, the procedure stops with success; if they are not, the same procedure is repeated recursively, with Z' now becoming the next "final state."

Of course, the programmer must also consider the possibility that the final state can never be reached from the initial state. The stopping criterion has to be formulated accordingly.

Control Questions

If you are unable to answer the following questions, return to the appropriate place in the preceding text.

- What is the difference between the computer's approach to planning and the human approach?
- Hand-simulate STRIPS' solution of the problem from Figure 5.1. By way of a simple exercise, do it not just for action a_1, but also for at least one of the remaining two.

5.5 Advanced Applications of AI Planning

The essence of the STRIPS's procedure was illustrated on toy blocks because this domain is so simple and easy to understand. However, its very simplicity may have sidestepped some of the approach's realistic challenges. Besides, trivial domains sometimes make the student ask the legitimate question, "what is it all good for?" To offer a broader picture, this section presents, and briefly discusses, some more advanced applications of planning.

Since many of these applications share certain common characteristics, it is customary to organize them in groups, each represented by a model that experts know how to deal with. Among the most popular of these models are the traveling salesman, the knapsack problem, and job-shop scheduling. Of course, a single chapter does not have enough space to detail their solutions—although the reader should by now be able to address them by using search techniques. Besides, recent experience indicates that they are

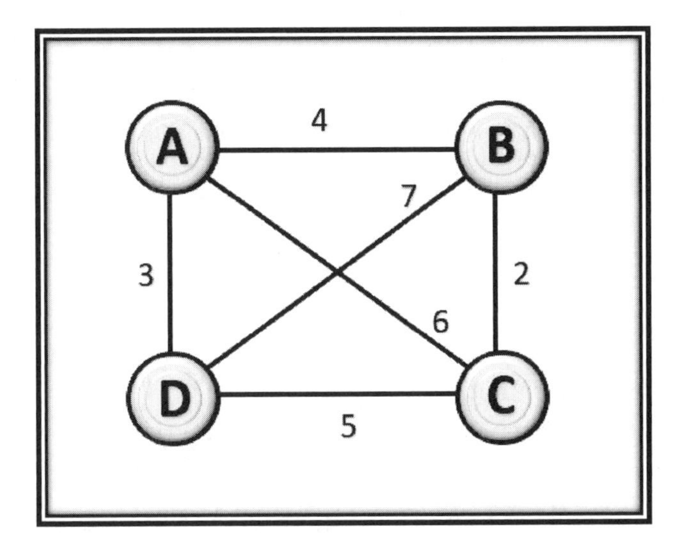

Figure 5.3 Traveling salesman. The nodes are cities, the edges are connections, and the integers are distances. The salesman wants to find the shortest path through all cities. This is computationally challenging if there are dozens of cities or more.

with much more efficiently solved by the techniques from the realm of *swarm intelligence* that will be the subject of Chapter 8. In this place, therefore, we will limit ourselves to just a brief outline of the nature of each of the model applications.

5.5.1 Traveling Salesman Problem

The principle is illustrated by Figure 5.3. The salesman is presented with a set of cities and a table of the distances between any two of them. The task is to find out in which order to visit all cities while minimizing the total distance covered by the salesman.

In the picture, we can see that if the salesman follows the sequence ABCDA, the total distance is $4 + 2 + 5 + 3 = 14$. If the sequence is ACDBA, the total distance is $6+5+7+4 = 22$. What about the other routes? In the case of just four cities, enumerating them all is feasible. But if there are *many* cities, the number of the permutations grows beyond all bounds of practicality; mechanical number crunching is destined to fail.

In artificial intelligence, as well as in some more traditional fields, quite a few methods to deal with this challenge have been worked out—some of them remarkably efficient, others less so. At this stage, the reader will have no difficulty suggesting a solution based on blind search.

5.5.2 Package Delivery and Packet Routing

Traveling salesman offers a general model to represent the needs of a company that delivers many packages to many destinations. For each individual trip, the driver is provided with a list of addresses to be visited in the shortest possible time and with minimum fuel. Each trip is different, and for each, the computer has to find the best path, and it has to be able to do so on a short notice. In the case of a sudden change (e.g., when a new address has been added at the last moment or some existing address dropped), an immediate update is needed.

The program has to do justice to the idiosyncrasies of the specific domain. Suppose a bicycle is used, instead of a car. If location X is in a valley, and Y on a hill, the distance from X to Y is perceived to be longer than in the opposite direction. Besides, the time to reach one city from another need not depend only on distance; what plays a role is also road quality and traffic density, the latter being subject to changes in time.

Problems of similar nature are faced also by internet routers. Each time a packet has arrived, the router has to decide where exactly to forward it. Diverse criteria have to be considered, and here optimization is far from trivial.

5.5.3 Ambulance Routing

Another interesting application is optimization of how ambulance vehicles are being dispatched. At each moment, many vehicles find themselves in different locations, some with patients, others empty. An ambulance with a patient may be close to an Emergency Room (ER), or may need a longer time to get there. The nearest ER may be overloaded (which may cause administrative delays) while a slightly more distant one may have a lot of spare capacity. Besides, hospitals may differ in the available equipment.

Once a call has been received, it is necessary to decide which of the vehicles to dispatch, and to which ER or hospital the patient is to be transported. The primary goal is to minimize time, but all the above-mentioned constraints have to be considered, too. In this sense, the ambulance-routing version of the traveling salesman is fairly complicated, and no easy solution exists.

5.5.4 Knapsack Problem

Consider a set of items, each with certain `weight` and `price`. You want to squeeze as many of these items in a knapsack of a limited size. Alternatively, the knapsack may permit only a certain maximum weight, and there is no way you can take everything because the total weight of all items would then exceed the limit. Either way, some compromises are necessary.

The task for AI planning is to identify a set of items that would maximize the total value of the items carried in the knapsack without violating the limit on their combined weight or size.

5.5.5 Job-Shop Scheduling

Figure 5.4 illustrates the principle of yet another model application. A manufacturer needs to complete four jobs, X, Y, Z, and W, each requiring the completion of certain tasks that have to be carried out in a pre-specified order. For instance, job X in the picture's upper-left corner consists of five tasks, A through E, and the flowchart informs us that A, B, and C have to be carried out in this concrete sequence, while the remaining two tasks, D and E, can be carried out independently of A, B, and C (in parallel with them), but E can only be started once D has been completed. The remaining three flowcharts specify the constraints involved in the other three jobs.

The work is to be carried out on three machines, M1, M2, and M3, each being able to handle only some of the tasks. For instance, the picture informs us that M1 can perform

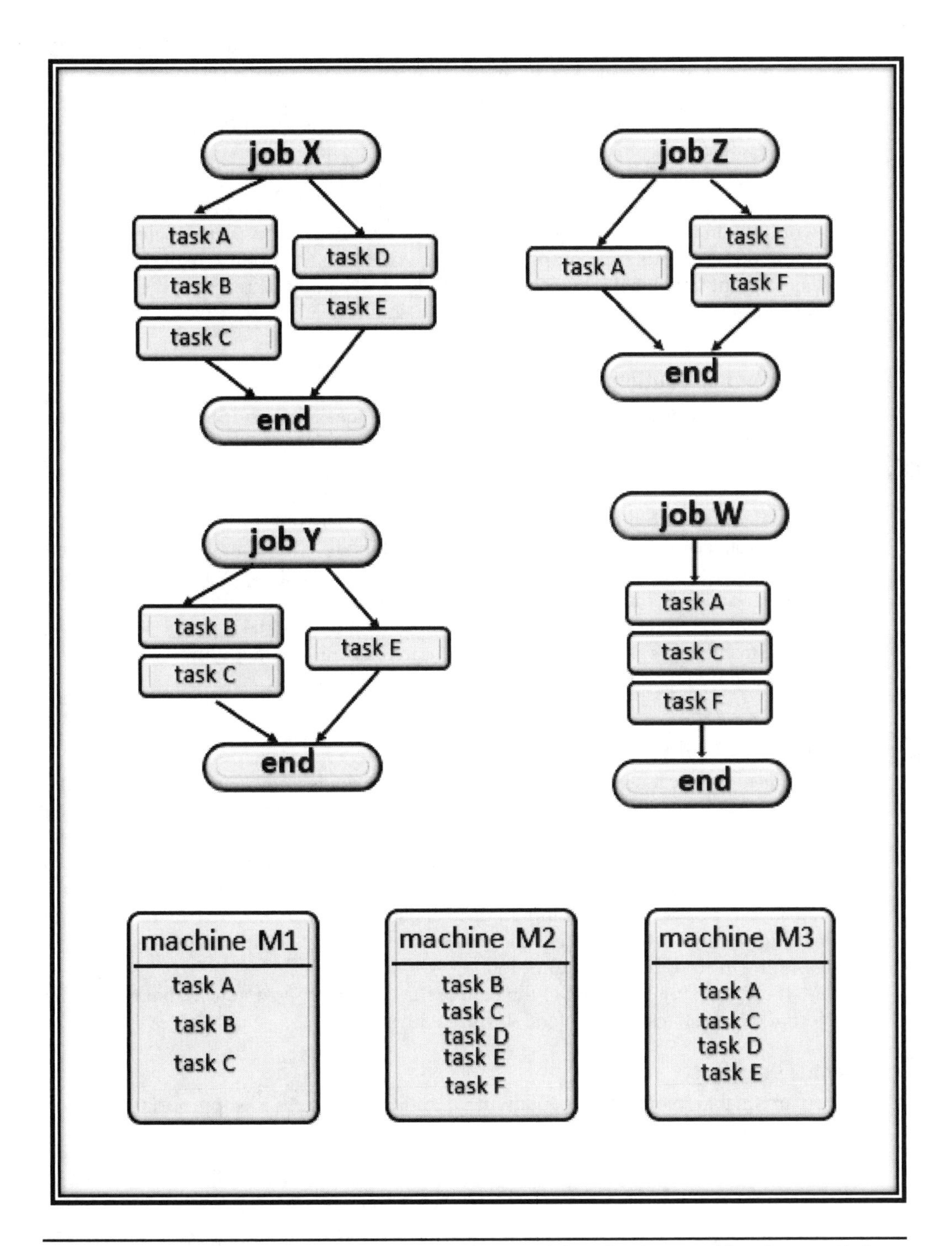

Figure 5.4 The goal is to assign the tasks needed for each job to the three machines in a way that minimizes overall time as well as the idleness of the machines.

tasks A, B, and C, but tasks D, E, and F have to be performed elsewhere: task F on M2, and task E either on M2 or on M3. Each machine can perform only one task at a time. The scheduler is to assign the tasks to the machines in a way that guarantees the completion of the jobs in the shortest possible time, while also minimizing the idle times experienced by the machines.

The scenario just described is known as *job-shop scheduling*. In its simplest version, each job consists of a predefined set of tasks that can be carried out in an arbitrary order (no workflow in the sense of Figure 5.4 has been specified). However, concrete circumstances and constraints can complicate the flowcharts to the point where the problem becomes all but unsolvable. Even if the absolute ideal cannot be reached, the industrial plant needs to find at least a "reasonably good" solution.

5.5.6 Word of Caution

In the early days of AI, many specialists believed that all these problems were best addressed by classical planning algorithms, being ideal targets for search techniques, heuristic or blind. Today we know better. Other approaches (some of them outside AI planning) can deal with these problems in computationally much more efficient ways. Traditional search techniques *are* useful here; but they may not be the best choice.

Apart from classical search, scholars have developed a whole family of modern techniques known as *swarm intelligence*. Their efficiency in the field of planning and scheduling is remarkable. At the very least, they have been shown to be much faster than, say, STRIPS. Moreover, they are easy to implement. No wonder that the new technology soon became so fashionable as to make traditional search techniques look almost obsolete. We will devote to it the entire Chapter 8.

5.5.7 Important Comment

Returning to classical search-based approaches, the reader now understands that the solution process, even in STRIPS, can be considerably expedited if the system is given the chance to benefit from some kind of background knowledge. For instance, the blocks-world problems may rely on the following rules: "start by putting down those blocks that in the final state will be on the table," or "moving blocks that in the final state are on top should be postponed." Human problem solvers subconsciously take advantage of many such rules. It thus makes sense to explore methods of implementing such ways of reasoning in computer programs, too.

Methods of doing so belong to the field of the *knowledge-based systems* that will be the subject of this book's later chapters, starting with Chapter 9.

Control Questions

If you are unable to answer the following questions, return to the appropriate place in the preceding text.

- Explain the principle of the traveling-salesman problem (TSP). Discuss some of its variations.
- Provide examples of real-world applications of TSP and point out the ways in which they digress from its baseline version.

- What is *job-shop scheduling* and which of its features make it particularly difficult? How is the problem related to TSP, and in what way it is different? Discuss some of its variations.

5.6 Practice Makes Perfect

To improve your understanding, take a chance with the following exercises, thought experiments, and computer assignments.

- Consider the domain from Figure 5.5. Suppose you want to write a planning program, based on STRIPS, that would create from these blocks various artefacts, such as a column, an arch, or a gate. What descriptors will help you characterize the problem's states? How will you define the actions? Note that, in this more general setting, some actions will be needed to rotate a block or to make it change its position from horizontal to vertical and vice versa.
- The main difference between the traveling salesman and job-shop scheduling is that the latter also needs to consider time relations. For instance, the job's tasks may have to be executed in a specific order, one after another, and each may need a certain time to complete. How will you modify the STRIPS technique so that it can do justice to this broader formulation?

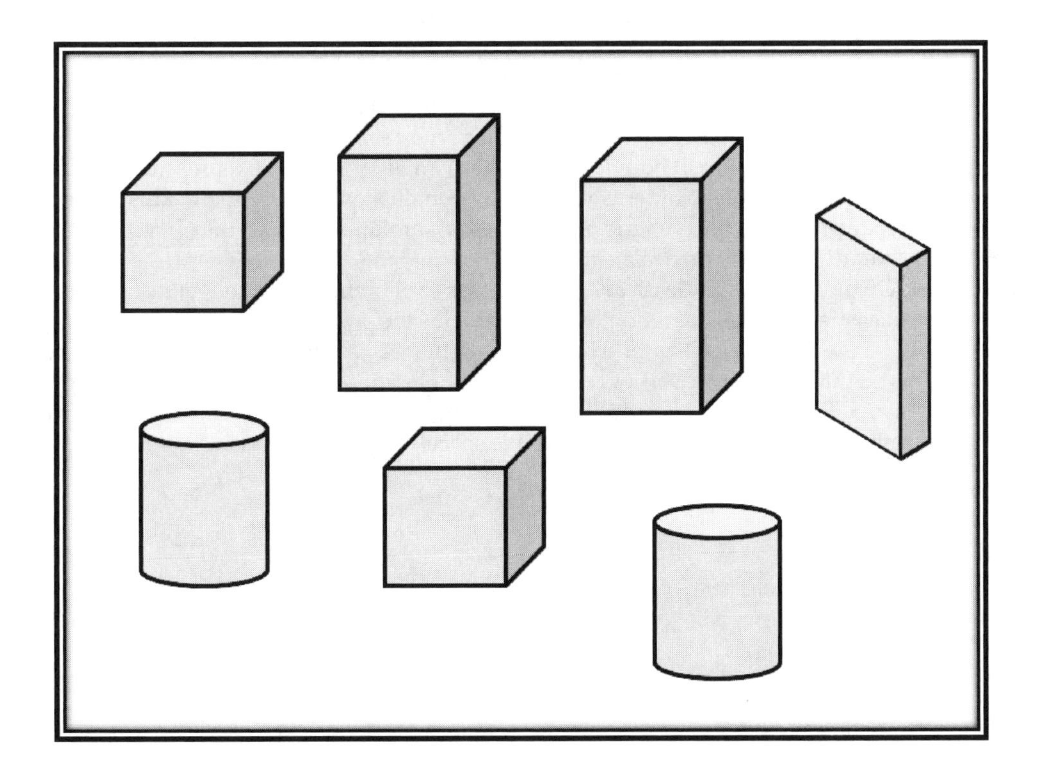

Figure 5.5 A more advanced blocks-world problem for practicing.

- Suggest some realistic applications that can be modeled by the TSP (but not those that were mentioned in this chapter). How will you address them by the STRIPS technique?
- Suggest realistic applications that can be modeled by the knapsack problem or by job-shop scheduling.
- Write a computer program implementing the STRIPS technique in a way that will enable it to deal with the simple blocks world. Run experiments with different initial and final states, and evaluate the computational costs of these experiments.

5.7 Concluding Remarks

Planning is a well-established discipline with many success stories to its credit. The first major milestone, STRIPS, was developed by Fikes and Nilsson (1971). The reason why their technique has been included here is that it demonstrates, in a very convincing manner, that one can indeed handle practical engineering problems by classical search. Besides, the STRIPS experience re-enforces in the reader the appreciation for the need to find a good representation for the problem's states.

Major applications of planning include such generic tasks as the traveling salesman, the knapsack problem, and job-shop scheduling—all of which this chapter described, even if only briefly. For more on the traveling salesman, see Applegate et al. (2006); for information about the knapsack problem, see Martello and Toth (1990); and for a good introduction into job-shop scheduling, see Chakraborty (2009).

These fascinating problems have served as useful test beds not only in artificial intelligence, but also in other fields of computer science and applied mathematics. The reason they have become so popular is that they can be shown to be representative of many practical applications. Students will greatly benefit from honing their skills of mapping concrete problems onto generic paradigms—assuming, of course, they have an idea how the paradigms are to be dealt with.

On the other hand, classical AI approaches to planning are no longer deemed as powerful as they were a generation or two ago. Thus the swarm-intelligence techniques that will be the subject of Chapter 8 have repeatedly been shown to be much more efficient. This, actually, is the reason why this chapter has presented just one famous representative, STRIPS, while ignoring the many alternatives that have been developed in the twentieth century. Swarm intelligence has largely superseded them.

CHAPTER 6

Genetic Algorithm

In domains whose evaluation functions are plagued by many local extremes, the power of classical AI search can be less than satisfactory. This is why alternatives have been sought. Among these, particularly popular is an approach that seeks to emulate the principles of Darwinian evolution. Having been shown to outperform heuristic search in many challenging domains, this remarkably flexible *genetic algorithm* has won the sympathies of theoreticians and practitioners alike.

The chapter explains the details of the genetic algorithm's building blocks, introduces some of its more advanced variations, and explores the secrets behind the technique's power. The chapter also presents ideas about how to implement the technique in a computer program and points out certain pitfalls that the cautious engineer wants to avoid. The potential of this paradigm is illustrated on the now-legendary test bed known as the *prisoner's dilemma*.

6.1 General Schema

The basic principle is captured by the endless loop shown in Figure 6.1. Throughout this chapter, we will refer to this loop by the acronym GA, standing for the *genetic algorithm*. Let us take a closer look at how it works—and why it works.

6.1.1 Imperfect Copies, Survival of the Fittest

Natural organisms are almost incapable of replicating flawlessly. Nearly always, children are nothing but imperfect copies of their parents, variations and combinations of their parents' genes. Some of these variations outperform others. The essence of Darwin's *survival of the fittest* is that individuals that are better adapted to their environment are more likely to survive and leave behind copious offspring that carry their parents' genes. In the long run, the population's level of adaptation to the environment is thus improving.

The nature of the process is probabilistic. A strong, fast, and smart individual can be unfortunate enough to perish under a landslide, whereas a weakling can make it to ripe old age simply by being lucky. In the billions of individuals, the law of large numbers prevails. Still, the occasional survival of a low-quality individual is beneficial. Its DNA may contain important genes that may prove invaluable in the future; it is a good thing that such genes are not destroyed prematurely.

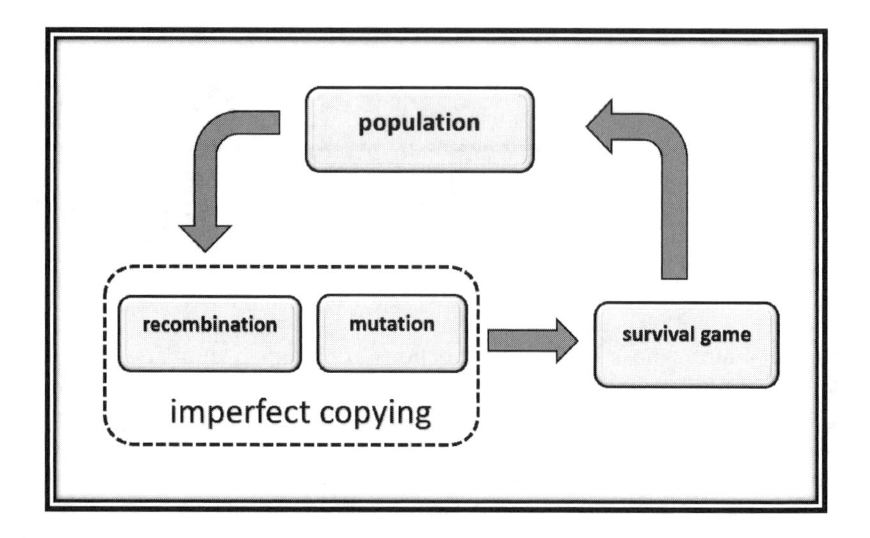

FIGURE 6.1 From the candidate solutions in a given *population*, GA creates imperfect copies. The subsequent *survival game* weeds out less promising solutions.

6.1.2 Individuals in GA Applications

Suppose we want to find a solution to some difficult problem, perhaps one of those mentioned earlier in this book. GA assumes that each candidate solution is encoded in terms of a *chromosome*. In the simplest implementation, the chromosome may be a string of bits, each bit standing for some feature or aspect of the solution. If the feature is present, the bit is set to 1; otherwise, the bit is set to 0. Any change in the chromosome thus implies a modification of the candidate solution. In simple applications, the chromosomes may consist of dozens of bits; in more ambitious projects, the bits will be counted in hundreds, thousands, and quite often even more than that.

Instead of bits, some applications work with chromosomes in the form of strings of integers, or strings of symbols, or even mixtures of bits, integers, and symbols. In more challenging domains, the chromosomes can be encoded using advanced data structures such as binary trees. Sometimes it is even beneficial to represent each candidate solution by two or more chromosomes. The flexibility of the paradigm is impressive.

6.1.3 Basic Loop

Figure 6.1 depicts an idea of how to cast the Darwinian principle in a computer program. The program maintains a population of individuals, each representing one candidate solution. From these individuals, children chromosomes are created by *recombining* the information contained in pairs of parental chromosomes, and by *mutation*. The population of parents and children then enters the *survival game*, a function meant to eliminate weaker individuals and retain those that are well-adjusted to momentary circumstances—those that are more likely than others to succeed in the engineering problem at hand.

So much for the general principle. Let us now take a closer look at some important details of its practical implementation.

6.1.4 Population

In a simple domain, a typical population will consist of a few dozen individuals, maybe a hundred or two. In more advanced applications, the population may comprise thousands of individuals. Most of the time, the population size remains the same in each generation. However, certain circumstances, such as the *premature degeneration* discussed in Section 6.4, can make it necessary that the population size be temporarily increased.

In certain domains, it is beneficial to work with multiple populations, each seeking to optimize a different aspect of the given problem. Inter-breeding among the populations then gives rise to individuals whose mix of the desired qualities is better than what could be achieved in the individual populations.

6.1.5 Survival of the Fittest

An important aspect of the GA is a function that accepts as input the description of an individual and returns this individual's value. In the context of GA, this function is usually called *fitness function*, but it plays essentially the same role as the evaluation function we know from the chapters on heuristic or adversary search.

The fitness function can take the form of a user-specified mathematical formula. For instance, if the chromosome has the form of a binary string representing integer x, then one such function can be $2.5 \times \sin(0.03x)$. Alternatively, the individual's fitness can be obtained from the result of some tournament-like competition among the population individuals (similarly to a soccer league). The motivation for the tournament is to put each individual's quality into the context of the other individuals in the population. This is why the fitness is sometimes called the individual's *survival edge*.

Just as in the case of heuristic search, the GA's ability to reach a solution in reasonable time will to a large extent depend on the quality of the fitness function. Poorly designed fitness function may lead to disappointing results.

6.1.6 How Many Generations?

Each cycle through the loop in the flowchart in Figure 6.1 is called one generation. GA typically runs through dozens or hundreds of generations, the concrete number depending on the complexity of the problem at hand, and on the stopping criteria specified by the programmer. The number of generations needed by the program to reach an acceptable solution also has a lot to do with the size of the population. Larger populations allow the program to find a good solution in fewer generations, though not necessarily with lower computational costs.

6.1.7 Stopping Criteria

Just like life itself, the GA loop is endless, and can run forever. Practically, though, the motivation for running the GA is not to simulate life, but to find a solution to a problem. The computer needs to be told how to recognize the situation where the best solution in the current generation is good enough to allow the termination of the process. Here are some popular possibilities:

1. Stop if the fitness of the best individual has reached a value whose interpretation indicates that the solution is good enough.

2. Stop if no major improvement in the fitness of the best individuals has been observed over a certain number of generations.
3. Stop after a predefined number of generations or after a certain pre-defined maximum time.

Control Questions

If you are unable to answer the following questions, return to the appropriate place in the preceding text.

- Summarize the fundamental aspects underlying Darwinian evolution. How are these aspects reflected in the basic loop of the GA?
- How does a typical implementation of the GA encode the individuals? What is a population? How will your GA program realize it is time to stop?

6.2 Imperfect Copies and Survival

The secret behind the GA's power is explained by two circumstances. First, individuals with higher fitness are more likely to survive and leave offspring, perpetuating their successful genes. Second, each child, being an imperfect copy of its parents, represents a minor modification of previously investigated solutions, something like a minor experiment of the Darwinian process. Let us take a look at how to implement these two aspects in a computer program.

6.2.1 Mating

In each generation of the GA process, the first task is to form pairs of mating individuals (future parents). The simplest mechanism will just pair them at random. A somewhat more sophisticated method will order the population according to the individuals' fitness values, from the highest to the lowest, and then pair the neighbors: the first with the second, the third with the fourth, and so on. Yet another strategy will select the mating pairs probabilistically, by a mechanism that will be explained later in this section (see the paragraphs dealing with the survival game and its use in mating).

6.2.2 Recombination

Once the mating pair has been formed, the next step creates their children by recombining their genetic information. For simplicity, let us suppose that each individual is described by a single chromosome in the form of a binary string. The simplest recombination technique is known as *one-point crossover*.

Figure 6.2 illustrates the principle. Suppose that the chromosomes have N bits. A random-number generator is asked to return an integer, $i \in [0, N-1]$. In the example from the picture, this random value, $i = 3$, determines the location of the little arrows under the chromosomes. Each arrow divides the binary string into a *leading part* and a *tail*. Each child combines the leading part of one parent with the tail of the other parent. For instance, the first child's chromosome consists of the leading part 11001 inherited from the first parent and the tail 001 inherited from the other parent. The other child combines the leading part of the second parent with the tail of the first.

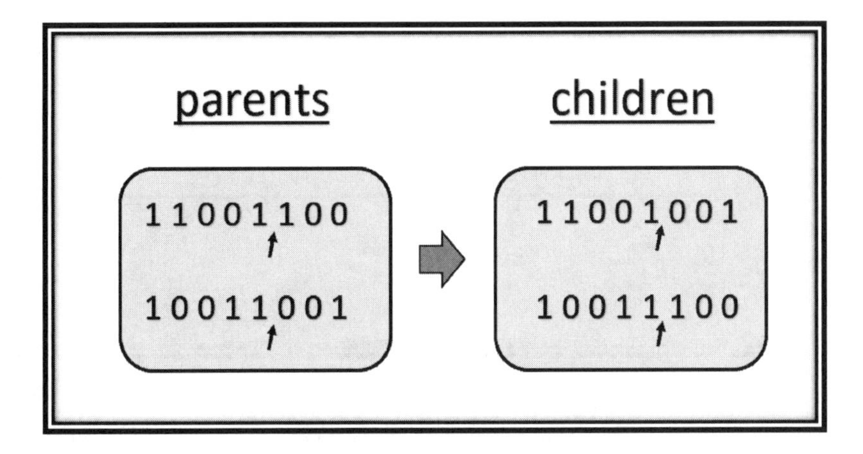

Figure 6.2 GA generates a random integer, here it is $i = 3$. The two children are created by swapping the last i bits of the parental chromosomes.

6.2.3 Mutation

Recombination is followed by mutation that considers one bit at a time and, with some small probability, flip-flops its value: if the bit is 0, then it is changed to 1; and vice versa. The probability of this happening is called *mutation rate*. Suppose that the mutation rate has been set to 2%, $p = 0.02$. For each bit, a random number, $x \in [0.00, 1.00]$, is generated. If $x < p$, the bit value is flip-flopped; otherwise, it is not.

Typical mutation rate will be somewhere between 1 and 3%. If the rate is lower, mutations are too rare to have any practical impact. If the rate is higher, mutations may destroy the GA's behavior because the bit-changes are then so frequent as to render the whole process too random to make sense. Usually, constant mutation rate is used, but special circumstances (such as the *premature degeneration* discussed in Section 6.4) may necessitate its temporal increase.

6.2.4 Implementing the Survival Game

Suppose, for simplicity, that the population consists of only five individuals with the following fitness values: 15, 15, 40, 20, and 10. Let each individual be represented by a line segment whose length is proportional to the individual's fitness, as indicated in Figure 6.3. This means that we have a sequence of five line segments, each for one individual. Suppose the fitness values have been normalized in a way that makes their sum equal to 100. If we generate a random number, $x \in [0.0, 100.0]$, its value will fall into one of the segments, thus pointing at the individual to be copied into the next generation. In the specific example illustrated by Figure 6.3, the third individual has thus been selected, the one whose fitness was 40.

For a population whose size is in each generation kept at N_p, this selection process is repeated N_p times. Note that a lower-valued individual has a small chance of making it into the next generation, but some high-valued individuals will have two copies, or even more.

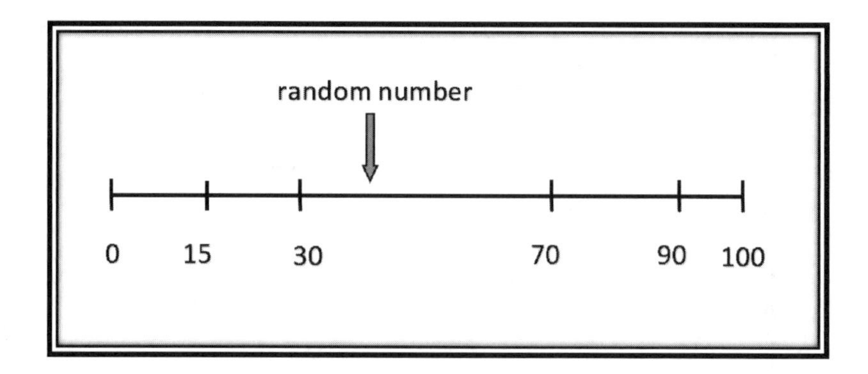

Figure 6.3 Each individual is represented by a line segment whose length is proportional to the individual's fitness. A randomly generated number then points at the line segment of the individual to be copied into the new generation.

6.2.5 Exploiting the Survival Mechanism for Mating

In this section, the first paragraph, Mating, suggested two simple mechanisms to form the pairs of parents whose chromosomes should be recombined to generate children.

Another way of doing so relies on the survival-game mechanism from Figure 6.3. First, the random-number generator chooses one individual to be copied into the next generation. Then, another individual is selected by the same random mechanism. This second individual, too, is copied into the next generation, and the two will become mating partners to recombine their chromosomes.

6.2.6 Commenting on the Survival Game

The world of biology knows only two possibilities: the individual either survives or perishes. The computational version of the GA is in this sense more flexible because it permits also the possibility that the next generation will contain two or more copies of the same high-valued individual.

The whole nature of the genetic search is probabilistic. A low-valued individual may, against all odds, survive by sheer good luck whereas a high-valued individual may not make it, simply because the random-number generator did not return the right number. This probabilistic aspect is very important. After all, the chromosome of a weak individual may still contain some useful sub-strings that may yet prove beneficial in the chromosome of a future child that inherits them from an otherwise weak parent.

6.2.7 Children Falling to Both Sides of Their Parents

Suppose we interpret each binary string as an integer. For instance, 011 will be interpreted as three and 1001 as nine. If the chromosome consists of 10 bits, then there are $2^{10} = 1{,}024$ different individuals, and they all can be arranged along the horizontal axis of a graph whose vertical axis represents the fitness function.

The example in Figure 6.4 illustrates what may happen when two parents swap, say, 3-bit tails. In this specific case, one parent's tail is 011 which means three; the other parent's tail is 000, which means zero. The child that obtains the leading part from the

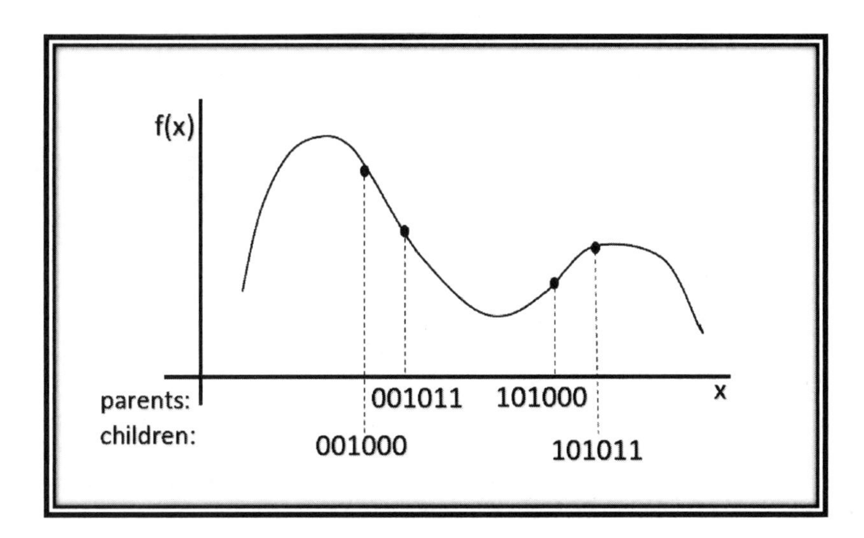

FIGURE 6.4 The two parents swapped their 3-bit tails. Since one tail is 000 and the other is 011, one child falls to the left of the first parent, and the other to the right of the second parent.

first parent and the tail from the second will therefore find itself, along the x-axis, to the left of the first parent (because zero is less than three). For a similar reason, the other child will find itself to the right of the second parent. This is what the graph illustrates.

Of course, with another pair of tails, the two children might find themselves on the opposite sides from their parents. Depending on the shape of the fitness function, some children may have higher fitness than their parents—but not necessarily so.

6.2.8 Simple Tasks for GA

Figure 6.4 also illustrates GA's most immediate application: to find the maximum (or minimum) of a function. Mathematicians know that local extremes are sometimes difficult to find by calculus. Whereas our graph shows a function of a single variable, the number of variables can be high, rendering an analytical search for global maximum tedious, especially if the function is complicated, and its analytical solution difficult.

Moreover, let us not forget that, in many technical applications, the exact shape of the function is unknown, and its values at individual points can only be established experimentally. This is the case of many optimization tasks in engineering, and we have touched upon something similar in the applications of the heuristic and adversary search techniques discussed in Chapters 3 and 4.

Finally, there exist (not only) engineering problems where the only objective way of evaluating candidate solutions is by comparing them with other solutions. In this event, the fitness can often be established by a "tournament," an open competition of alternative solutions. We will meet this last case in the analysis of the *prisoner's dilemma* in Section 6.7.

6.2.9 Exploration of the Parents' Neighborhood

How far the children fall from their parents (e.g., along the x-axis in Figure 6.4) depends to a great extent on the randomly generated length of the tails that the parents swap. If it is just one bit, then the children are close to the parents; if both parents happen to have the same value of this least-significant bit (the same tail), then the children will only be copies of the parents.

The longer the tail, however, the greater the potential difference between the children and their parents. Thus in the case of 4-bit tails, it may be that the two tails have values 1111 and 0000, respectively, which amounts to parent-to-child distance of 15. But then again, in some parents the 4-bit tails can be 1110 versus 1111, in which case the distance is only 1. We see that the parent-to-child distance depends on the length of the tail only probabilistically.

6.2.10 Recombination versus Mutation

Pondering on what we learned in the previous paragraph, we realize that the one-point crossover operator helps the GA explore the parents' neighborhoods. We now also understand that close neighborhoods are likely to receive more attention than more distant ones.

Mutation, on the other hand, can institute more significant jumps along the x-axis. Each bit in the chromosome faces here the same chance of being inverted from 1 to 0 or vice versa. If the least significant bit is affected, the chromosome after mutation will be located, along the x-axis, next to the original chromosome. If, however, the leftmost bit is inverted, a much bigger "jump" is made.

In this sense, mutation neatly complements recombination in the GA's efforts to explore the search space.

6.2.11 Why the Algorithm Works

The fact that the child is located to the left or to the right of the parent does not yet mean that the child's fitness will exceed the parent's fitness. What matters is the shape of the fitness function. In the specific case of Figure 6.4, each child's fitness is higher than that of the nearest parent; with a different shape of the function, the children's values can be lower than the parents' values.

In the whole population, some children will exceed the qualities of their parents, and other children will not—but it is the survival game that has the final say: higher-valued children stand a higher chance of making it into the next generation. At the end of the day, then, later generations tend to be populated by individuals whose fitness values have on average increased.

And yet, there is no guarantee. Some exceptions to the general tendency in the GA's behavior will be discussed in Section 6.4.

Control Questions

If you are unable to answer the following questions, return to the appropriate place in the preceding text.

- Discuss the mating principles employed by the GA. Explain how the simple one-point crossover recombines the information from the parents' chromosomes.
- How will you implement mutation? What is the impact of mutation rate?

- Explain the mechanism that implements the survival game. How will you apply the same mechanism when implementing the mating strategy?
- Recombination and mutation affect the GA's search-space exploration in different ways. Explain how they complement each other.
- What constitutes a typical problem to be addressed by the GA? If the goal is to find a maximum of a fitness function, why not prefer calculus?

6.3 Alternative GA Operators

To simplify our analyses, we have so far considered only the simplest GA operators. These, however, are rarely used. Let us take a look at some more common alternatives.

6.3.1 Two-Point Crossover

Perhaps the most popular recombination operator is the two-point crossover illustrated by Figure 6.5. For a binary string consisting of N bits, the random-number generator provides two integers from the interval $[0, N-1]$. These two numbers then define a "middle part" to be swapped between the two parents. In the picture, the locations of the two integers are indicated by the little arrows under the strings. The two bits between the arrows are those that this operator swaps.

In the last section, Figure 6.4 explained only the behavior of the one-point crossover operator, but the observations will essentially be the same even when the two-point crossover is used. To see why this is the case, just recall that the order of the bits in the chromosome is arbitrary, each bit representing one of a set of independent features. If we rearrange the features in a way that shifts the middle part to the chromosome's tail, we will get the one-point crossover.

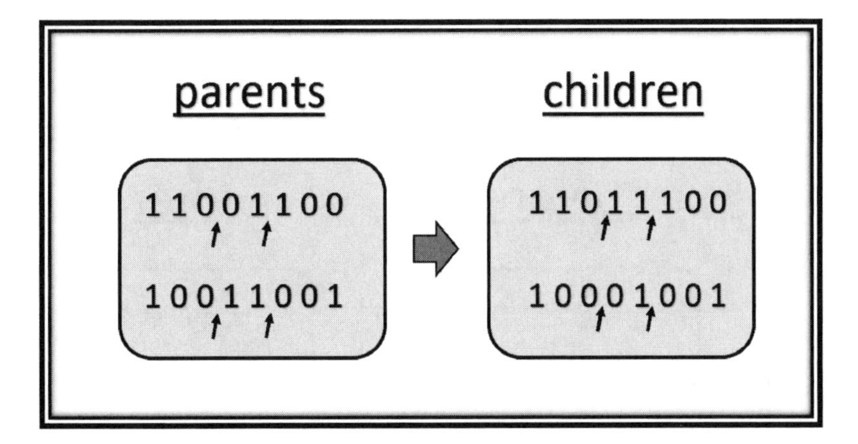

FIGURE 6.5 Two-point crossover: GA generates two random integers. The two children are created by swapping the bits between the locations defined by the two integers.

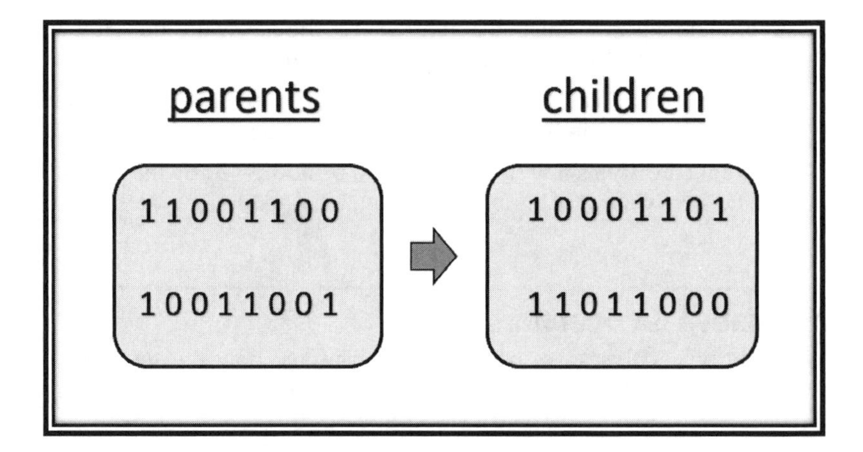

Figure 6.6 Random bit exchange. GA generates a set of random integers. The children are created by swapping the bits pointed to by these integers. Here, the integers were 0 and 6. This is why the rightmost bit and the seventh bit from the right were swapped.

6.3.2 Random Bit Exchange

Yet another possibility is illustrated by Figure 6.6. Here, the random-number generator returns a few integers from $[0, N-1]$. These are then interpreted as the indexes (locations) of the bits that are to be swapped between the parents. The index of the rightmost bit is 0. In the specific case shown in the picture, the random integers were 0 and 6. This is why the rightmost bit and the seventh bit from the right were swapped.

6.3.3 Inversion

Whereas the recombination operators recombine the genetic information of a *pair* of chromosomes (parents), *inversion* operates on a single chromosome. Two random integers are generated, and the sub-string between the locations they define is inverted—the order of the bits between the two locations is reversed. In the case from Figure 6.7, the

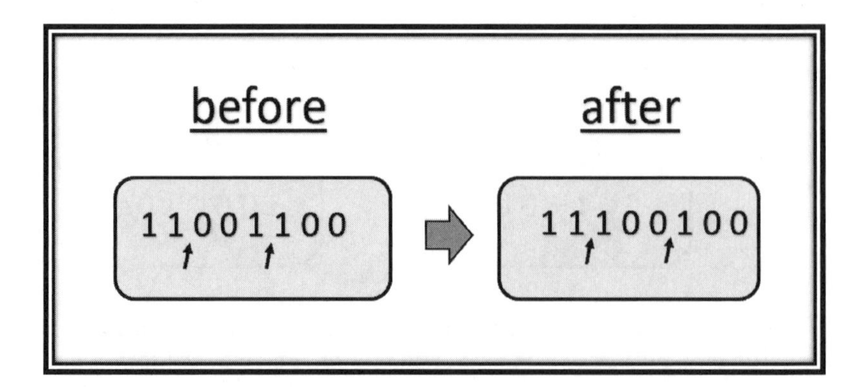

Figure 6.7 Inversion operator: GA generates two random integers. The sub-string they define is then inverted.

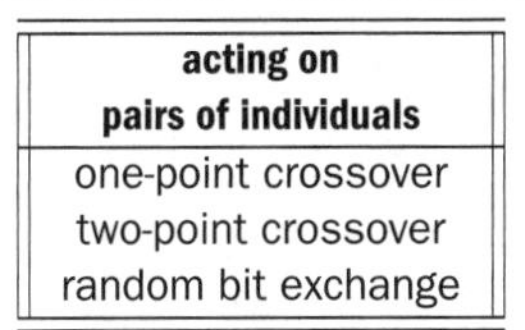

acting on pairs of individuals
one-point crossover
two-point crossover
random bit exchange

acting on single individuals
mutation
inversion

Table 6.1 Typical Search Operators Used by the Genetic Algorithm

two randomly generated locations are indicated by arrows. The sub-string contains 001; inverting it, we obtain 100 which is what it is in the bit string on the right.

6.3.4 Programmer's Ways of Controlling the Process

All in all, programmers wishing to implement the GA have at their disposal the five basic operators listed in Table 6.1. Note that three of them operate on pairs of chromosomes, and two (mutation and inversion) operate on single chromosomes.

In professional implementations, the one-point crossover is rare. Most typical are various combinations of the remaining four. For instance, the programmer may decide that 70% of the mating partners should undergo recombination by the two-point crossover, and the remaining 30% should use random bit exchange. After this, the chromosomes should be subjected to, say 2% mutation rate, and 3% of the individuals should be subject to sub-string inversion. The concrete mixture will depend on the engineer's experience, and can vary in the course of the GA search. For instance, the frequency of sub-string inversion may have to be increased when the population is to be rescued from premature degeneration (see Section 6.4).

Control Questions

If you are unable to answer the following questions, return to the appropriate place in the preceding text.

- Explain the principle of the two-point crossover operator. Which of its aspects are controlled by the random number generator?
- Explain the principle of random bit exchange. Explain the principle of the inversion operator.
- Discuss the "mixtures" available to the programmer.

6.4 Potential Problems

The principles of the GA are now clear, and so are the diverse possibilities of their practical implementations. To deepen our understanding and to increase our chances of success when employing the technique in engineering practice, let us now take a brief look at some typical situations that may cause the GA to fail.

ID	binary string
a	1 0 0 1 0 0
b	1 0 0 1 0 0
c	1 0 0 1 0 1
d	1 0 0 1 0 1

Table 6.2 An Example of a Degenerated Population

6.4.1 Degenerated Population

The power of the GA depends on the population's diversity. The greater the variance among the individuals, the greater the power of the recombination operators and the greater the chance of hitting on some interesting solution soon. Conversely, if the diversity is inadequate, not much can be discovered. In the extreme, the population can degenerate into a state where the GA's operators are essentially incapable of creating individuals that would differ from their parents.

To illustrate the last point, Table 6.2 shows a very small population consisting of only four individuals. A closer look reveals that recombination is incapable of creating new chromosomes. For instance, individuals **a** and **b** are identical and nothing can be gained by any crossover operator. Individuals **b** and **c** are *not* identical, but the reader will easily verify that crossover will never result in binary strings that differ from those already present in the population. We say that the population is *degenerated*.

6.4.2 Harmless Degeneration versus Premature Degeneration

There is no need to complain that the population has degenerated if it already contains a good solution to the given problem.

Sometimes, however, the population degenerates way too early. The individuals' fitness values are still low, and yet the GA does not seem to be getting anywhere. It is in this sense that the degeneration is regarded as *premature*. A well-written GA program should identify this harmful situation when it occurs and take a corrective action.

6.4.3 Recognizing a Degenerated State

The first indication that the population has degenerated is the observation that, over the last N generations, no improvement in the value of the population's best individual has been seen.

The suspicion can be corroborated by a program module that evaluates the population's diversity. A simple way of doing so is to calculate for each pair of individuals their Hamming distance: the number of bits on which the two binary strings differ. For instance, the Hamming distance between 00011100 and 10011101 is two because the two strings differ in two bits, the first and the last.

Once the degeneration has been detected, the program informs the user about the quality of the currently best solution. If this best solution appears acceptable, the program may stop with success. If the solution is still poor, measures have to be taken to extricate the population from the degenerated state.

6.4.4 Getting Out of the Degenerated State

The population from Table 6.2 cannot be modified by recombination. Mutation *can* modify it, but very slowly. With mutation rate of 2%, it may take quite a few generations before the number of bit changes is sufficient for the recombination to become functional again. Of course, the programmer may decide to increase mutation rate dramatically (say, to 20–30%) for the duration of a few generations, and then reduce it again. This will improve the population's diversity quickly enough; however, such drastic measure may destroy some chromosomes that have already been shown to be quite good, and as such do not deserve to be corrupted in mutation.

Another possibility is simply to create some randomly generated new individuals and insert them into the degenerated population. This addition increases diversity, and thus provides useful impetus to the evolution process to restart. Diversity can also be improved by increasing the frequency with which the inversion operator is employed. By way of a simple exercise, the reader is encouraged to try and see how inversion, applied to the first individual in Table 6.2, may (or may not) result in a new individual that is sufficiently different from the rest of the population: All depends on which sub-string is inverted.

6.4.5 Poorly Designed Fitness Functions

Another source of the GA's disappointing performance can be traced to an inappropriate fitness function. Figure 6.8 illustrates two poor choices. The flat one on the left hardly distinguishes good individuals from bad. The survival-of-the-fittest mechanism will then drive the subsequent generations to the function's maximum only very slowly. Note that this weakness can be remedied by simply replacing the flat function $f(x)$ with something "steeper," such as $f^2(x)$.

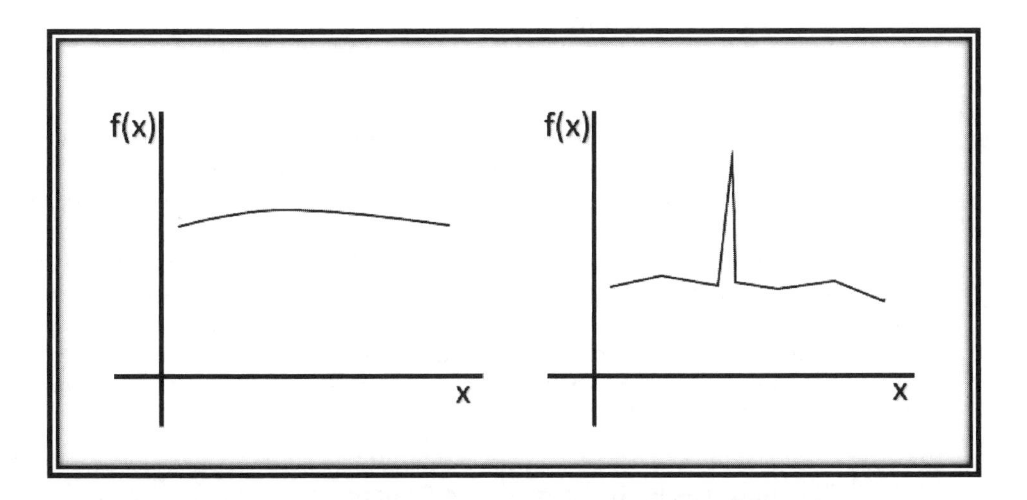

FIGURE 6.8 Both of these fitness functions are poorly designed. The one on the left is too flat, the one on the right does not allow for *gradual* improvement.

Another harmful situation is depicted on the right. Here, the peak is too narrow, and it may take quite some time before the GA generates individuals that fall in that narrow area—unless such an individual is created right at the beginning by the random-number generator, or soon afterwards by a lucky mutation.

6.4.6 Fitness Functions Not Reflecting the GA's Goal

Of course, the engineer must make sure that the fitness function, besides having the right shape, also reflects the needs and goals of the given application. Here, the same considerations apply as in the evaluation functions used by heuristic or adversary search.

Control Questions

If you are unable to answer the following questions, return to the appropriate place in the preceding text.

- What is *degenerated population*? Is it always harmful?
- How can the GA recognize that the population has degenerated, and what can be done to re-introduce diversity?
- What should the fitness function look like? Discuss examples of poorly designed fitness functions.

6.5 Advanced Variations

The baseline version of the GA can be improved in a number of ways. One possibility relies on multiple inter-breeding populations, another seeks to emulate the principles of Lamarckian evolution. As for the chromosomes, they can be more sophisticated than the plain bit strings from the previous sections.

6.5.1 Numeric Chromosomes

In domains where the individuals are described by numeric chromosomes (vectors of integers instead of bits), the same recombination operators can be used: one-point crossover, two-point crossover, and swapping the values in randomly selected locations of the vectors.

As for mutation, perhaps the simplest method is illustrated in Figure 6.9. The idea is to select a certain percentage of locations in the numeric vector, and then add to them a randomly generated number. Sometimes (as in this particular illustration) the programmer needs to prevent the values from exceeding a certain maximum. In that event, the modulo function is used. For instance, if the maximum permitted value is 20, then the added noise may result in $19 + 3 = 2$.

6.5.2 Chromosomes in the Form of Tree Structures

In some domains, bit strings or vectors of numbers are not enough; more advanced data structures are needed. For instance, mathematical expressions can be represented by binary trees. To represent expression $\sin x + \cos y$, the symbol "+" is placed at the root from

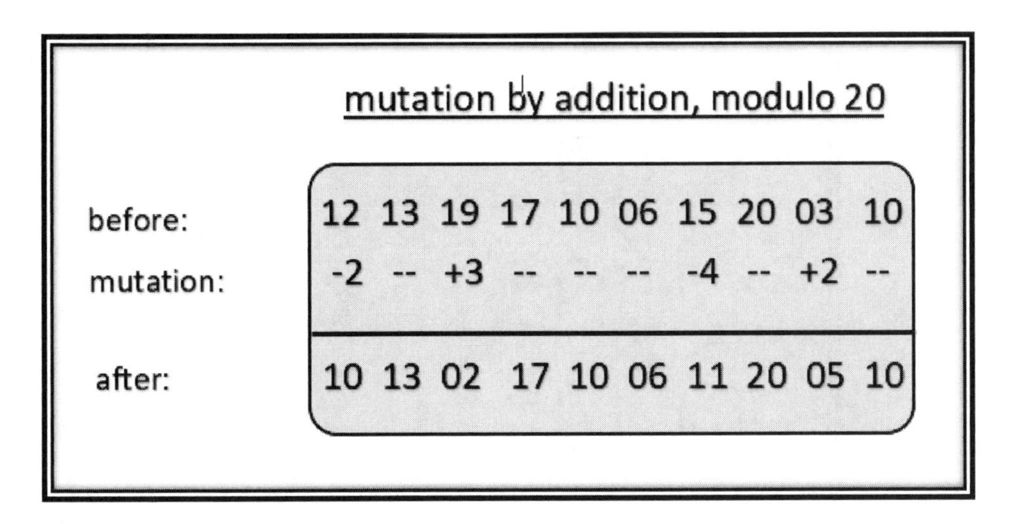

Figure 6.9 In numeric chromosomes, mutation can be modeled by adding a randomly generated integer, modulo the permitted maximum (so that $19 + 3 = 2$).

which one edge points to sin and the other to cos. From each of these, only one edge is emanating: from sin to x and from cos to y.

Figure 6.10 shows us how to recombine these tree-structure chromosomes. In each of the parent chromosomes, one edge is selected at random; the children are then generated by swapping the subtrees defined by these edges.

As for mutation, the program can randomly modify the contents of some nodes in the tree. Sometimes, only the terminal nodes are subjected to mutation. For instance, in the mathematical expressions mentioned above, the terminal nodes contain variables; the programmer may decide that only these variables be mutated, say, by changing x to y.

6.5.3 Multiple Populations and Multiple Goals

Textbook examples usually assume that there is only one fitness function to optimize. In reality, this may not be enough. Quite often, one needs to optimize two or more aspects that may even contradict each other. Taking care of all of them is sometimes easy—and sometimes difficult. If we want to maximize the quality of a product while minimizing its price, it may be possible to combine the two aspects in one carefully crafted function; for instance, $fitness = quality/price$. In other domains, such attempts would be awkward, and may even mislead the GA.

In this event, one successful strategy relies on multiple populations, each optimizing a different fitness function that represents a population-specific aspect. Occasional *interbreeding* (when each parent comes from a different population) may give rise to individuals that satisfy competing aspects.

6.5.4 Lamarckian Approach

In Darwinian evolution, an individual's genetic information does not change during this individual's life. Changes occur only during the transition from parents to children: by

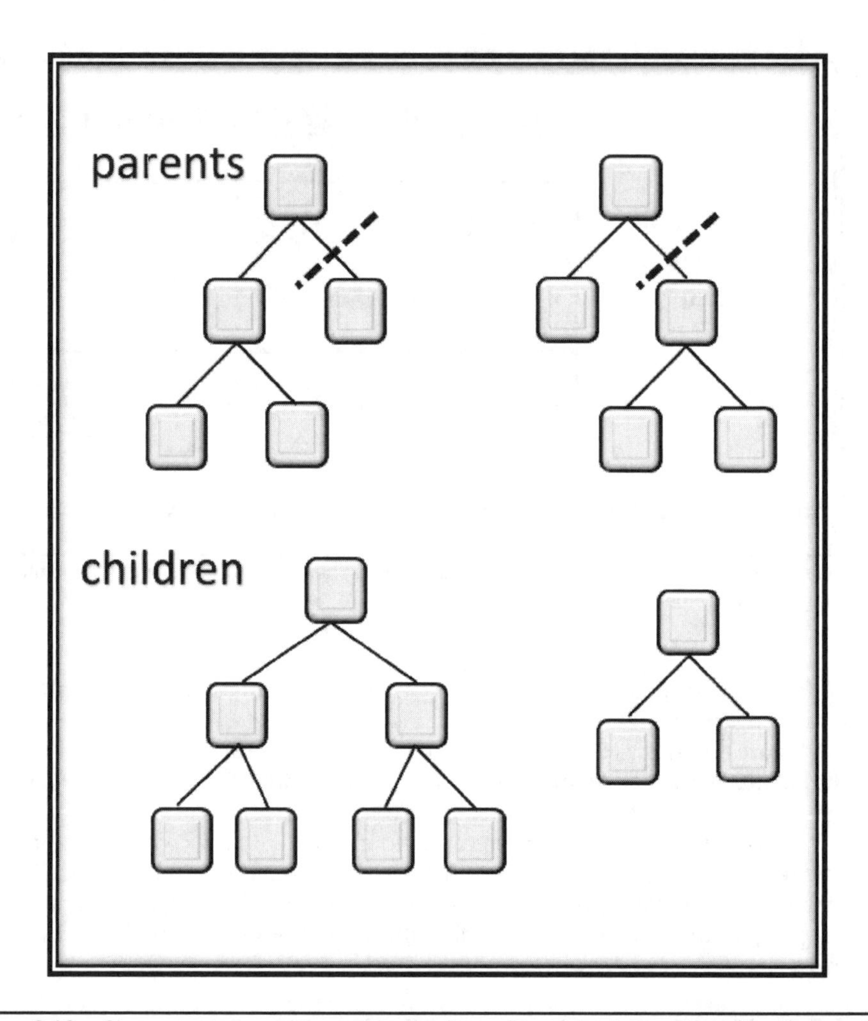

Figure 6.10 Chromosomes sometimes take the form of tree structures. Recombination can then be implemented as the exchange of randomly selected subtrees. In this picture, each of the parent trees has four edges. In each tree, one edge is randomly selected for the subtree exchange (note the dashed lines).

mutation and recombination. Lamarck, a scholar preceding Darwin by one or two generations, held a different opinion. In his view, the change *could* occur in the course of an individual's life.

Today we know that Lamarck was mistaken in the context of the genetic information in DNA. However, one can argue that the principle he proposed is observed in other aspects of evolution. For instance, a professor may pass to the students some knowledge acquired in the course of his or her own studies. This kind of evolution is faster than the classical Darwinian process. It is thus legitimate to ask how to implement this idea in the GA.

One possibility is to place the "Lamarckian operator" somewhere in the schema from Figure 6.1, perhaps after the *survival game*. This operator then tries to improve each individual's fitness by tweaking its parameters—for instance, coefficients or added constants in applications that seek to find a mathematical expression that best models a certain process.

Control Questions

If you are unable to answer the following questions, return to the appropriate place in the preceding text.

- How will you implement the recombination and mutation operators in a domain where the individuals are described by numeric vectors instead of binary strings?
- How will you implement the recombination and mutation operators in a domain where the individuals are described by tree structures?
- Explain the multi-population implementation, including its motivation and the benefits of interbreeding.
- What is the difference between Darwinian and Lamarckian evolution?

6.6 GA and the Knapsack Problem

To illustrate the benefits of the GA in the context of a more advanced application, let us take a look at how we might use the technique to address one of the model tasks from Chapter 5, the *knapsack problem*.

6.6.1 Knapsack's Rules (Revision)

The agent has at its disposal a set of objects. Suppose the i-th object is characterized by its price, P_i, and weight, W_i. The goal is to find a subset of objects that maximizes the sum of their prices, $\max_i \Sigma P_i$, under the constraint that the sum of their weights, ΣW_i, is not allowed to exceed a user-specified maximum, MAX.

A set of N objects has 2^N subsets. Which of them represents the best combination of objects to be packed in the knapsack? To evaluate the price and weight of an exponential number of subsets is unrealistic. Intuitive approaches do not seem to work, either. Should the agent pack as many light objects as possible, or should it rather focus on a few high-valued objects? Also heuristic search tends to disappoint because we may not know how to design the evaluation function.

6.6.2 Encoding the Problem by Binary Strings

Here is a simple way of addressing the problem by the GA. If the total number of available items is N, we will represent the knapsack contents by a binary string of N bits, where the i-th bit is set to 1 if the i-th item is to be placed in the knapsack, and to 0 if not. The population then consists of these binary strings.

The fitness function sums the values of all items for which the given chromosome contains 1. At the same time, the function watches the sum of the items' weights, and if this sum exceeds MAX, the returned fitness is 0, the lowest value possible. An even better fitness function will reflect also *how much* the maximum has been exceeded, and it will penalize the failure in proportion to the excess weight.

6.6.3 Running the Program

The GA process begins with a randomly generated population. Mating is either random or controlled by the survival game; recombination relies on the two-point crossover and random bit exchange, and mutation is set to a small value, say, 2%.

The engineer can be very creative when it comes to the stopping criterion. Here is one line of reasoning. When the best individuals' fitness values get close to the problem's best possible solution, the population tends to be marked by an increased frequency of individuals whose total weight exceeds the permitted maximum, $\Sigma W_i > \text{MAX}$, for which they receive fitness 0. In view of the observation, the programmer may decide to instruct the program to stop if the proportion of chromosomes with zero fitness reaches, say, 10%.

As for the population size, it should be large enough to minimize the danger of premature degeneration. The size should also reflect the length of the chromosomes: the longer the chromosomes, the larger the population size.

6.6.4 Does the GA Find the Best Solution?

For high values of N, the GA will rarely identify the absolutely best solution. More often than not, however, the practically minded engineer is content with finding a combination that is "reasonably good."

6.6.5 Observation: Implicit Parallelism

Experience shows that the GA finds a good solution to the knapsack problem in a relatively small number of generations. Beginners are often surprised by how quickly the solution is found. Suppose we denote the items by $I_0, \ldots I_{N-1}$. The number of all possible subsets is then 2^N. To consider all of them is in a realistic application virtually impossible. And yet, if the GA has succeeded in, say, 20 generations, then only $20M$ combinations have been investigated where M is the population size.

The secret of this staggering efficiency is hidden in what we know as *implicit parallelism*. Suppose, for simplicity, that $N = 8$, which means that the problem can be represented by 8-bit strings. Consider the following chromosome:

```
0 0 1 1 0 1 1 0
```

If the leftmost bit represents the zeroth item, I_0, the chromosome is telling us that the knapsack contains items I_2, I_3, I_5, and I_6. At the same time, however, the same chromosome subsumes any subset of these four items. If it is beneficial that, say, I_5 should be combined with I_6, the combination of these two is already contained in this chromosome, and the pair is likely to survive recombination, mutation, and the survival game—and as such will make it to the next generation.

In this sense, each chromosome represents a set of (possibly many) different combinations. In other words, processing a single chromosome means the evaluation of a great many alternative item combinations.

6.6.6 Encoding Knapsack Contents with a Numeric String

If the number of items, N, is very high, say, tens of thousands, then the bit strings may prove impractically long; at the very least, they would then call for very large populations, and this can entail prohibitive computational costs. In this event, the engineer may prefer to encode the knapsack contents in terms of a list of integers.

The idea is to represent each item by its index, an integer from $[0, N-1]$. The chromosome then contains only the indices of those items that are going to be packed. For instance, a knapsack containing the second, tenth, and twenty-first item will be represented by $\{2, 10, 21\}$. Note that this means that each chromosome can have a different length: A knapsack packed with 100 items will be represented by a chromosome containing 100 integers, and a knapsack packed with 23 items will be represented by a chromosome of length 23.

6.6.7 Mutation and Recombination in Numeric Strings

Suppose the chromosomes take the form explained in the last paragraph. Mutation can be implemented in the way described in Section 6.5 (see also Figure 6.9). Recombination may permit swapping sub-strings of different lengths. Thus in the case illustrated by Figure 6.11, the random-number generator has chosen for the first parent the sub-string between the first and the fifth integer from the left; and for the second, the sub-string between the first and the fourth. The children are obtained by swapping the contents marked by the ovals.

Note that this way of encoding the chromosomes may call for some additional programming. For instance, unless we want to permit multiple copies of the same item, the program has to make sure that each item is present in the chromosome only once. Also, this type of recombination tends to hurt the neatly ordered sequence of items in the chromosome; the programmer may prefer to make sure that proper order has been maintained (this, of course, may not be necessary). Such details will reflect the subtleties of the concrete application, and also the programmer's personal experience and preferences.

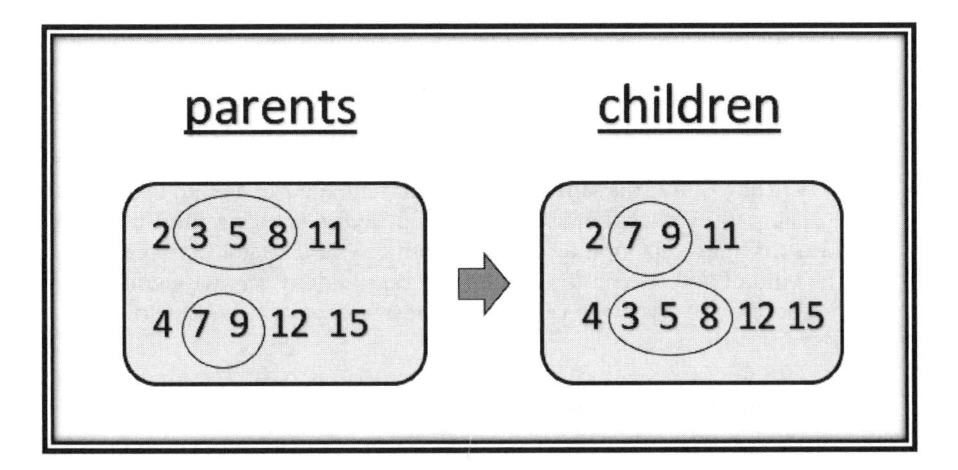

Figure 6.11 Recombination in these numeric chromosomes is allowed to swap sub-strings of different lengths.

6.6.8 Summary

The domain briefly discussed in this section has taught us an important lesson: Even a very simple problem may offer a broad range of alternative solutions. We have seen two different ways of representing the solutions by chromosomes. We also understand that the fitness function can be designed in many different ways. When implementing mutation and recombination, and even when formulating a useful stopping criterion, the engineer has ample opportunity to experiment with quite a few alternative approaches.

Such remarkable flexibility makes the GA very attractive for engineers who like to be creative. No wonder that many of them prefer the GA over classical search techniques.

Control Questions

If you are unable to answer the following questions, return to the appropriate place in the preceding text.

- Summarize the rules of the *knapsack* test bed.
- Discuss the two alternative ways of encoding candidate solutions in chromosomes.
- How will you implement here the fitness function, mutation, recombination, and the stopping criterion?

6.7 GA and the Prisoner's Dilemma

The knapsack problem was arguably simple and easy to address by the GA. Let us take a look at something more challenging: a popular problem known as the *prisoner's dilemma.*

6.7.1 To Be Tough or to Squeal?

Suppose that you and a friend of yours have robbed a bank. Later, you got arrested. At the moment, both of you are being kept in separate cells without any possibility of communicating with each other and synchronizing your future strategies.

As it turns out, evidence against you is flimsy, and so the police offer you a deal. If you confess and testify against your friend, they will let you go as a collaborating witness. whereas your friend will then face a five-year sentence. Since you naturally assume they are likely to have made the same offer to your friend, you are bound to ask: suppose both of us collaborate? Well, having two confessions reduces the incentive for rewarding any of you as a single collaborating witness; both of you thus get four years. On the other hand, if both of you remain tough, and stubbornly deny any wrongdoing, you should both expect only two years, on account of the evidence being so shallow.

6.7.2 Practical Observations

Table 6.3 summarises the situation. Each of you has two options, to be tough and to squeal. For each of the four combinations, the corresponding field in the table contains two numbers: the first is your sentence and the second is your friend's sentence. For instance, if you are tough and he squeals, you will get five years and he gets zero.

		your friend	
		tough	squeals
you	tough	2,2	5,0
	squeals	0,5	4,4

The fields in the table specify the expected prison terms. For instance, if you are tough and your friend squeals, you will get five years and he will be set free.

Table 6.3 Penalties in the Prisoner's Dilemma

The table indicates that the lowest average sentence is expected if both of you refuse to cooperate with the police. In that event, it is two years each. But then, do you really know what your friend is going to do? Not being sure, you realize that you are better off giving him up because then the worst sentence is four years, and there is still a chance of being set free. If you remain tough and he squeals, it is going to be five years. Obviously, a lot depends on how much you two trust each other.

6.7.3 Strategies for Repeated Events

The problem becomes more interesting if we allow for the same situation to be repeated many times over. In this scenario, you no longer rely on trust alone because you can take into consideration your experience with what happened previously. In the end, you will probably develop a rule such as, "I will trust him in the first round; but if he betrays me, I will retaliate the next time around." Many such strategies can be designed. Which of them will be best in the long run? Which will result in the lowest average sentence?

Puzzles of this kind used to be studied by a discipline called *game theory*. Understanding their nature was vital in the days of Cold War when the two superpowers could expect to find themselves in a situation very much like the one described above. The cardinal question was, "should we launch a preemptive nuclear strike?" Consequences were bound to be disastrous, but perhaps still better than what would happen if the other side were faster.

Scientists developed dozens of strategies for handling the dilemma, and then compared them in a simulated tournament where different strategies were pitted against each other in a way reminiscent of a soccer league. Somewhat surprisingly, the tournament was won by the simplistic *tit-for-tat* approach: be tough the first time around, and then copy your comrade's latest behavior.

6.7.4 Encoding the Strategy in a Chromosome

In each round, the agent has two actions to choose from: to be tough (*T*) or to squeal (*S*). Let us begin with something simple: The agent will base its decision just on what happened in the previous round. There are four possibilities: TT, TS, ST, and SS. The first letter denotes the agent's action, and the second letter denotes its comrade's action (in the previous round). For instance, TS means that the agent was tough but the comrade squealed.

The agent's strategy consists in answering the following question: "What should I do in any of these four situations?" For this, a 4-bit chromosome is enough: The first bit represents the action in the TT case, the second in the TS case, the third in the ST case, and the fourth in the SS case. If the bit is set to 1, the agent is tough; if the bit is set to 0, the agent squeals. For instance, strategy `[0 1 0 0]` is interpreted as telling the agent, "be tough if the previous round saw TS (you were tough and your friend squealed); squeal in all the other situations."

In a famous experiment, the agent was to base the decision not on what happened in the last round, but on the history of the last three rounds. In each round, the same four situations could occur, which amounts to the total of $4^3 = 64$ different possibilities. Below, each of these 64 possibilities is represented by a string of six letters, grouped in pairs, each pair representing one of the last three rounds:

1. TT TT TT
2. TT TT TS

⋮

64. SS SS SS

A chromosome defining a concrete strategy is a string of 64 bits, each bit corresponding to one of the 64 different situations. If the i-th bit in the chromosome is set to 1, then the strategy tells the agent to be tough in the i-th situation; conversely, 0 tells it to squeal. Since there are 64 bits, the total number of different strategies is 2^{64} which is a number close to 20×10^{18}. To find the best strategy (or at least an almost-best strategy) in a space so vast is an ambitious goal to say the least.

6.7.5 Early Rounds

When writing the computer program, the engineer has to pay attention to details that, for simplicity, the previous paragraphs neglected. For instance, since the chromosomes define the concrete action for each situation in the last three rounds, the program needs to be told what to do at the beginning of the game, before the first three rounds have been completed.

Let us suggest a simple solution. For the first round, the action is fixed; for instance, "be tough in the first round." For the second round, the strategy is based on what happened in the first round. For the third round, the strategy is based on what happened in the first two rounds. From the programmer's perspective, one possibility is to add to the chromosome some extra bits that address the game's beginning.

6.7.6 Tournament

Here is how the whole procedure was organized in a famous experiment that helped popularize the GA's potential in the 1980s.[1] A random-number generator creates an initial population of, say, 20 chromosomes, each a binary string of 64 bits (plus a few

---

[1] Somewhat unexpectedly, the experiments were conducted not by an engineer or a computer scientist, but by a political scientist, Robert Axelrod (1984).

more bits representing special situations such as the one from the last paragraph). The fitness of each individual is evaluated by a method called *tournament*, organized as follows.

X is paired with another individual, and they become the two bank-robbing comrades. They go through, say, 60 rounds of bank robbing, and each chooses its action according to its chromosome. Each round results in a certain sentence for X, and the sum of the 60 sentences is calculated. Once this is finished, X is paired with another individual, then yet another, and so on, each time going through the same 60-round experiment. In the end, X's fitness is defined as the average sum of sentences from all pairings.

6.7.7 Performance

At the time when this study was conducted, about 60 strategies had been known to mathematicians and game theorists. In one experiment, the fitness of the chromosomes was evaluated by measuring its average performance against all of these. In another study, the *tournament* of randomly generated strategies was arranged.

Two observations were made. First, the best strategy discovered by the GA outperformed any of the previously known manually created strategies. Second, to find this powerful strategy, the GA needed only 50 generations, always with the population size fixed at 20 individuals. This means that to find a good solution, the GA needed to evaluate only $50 \times 20 = 1,000$ chromosomes out of the total of 2^{64} possibilities. In other words, not only did GA outperform anything discovered during the previous decades of scientific studies; it did so with remarkable efficiency.

The fact that only a small percentage of all strategies needed to be explored is explained by the implicit parallelism discussed in Section 6.6.

6.7.8 Summary

One of the reasons for including here the problem of the *prisoner's dilemma* is that it illustrates the tournament approach to fitness evaluation. In a sense, this is what is going on in the world of biology. Besides, the tournament can be used if we want to see how the population adapts itself to changed circumstances. Consider a scenario in which the prison terms from Table 6.3 are occasionally modified. Under the new circumstances, the successful individuals may have to follow a different strategy.

This said, one has to keep in mind that the computational expenses of the tournament may force the engineer to keep the population small: The tournament table for M individuals will contain the results of $M(M-1)/1$ encounters which is 1,770 in the case of, say, $M = 60$. This has to be multiplied by the number of rounds involved in each encounter. The prisoner's dilemma is still a simple domain where the 60-round experiment can be completed in less than a second. However, the engineer may face difficulties in a domain where each individual-to-individual game is expensive.

Finally, do not forget that the small population of just twenty individuals may imply the danger of premature degeneration (Section 6.4). In some domains, the chromosomes have to be long. For instance, if the *prisoner's dilemma* considers the last four rounds instead of three, the chromosomes will have to consist of at least $4^4 = 256$ bits. The reader already understands what this means for the necessary population size, for the involved computational costs, and for the chances of premature degeneration.

Control Questions

If you are unable to answer the following questions, return to the appropriate place in the preceding text.

- Summarize the nature of the problem known as the *prisoner's dilemma*. For What application domain was it originally intended?
- Explain how the agent's strategy was here represented by the binary-string chromosomes. How will you handle the game's initial rounds?
- Explain the essence of the *tournament* approach to fitness evaluation. Comment on its computational costs.
- Why is the GA so efficient in finding a good solution?

6.8 Practice Makes Perfect

To improve your understanding, take a chance with the following exercises, thought experiments, and computer assignments.

- Write a program implementing the general schema of the GA from Figure 6.1. For recombination, use two-point crossover.
- Design a multi-peak function, perhaps a high-order polynomial or a goniometric function such as $x + 2\sin(0.02x)$. The more complicated it is, the better. Using the GA program implemented in the previous task, find this function's global maximum for a given range of the values of x.
- How would you implement the GA if you wanted to find the maximum of a multivariate function?
- Implement a function that accepts as input a population, and returns *true* if the population is degenerated and 0 otherwise. Write another function that increases the diversity of a degenerated population by inserting randomly generated new individuals. What other mechanisms to increase the population's diversity do you know?
- The GA solution suggested in this chapter for the knapsack problem was relatively simple, which was necessary for educational purposes. Can you suggest a more sophisticated (and more efficient) way of handling the task?
- Write a program implementing the *prisoner's dilemma*, then test its behavior for different sizes of the population. Large populations are less likely to degenerate, but they may prove computationally expensive. Run a series of experiments to explore this trade-off.
- *Prisoner's dilemma* has been chosen here because it neatly illustrates the *tournament* approach to fitness calculation. Can you think of another problem where the tournament approach is likely to be beneficial?
- The knapsack problem and the *prisoner's dilemma* are still relatively simple domains. How would you employ the GA if you needed to address something more complicated, say, the travelling salesman or the job-shop scheduling from Chapter 5?

6.9 Concluding Remarks

The revolutionary book by Holland (1975) popularized the idea that biological evolution can be described in terms of a computer algorithm. The work was original, but its author was unaware that similar ideas had been formulated earlier by Rechenberg (1973) whose book, however, was written in German and thus lacked the impact it deserved. Prisoner's dilemma was explored at length by Axelrod (1984). The mechanism that encodes the knapsack problem by way of numeric chromosomes (plus its impact on recombination) was employed in a different context by Roszypal and Kubat (2001), but it is more than likely that other authors had employed the same idea even earlier.

The GA is less intuitive than heuristic search, but it is known to be more powerful than heuristic search in advanced applications with complicated fitness functions. In simple domains, hill climbing and best-first search are often good enough. The engineer should always choose a tool that is most appropriate for the given problem.

One of the advantages of the GA is its ability to react to changed circumstances. Thus in the *prisoner's dilemma*, the authorities may decide radically to modify the penalties; and in the knapsack problem, the values and weights of some items may also vary in time. Many engineering problems are in this sense non-stationary. From the programmer's perspective, these time-varying aspects may necessitate modifications of the way fitness is evaluated.

Let us mention in passing that, in the 1990s, the successes of the GA raised a lot of interest in *genetic programming*, a discipline that explores ways of developing *Lisp* programs by GA-based evolution. These efforts, however, are outside the scope of this textbook.

The GA was the first biologically inspired approach to artificial intelligence. Later, other successful attempts that sought inspiration in nature were made. Some of the most successful will be the subject of Chapter 8.

CHAPTER 7
Artificial Life

Quite a few approaches to artificial intelligence have been inspired by studies of biological systems. Apart from the genetic algorithm from the previous chapter, this is the case of *Artificial Life*, a scientific discipline that explores the possibilities of making a computer program appear to be alive. Strictly speaking, this field does not address this book's primary interests (problem solving and automated reasoning). Still, it deserves at least a brief exposition because it helps us appreciate the potential of *emergent properties*, a concept that plays a critical role in the swarm-intelligence techniques that are the subject of Chapter 8.

It is thus mainly for illustrative purposes that we outline here the principles of some of the most typical examples of artificial life: L-systems, cellular automata, and Conway's game of life.

7.1 Emergent Properties

The concept of *emergent properties* may appear rather philosophical. The reason we are investigating it here is that it has proved invaluable in some recent breakthroughs in the studies of artificial intelligence.

7.1.1 From Atoms to Proteins

An atom's behavior follows quantum-physics laws that explain why and how it combines with other atoms, thus forming molecules such as amino acids. An amino acid exhibits properties that at the level of atoms do not make any sense. For example, each of these molecules has a specific shape, contains a different mix of atoms, tends to attach itself to other amino acids, thus forming proteins. When studying proteins, biologists rely on features that are absent not only in atoms, but also in amino acids. A protein fulfills a specific task in the cell, consumes a certain amount of energy, and so on.

We observe that higher levels of organization are marked by traits and properties that are not found at lower levels. We say that these new traits and properties *emerge* only at this higher level, which is why they are referred to as *emergent properties*. The higher we climb the ladder of complexity, the more interesting these emergent properties become.

7.1.2 From Molecules to Society

A living cell is characterized by its way of locomotion, by the mechanism it uses to ward off bacteria, by its ability to multiply, and by many other such properties, most of which would never be mentioned in a discussion of a protein.

A great many cells may form an animal. The animal can be carnivorous or herbivorous, it has a specific nervous system, it may be covered by fur, it may be a predator hunting in packs, it prefers warmer or colder climates, it hibernates during winters, and so on. These, again, are properties that have emerged only at this level of complexity. There is no sense in mentioning these properties when talking about proteins or atoms.

7.1.3 From Letters to Poetry

An individual letter represents a phoneme. A specific group of letters put together becomes a noun or a verb. In connection with these, a grammarian will use terms such as conjugation or inflexion or tense. A word can be of foreign origin, it can be archaic or literary or colloquial, and it may exhibit many other aspects that we would never even consider when talking about isolated letters.

A sequence of words may form a poem which, in its turn, is marked by rhythm and rhyme and metaphor, and may convey meanings much richer than any isolated word. Poems can be classified as lyrics or epics, they may be profound or shallow, classical or modern—and all this constitutes yet another set of properties that have emerged with the transition to a higher level or organisation.

7.1.4 A Road to Artificial Life

With elevated complexity, new properties emerge. At a relatively high level, life emerges, exhibiting behavioral aspects denied to the lower levels. This observation has motivated a stunning question: Can we write a computer program that behaves in ways that make it look alive? Attempts to answer the question have given rise to a research field known as *artificial life*. The term is perhaps an overstatement, even exaggeration. Still, it is fascinating to realize that a computer program can move, replicate, even be capable of evolution, and that it can achieve these aspects with fairly simple means. This chapter will take a look at some of the best-known ideas.

By doing so, it will prepare the soil for the next chapter which will show how *swarm intelligence* can be achieved as an emergent property characteristic of a group of simple agents. Computational techniques based on this principle often outperform classical search techniques on a wide range of tasks.

Control Questions

If you are unable to answer the following questions, return to the appropriate place in the preceding text.

- What do we understand by the term, *emergent property*?
- Discuss this section's examples of emergent properties. Can you find other examples? If yes, explain which new properties emerge at each new level of complexity.
- What lesson can AI draw from these observations?

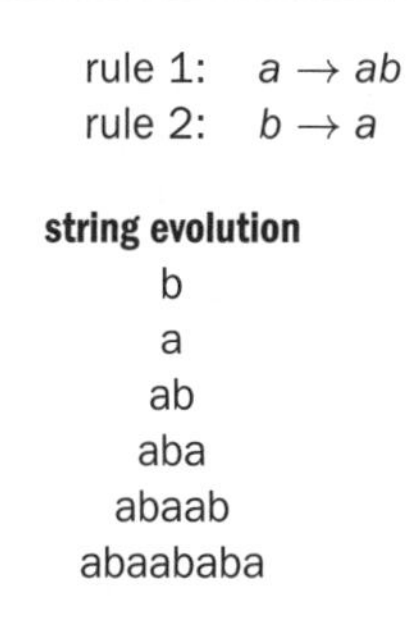

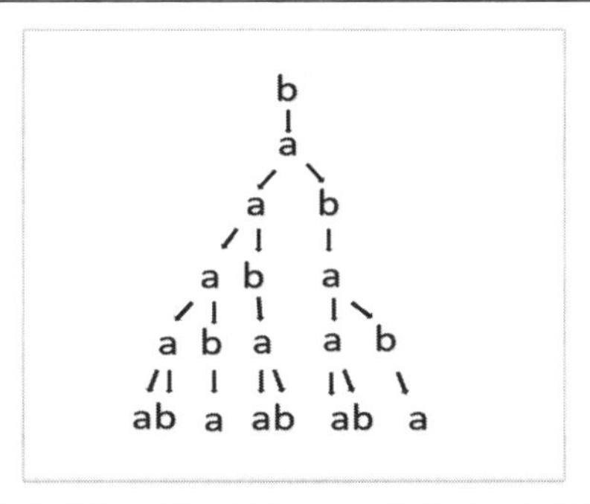

Table 7.1 Evolution of a String of Symbols in the Basic Version of the L-system

7.2 L-Systems

In the 1960s, Aristid Lindenmayer proposed his L-system, a formal language with which he intended to model the behaviors of diverse cells or plants. This may have been the first time that humanity came to recognize the phenomenon of emergent properties—and its potential (at first just theoretical) in computer programs.

7.2.1 Original L-System's Rules

This is perhaps the simplest example of artificial life. To appreciate the principle, let us constrain ourselves to an almost ridiculously simple domain: gradually evolving sequences of two letters, *a* and *b*. A string of these two letters is formed, and then converted to another string, by the systematic application of the following two rules:

rule 1: $a \rightarrow ab$
rule 2: $b \rightarrow a$

Suppose the evolution begins with a single letter, *b*. Rule 2 says that, in the next generation, this letter should be replaced with *a*. In yet another generation, rule 1 replaces *a* with the pair *ab*. Then, the first of these letters is replaced with another *ab* (rule 1), and the second letter is replaced with *a* (rule 2). Table 7.1 shows how the sequence evolves over a few generations. In this table, it is easy to notice a specific pattern that the changes adhere to.[1] Note also that the lengths of the successive strings follow Fibonacci sequence: 1, 1, 2, 3, 5, 8, ... We have just experienced that two very simple rules are capable of generating relatively complex patterns marked by remarkable regularity.

[1] Lindenmayer was a biologist. He argued that these ever-changing sequences can be used as a model of the growth of algae as well as certain branching structures in plants.

7.2.2 Another Example: Cantor Set

Over time, quite a few other pattern-generating rules of similar nature were found and experimented with. Many of them were meant to model specific aspects of the world of nature. Perhaps the most famous one, known as *Cantor set*, relied on the following two rules:

rule 1: $a \rightarrow aba$
rule 1: $b \rightarrow bbb$

The reader will easily verify that if we start with a, the procedure generates the following sequence (again, the pattern is easy to notice):

aba
aba bbb aba
aba bbb aba bbb bbb bbb aba bbb aba

7.2.3 Lesson

We have convinced ourselves that even a trivial language (consisting of just two symbols), and some very simple rules, can give rise to interesting sequences reminiscent of growth patterns.

Control Questions

If you are unable to answer the following questions, return to the appropriate place in the preceding text.

- What two rules are used in the L-system? What kinds of regularity do they generate in strings of symbols?
- Do you know another mechanism of a similar nature?

7.3 Cellular Automata

L-systems were mentioned here mainly for historical reasons, as the oldest examples of artificially designed emergent properties. Somewhat more advanced are *cellular automata*. These, too, typically operate with binary sequences, but they rely on more sophisticated rules.

7.3.1 Simple Example

Consider an automaton consisting of 10 cells arranged in a line. Each cell finds itself in one out of two possible states: alive or dead. In Figure 7.1, the live cells are filled with gray squares, and dead cells are empty. For convenience, individual cells will be referred to by integers from 1 through 10 (from left to right). At the top of the picture is the automaton's randomly initialized zeroth generation that has four live cells.

For each cell, we will consider a neighborhood that stretches two cells to either side (and includes this cell). For instance, the neighborhood of cell 4 consists of cells 2, 3, 4, 5,

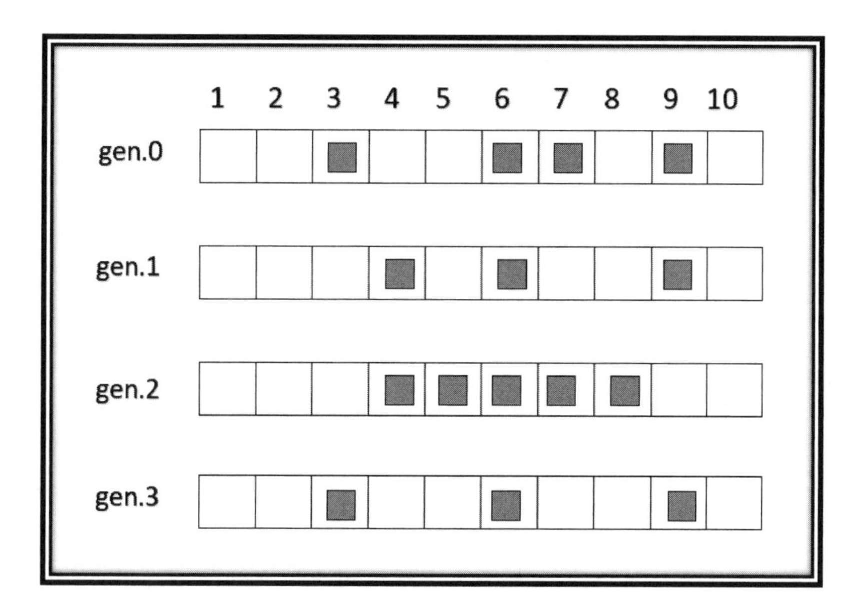

FIGURE 7.1 Cellular automaton. Live cells are filled with gray squares, dead cells are empty. For each cell, consider its value, and also the value of its four nearest neighbors. If exactly 2 or exactly 5 cells in the neighborhood are alive, the cell will be alive in the next generation; otherwise, it will be dead.

and 6. If the cell is close to one of the automaton's boundaries, the neighborhood is shortened accordingly; it cannot reach beyond the boundary. Thus the neighborhood of cell 2 consists of cells 1, 2, 3, and 4.

The contents of the automaton will change from one generation to the next according to the following simple rule:

> "A cell whose neighborhood contains exactly 2 or exactly 5 live cells will be alive in the next generation. If its neighborhood contains a different number of live cells, it will be dead in the next generation."

For instance, in the zeroth generation, cell 3 is alive, and we can see that its neighborhood contains only this one live cell. The above rule therefore ensures that in the next generation, cell 3 will be dead. The same rule is applied to each of the 10 cells, which converts the automaton of generation 0 into the one labeled in the picture as generation 1. This is then continued for two more generations. It can happen that at a certain generation, all cells in the automaton are dead, in which case further changes are no longer possible, and the automaton's evolution stops.

7.3.2 Variations

Whether a cell will in the next generation be dead or alive depends on its 5-cell neighborhood. In our case, the cell was bound to be alive if exactly 2 or exactly 5 cells in the neighborhood were alive. Another rule may command that the cell will be alive if exactly

1 or 3 cells in the neighborhood were alive—or any other combination. Besides, the size of the neighborhood does not have to be five; it can be three or seven or any other number. Each of these rules will result in a different way the automaton evolves.

7.3.3 Adding Another Dimension

The cell automaton considered in this section is linear, consisting as it does of a single dimension. This is not mandatory. The next section will show how adding another dimension can lead to a two-dimensional automaton capable of exhibiting much more interesting behaviors.

Control Questions

If you are unable to answer the following questions, return to the appropriate place in the preceding text.

- What is the nature of the one-dimensional automaton from this section?
- What variations are possible?

7.4 Conways' Game of Life

Perhaps the most impressive—and certainly the best known—of all artificial-life adventures is *Conway's game of life.* We can regard it as a more sophisticated version of the cellular automaton from the previous section. Its rules are more complicated and they result in more interesting patterns of behavior.

7.4.1 Board and Its Cells

Consider a two-dimensional grid such as the one in Figure 7.2. Taking inspiration from chess, we will refer to the individual cells by their coordinates, using letters for columns

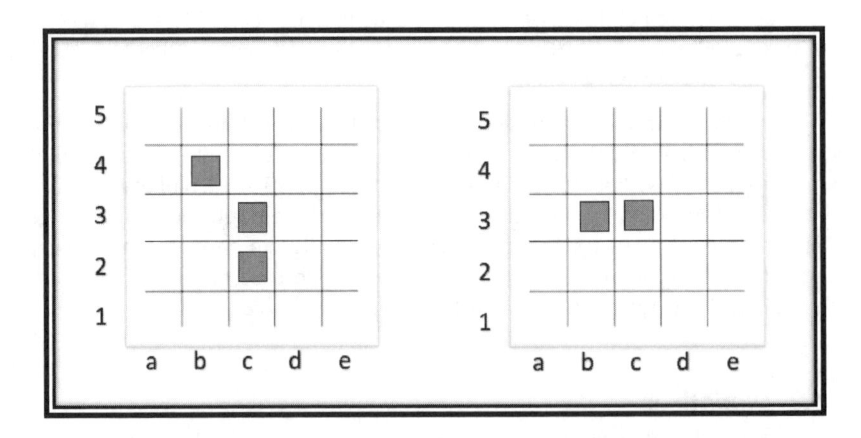

FIGURE 7.2 Applying the four rules to the pattern on the left will cause it to change to the pattern on the right.

1. A live cell with two or three live neighbors stays alive.
2. A live cell with more than three live neighbors dies of overcrowding.
3. A live cell with less than two live neighbors dies of loneliness.
4. A dead cell with exactly three live neighbors becomes a live cell.

Table 7.2 The Four Rules of Conway's Game of Life

and integers for rows. For instance, the cell in the upper-left corner is a 5 and the cell in the bottom-right corner is e1. For simplicity, the picture shows only a 5-by-5 board; in reality, the board will be much larger.

Just as in cellular automata, each square finds itself in one out of two possible states, dead or alive. In Figure 7.2, the live cells are marked by gray squares and the dead cells are empty. In the board depicted on the picture's left, we can see three live cells; all the other cells being dead.

7.4.2 Rules

Whether a dead cell is converted into a live cell, or a live cell into a dead one, is decided by the four rules summarized in Table 7.2. Similarly to cellular automata, the rules operate over neighborhoods. Here, however, the neighborhood comprises only the immediate neighbors, and does not include the cell itself. Thus in Figure 7.2, the neighborhood of a1 contains a2, b1, and b2. The neighborhood of c3 contains eight cells.

In the left part of Figure 7.2, cell b4 is alive, but since it has only one live neighbor, it dies of loneliness. The same happens to cell c2: having only one live neighbor, it dies of loneliness. This is why b4 and c2 in the right part of the picture are both empty. Cell c3 on the left is live, and since it has two live neighbors, it survives. Cell b3 is dead, but since it has three live neighbors, it becomes a live cell itself. The reader is encouraged to verify that systematic application of the four rules converts the pattern on the left to the pattern on the right.

7.4.3 More Interesting Example

The upper-left corner of Figure 7.3 contains a simple pattern which, when subjected to our four rules, leads to some interesting behavior. Next to the original pattern, the picture shows the following two generations, the second being the one in the upper-right corner. If we continue in this manner (see the two boards in the second row), we realize that in a few generations, the pattern from the upper-right corner has reappeared, but not in the same location on the board; it has been shifted by one column and by one row.

It is easy to implement the four rules in a computer program, and to visualize the board positions on the monitor. If we do so, we will observe that, throughout the follow-up series of generations, the pattern keeps reappearing at regular time intervals, always shifted by one square down and by one square to the right. The computer animation then leaves the impression that the pattern is "gliding" from the board's upper-left corner toward the board's bottom-right corner. This is why literature refers to this pattern as a *glider*.

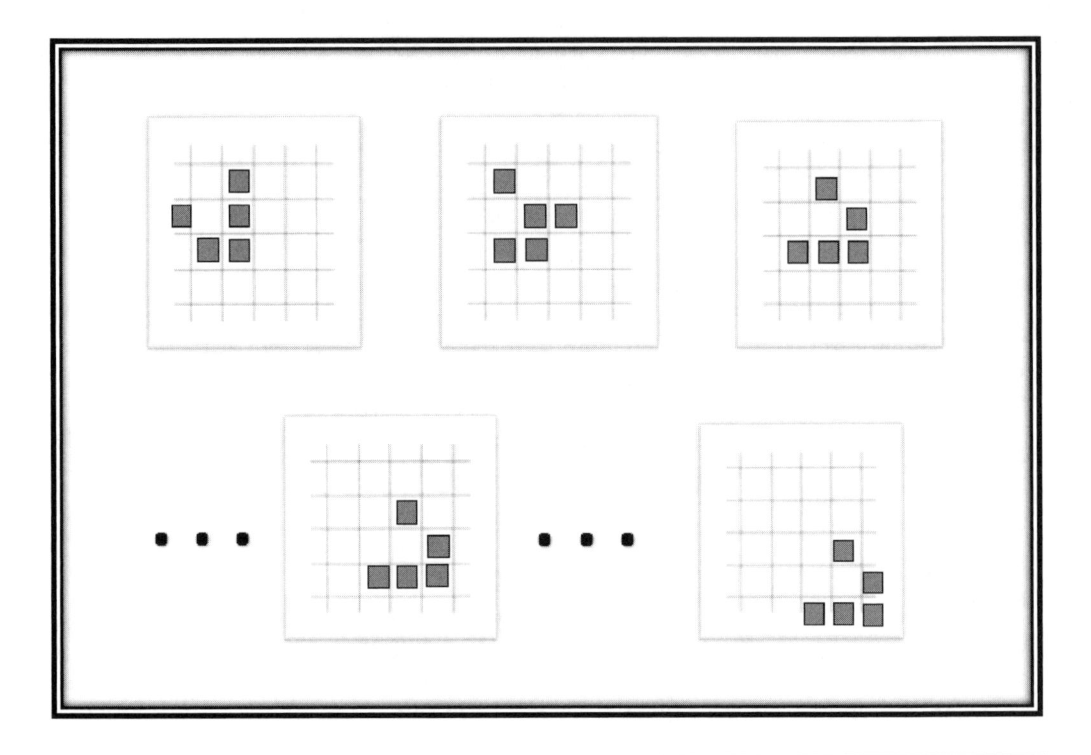

Figure 7.3 Look at the pattern formed in the third generation. The same pattern reappears a few generations later, only that it has now been shifted down and to the right. Later, it again reappears, again shifted. The pattern is known as *glider*.

7.4.4 Typical Behaviors

Depending on the pattern's shape at the games' beginning, different observations are made. Some patterns have a way of recurring: They keep changing until, after a few generations, the same initial pattern reappears. Other initial patterns are called *oscillators* because the game keeps switching between two patterns. Yet others result in *movements* that create an impression of something alive, even reproducing itself. Finally, some patterns are static in the sense that they do not change at all. No wonder that this algorithm has been given the name, *game of life*.[2]

The game's variability knows no bounds, which is what makes it so popular. The fact that any student can easily write a program that generates new patterns based on a few user-supplied rules has encouraged a lot of experimentation. Fans have come up with fancy names for specific patterns: glider gun, spaceship, toad, pulsar, and others.

[2] The reader will find two additional examples in Section 7.5.

7.4.5 Summary

A more detailed study of these patterns, and of the diverse behaviors generated by different sets of rules, does not belong to the field of artificial intelligence. The reason why Conway's intriguing algorithm has been mentioned here is that it illustrates the idea of emergent behavior discussed at the beginning of this chapter. At the lowest level, a cell is just alive or dead, and nothing more can be said about it. At higher levels, we observe groups of cells arranged in interesting shapes undergoing specific changes, exhibiting strange new properties that include movement, replication, and even multiplication.

Over the last decades, emergent behavior and emerging properties have become popular topics among AI theoreticians and practitioners alike—and has transformed their view of what strategies are likely to become AI's center of gravity in the next generation or so.

Control Questions

If you are unable to answer the following questions, return to the appropriate place in the preceding text.

- List the four rules employed in Conway's game of life, and show examples of their application to the board's individual cells.
- Why this simple automaton is called a "game of life"? What typical behavior does it exhibit?

7.5 Practice Makes Perfect

To improve your understanding, take a chance with the following exercises, thought experiments, and computer assignments.

- Return to the L-system's example from Table 7.1 and use its two rules to generate a few more letter strings. Then create an initial string of your own (different from those in Table 7.1) and see what happens if you subject it to the same rules.
- Write a program that accepts as input a string of zeros and ones, and employs rules such as those from Section 7.3. Use computer animation to illustrate this cellular automaton's behavior.
- Invent an interesting *L*-system of your own. Invent a cellular automaton of your own. Write computer programs visualizing their behaviors.
- Take a look at the two patterns shown in Figure 7.4. Hand-simulate Conway's algorithm and comment on the observations you have made. What happens with these patterns over a series of generations?
- Write a program that will automate the exercise from the previous task. The input is a specific pattern. Computer simulation will show how this pattern evolves over a series of generations.

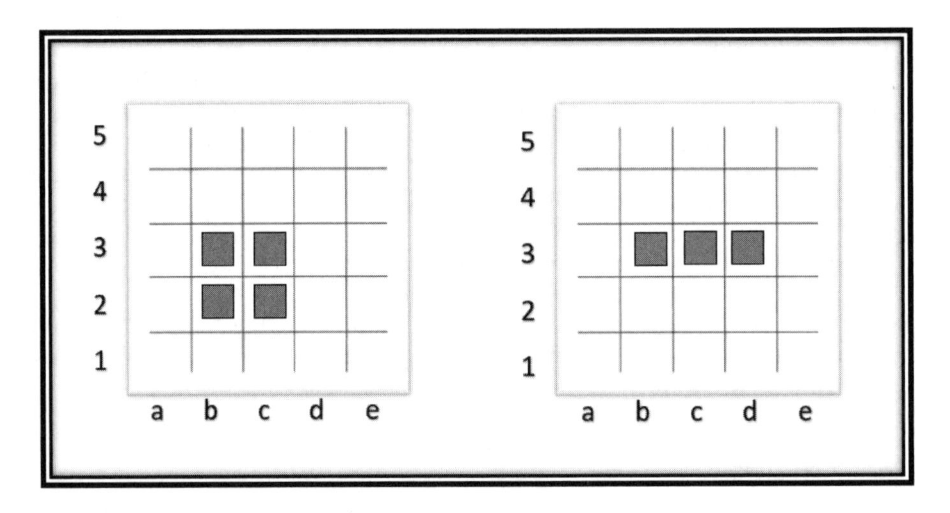

Figure 7.4 Examples of Conway's game of life for practicing.

7.6 Concluding Remarks

The question of artificial life was for the first time raised even before computer programming was invented. As early as during WW II, John von Neuman (one of the fathers of the digital computer) came up with his own definition of life, listing its characteristic features such as the ability to reproduce. He even considered the possibilities of implementing such self-reproducing entities in hardware. Unfortunately, the technology of the day lacked the sophistication needed for the construction of a working prototype. Besides, war time had other challenges to cope with.

Later generations realized that artificial life would be more easily implemented in software. This philosophy led to the invention of the cellular automata introduced by Ulam (1962), a groundbreaking paper that raised a lot of interest. The fresh new ideas and suggestions caught the scientists' imagination, and it is almost to be regretted that this book cannot afford to devote more space to them.

Closely related to Ulam's line of thought was the invention of what came to be known as L-systems. Their principles were first published by Lindenmayer (1968). True, he meant them to be models of natural systems, tools to help us improve our understanding of these systems. But while his primary interests lay outside computer science, the ideas that he pioneered surely influenced subsequent studies of artificial life.

The game of life was invented by John Horton Conway. He did not publish the idea himself, only suggested it in a letter to Martin Gardner who then wrote a paper for *Scientific American* (Gardner, 1970), giving full credit to its originator. Gardner's article was a sensation. The experience that a few simple rules can give rise to entities that appear to be alive inspired quite some following.

From the perspective of this textbook, the most important aspect of artificial life is that it illustrates the unexpected consequences of emergent behavior and emergent properties. The experience paves the way to the more advanced (and more practical) principles of *swarm intelligence* that are the subject of Chapter 8.

CHAPTER 8
Emergent Properties and Swarm Intelligence

For decades, the main thrust of artificial intelligence's (AI's) approach to problem solving focused on search. In the end, relevant techniques became so sophisticated and well-understood that no further work along these lines seemed possible, or even needed. Then came the revolution: The *swarm intelligence* that attacked AI problems by groups of small agents, each executing an action so simple as to seem almost trivial. The sum of these actions proved capable of remarkable feats, thanks to the phenomenon of the emergent properties we know from Chapter 7.

This decentralized approach is very efficient; quite often, it finds a solution at a fraction of the time that would be needed by more traditional approaches. No wonder that swarm intelligence has come to dominate the field to the point where some experts have labeled heuristic search as dead and buried. True, this may be going too far, but the new paradigm certainly does merit any AI student's undivided attention.

Among the relevant multi-agent techniques, the following three answer this book's needs best: ant-colony optimization, particle-swarm optimization, and the honey bees' algorithm. The first deals with discrete decision-making, the latter two target continuous-valued domains.

8.1 Ant-Colony Optimization

Many AI problems, including those from Chapter 5, can be cast as graph analysis. Thus the traveling salesman needs to find the shortest path through a set of cities, and many puzzle-solving approaches traverse a search tree—which, again, can be seen as a graph-related exercise. Problems of this kind are successfully handled by a technique known as *ants-colony optimization, ACO.*

8.1.1 Trivial Formulation

Let us begin with an extreme simplification of the basic problem statement. The point is to get acquainted with the basic terminology and with the technique's overall philosophy. ACO's full version will then be the subject of Section 8.2 that will also explain how to use it when dealing with practical problems such as the traveling salesman.

8.1.2 Ant's Choice

The top of Figure 8.1 shows an ant facing a conundrum: Which of the three roads to follow? One leads to a doughnut, another to a cheese, and the last to a pie. Two criteria

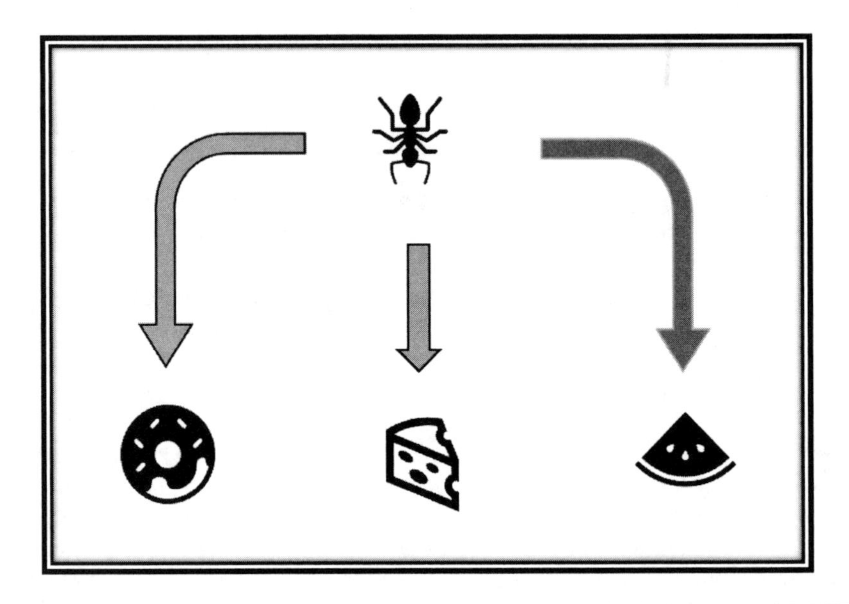

FIGURE 8.1 When choosing the most attractive destination (doughnut, cheese, or pie), the ant considers two criteria: distance and food quality, inferring the latter from its popularity among other ants.

spring to mind: distance and attractiveness. Distance is obvious. The ant is hungry and wants to reach the food cache fast.

As for attractiveness, however, this will only be known once the site has been reached. Fortunately, an indirect clue can be used: Suppose that many other ants have chosen, say, the pie. If so, they surely have done so for a reason. Maybe they liked the food's taste, maybe its amount, maybe something else. At any rate, if many have found the choice so attractive, why not just follow suit?

8.1.3 Pheromone Trail

Reality is more complicated than Figure 8.1 indicates. For instance, the ant may have to face *many* options, not just three. Besides, not each path leads to a food cache. Far from it. Food is scarce. It does exist, but no one knows where, and it is thus necessary to look for it. This is why the ants engage in *foraging*.

At the beginning, the foraging ants just wander around aimlessly, not knowing where to turn. Sooner or later, though, one of them hits on food source that others may take advantage of, too. To show them the way, the successful individual leaves, on its way home, a trail of a chemical known as *pheromone*. In this way, the path is highlighted for others to follow. If the number of foraging ants is high, there may be quite a few such trails to choose from, some of them rich in pheromone, others less so.

8.1.4 Choosing the Path

From this moment on, the search is no longer blind. The marked trail points the way. The rule is, *prefer* the path covered by pheromone. Note the emphasis on the word, "prefer." The ant is attracted by the scent, but is under no obligation to pursue it; the decision is only probabilistic. Those ants that have followed the indicated direction will find food, too, and having thus been successful themselves, they will add their own dose of pheromone. Those paths that have been traveled by many successful ants collect a growing amount of the deposited chemical.

The ant's world consists of a network of trails, some with lots of pheromone, others with less, or none at all. How attractive each path is depends on how much of the chemical covers it. The more of it, the higher the chance that the individual will choose it.

8.1.5 Evaporation versus Additions

The chemical being unstable, it gradually evaporates. Given enough time, it will disappear altogether. But if successful ants have added more pheromone than what has been lost by evaporation, the density increases. Conversely, if the reinforcements are rare, evaporation gets the upper hand and the thread gets thinner.

8.1.6 Programmer's Perspective

Let us take a look at how to implement the path-choosing process in a computer program. Here are the criteria the ant wants to consider: the goal's proximity and the path's popularity. The former is the reciprocal of distance. The latter is proportional to pheromone density. The desirability of any given path is the product of these two.

In Figure 8.1 the ant faced three options: doughnut, cheese, and pie. For each, the first numeric column in Table 8.1 provides an example distance, d_i. From the distance, the goal's proximity is obtained as $\eta_i = 1/d_i$ (next column).

Then comes the pheromone density, τ_i. Later, we will learn how this density evolves over time. For the time being, however, let us simply assume that the values are those given in the next column. Finally, the table's last column contains the *desirability* of each of the three paths, calculated as the product of pheromone density and proximity, $\delta_i = \tau_i \cdot \eta_i = \tau_i \cdot \frac{1}{d_i}$.

The values of distance and pheromone density for each target are given. Proximity is the reciprocal of distance. *Desirability* is the product of proximity and pheromone density.

Goal	Distance d_i	Proximity $\eta_i = 1/d_i$	Pheromone density, τ_i	Desirability $\delta_i = \tau_i \cdot \eta_i = \tau_i/d_i$
Doughnut	1	1.00	1	1.00
Cheese	2	0.50	1	0.50
Pie	3	0.33	2	0.66

The probability of choosing a concrete path is calculated as follows:

$$P_d = \frac{1}{1+0.50+0.66} = \frac{1}{2.16} = 0.46$$

$$P_c = \frac{0.5}{1+0.50+0.66} = \frac{0.5}{2.16} = 0.23$$

$$P_p = \frac{0.66}{1+0.50+0.66} = \frac{0.66}{2.16} = 0.31$$

TABLE 8.1 Numeric Example: An Ant Choosing a Path

8.1.7 Probability of Choosing a Concrete Path

The likelihood of a path being selected is determined by the desirabilities thus obtained. The idea is to employ the *soft-max* function that converts these desirabilities to probabilities. Here is how the probability of the *i*-th path is calculated (the proportion of δ_i in the sum of all desirabilities):

$$P_i = \frac{\delta_i}{\Sigma\delta_i} \tag{8.1}$$

At the bottom of Table 8.1, we can see how these numbers are calculated for the three paths in Figure 8.1. From the results, we conclude that if 100 ants face the same choice, then on average 46 will head for the doughnut, 23 for the cheese, and 31 for the pie.

Let us reemphasize the probabilistic nature of the whole undertaking. ACO's underlying philosophy assumes that all ants are subject to the same decision-making rules. As a result, the majority will take the decisions that have proved most attractive in recent history. Every now and then, however, an individual will pursue a path that has so far been neglected. A theoretician will say that an essentially greedy policy (one that always strives for the maximum benefit) is thus complemented by occasional exploration.

8.1.8 Path-Selecting Mechanism

Once we know the probabilities of the individual paths, the concrete choice is determined by a random-number generator using the same mechanism as the one that Chapter 6 recommended for the genetic algorithm's survival game. Figure 8.2 illustrates the point. In the interval $[0.0, 1.0]$, each option receives a segment whose length is proportional to its probability, P_i (the probabilities are calculated at the bottom of Table 8.1). The random-number generator returns a number from $[0.0, 1.0]$. The number's location along the segments decides which goal to pursue.

8.1.9 Adding Pheromone

An earlier paragraph explained that each ant, on its way back from the food cache, leaves on the trail a certain amount of pheromone. Here is a popular way of implementing this in a computer program.

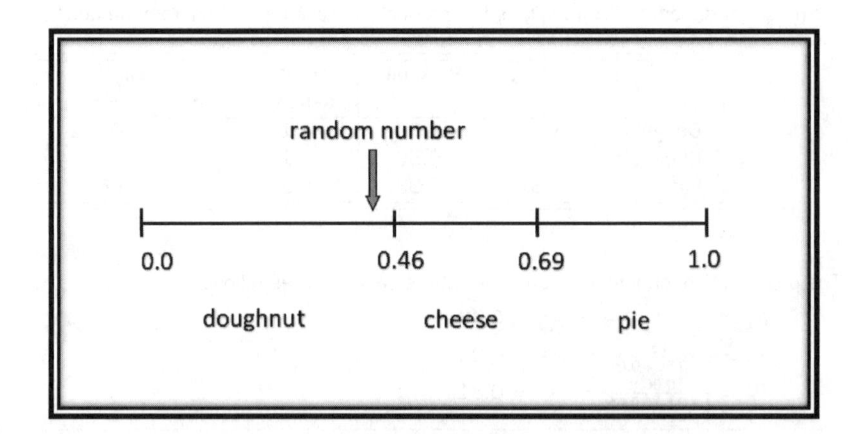

FIGURE 8.2 ACO's probabilistic decision-making. Each goal is assigned a line segment whose length is proportional to desirability. A random number then falls into one of these segments.

Suppose there is not just one ant, but many of them, and each has the same amount of pheromone to use. If the path is long, the added chemical has to be spread thinly. The shorter the path, the higher the density of the pheromone the path receives. If the length is 10, then twice as much pheromone is added per unit of distance than in the case of length 20. Put more technically, the amount of added pheromone is inversely proportional to the distance.

Note that this means that the amount of pheromone along a path informs the decision-maker not only about the food's popularity, but indirectly also about its distance. This may not be exactly what is going on in nature, but it is okay for a computer implementation.

8.1.10 Pheromone Evaporation

Pheromone is not only being added; it also keeps evaporating. How quickly it evaporates is controlled by the *evaporation coefficient*, ρ, whose value is provided by the programmer. At fixed time intervals, say, each five seconds, the pheromone density along each of the three edges is reduced according to the following formula:

$$\tau = \tau(1 - \rho) \tag{8.2}$$

Suppose that the programmer has set the value of the evaporation coefficient to $\rho = 0.1$. If the pheromone density on a certain path at time t is $\tau_t = 10$, then, unless some more pheromone has meanwhile been added, the density at time $t + 1$ drops to $\tau_{t+1} = 10 \times (1 - 0.1) = 9.0$.

8.1.11 Non-stationary Tasks

The probabilistic approach just described gives preference to edges that are short and popular. This said, some of the longer and less frequently visited edges are occasionally explored, too, especially if there are a great many ants. This exploration is important in *non-stationary* domains whose characteristics are prone to change in time. For instance, some of the food sources have been depleted, others forcefully removed, and yet others added. Domains of this kind are called *non-stationary*.

In these non-stationary domains, ACO is flexible enough to adjust to the changed circumstances. If food is no longer found, at a certain location, ants will cease to add pheromone to the path leading to it, and the chemical will gradually evaporate. Conversely, a newly emerged food source may cause a previously "unattractive" path to gain in popularity. At the beginning, one of the ants hits on the cache by sheer good luck. Having discovered food, it will deposit pheromone—which will cause other ants to pursue the same path.

Note that ACO's ability to adjust to these changes has been made possible by the probabilistic nature of the underlying decision-making process.

Control Questions

If you are unable to answer the following questions, return to the appropriate place in the preceding text.

- What are the two fundamental criteria that guide the ant's choice of a path? Write down the probabilistic formula that underlies the decision where the ant should go.

- Describe how pheromone density is increased by the ant returning from a food cache, and how pheromone evaporation reduces this density.
- Comment on how ACO's flexibility helps the paradigm to adjust to changed circumstances in non-stationary domains.

8.2 ACO Addressing the Traveling Salesman

Section 8.1 illustrated ACO's fundamentals on an extremely simplified domain: Each ant was to choose one out of three paths, and there was food at the end of each path. Reality is more interesting than that. Instead of choosing a single path, the ants have to wander around an entire graph, sometimes a very large one. The mechanisms that update the pheromone density along each of the graph's edges are in this case somewhat more complicated.

The best way to explain the principles is by considering a concrete application. A reasonable choice is here the traveling salesman because it serves as a model of many engineering problems. Anybody who understands how to address this particular problem will know how to handle other domains of a similar nature. Besides, the traveling salesman's essence makes it a convenient target for an ant colony.

8.2.1 Ants and Agents

Swarm-intelligence techniques, such as ACO, operate with sets of agents, sometimes a great many agents. In this section, each agent mimics the behavior of a single ant. It thus makes no difference whether we talk about an *ant* or an *agent*. In what follows, we will use the two terms interchangeably.

8.2.2 ACO's View of the TSP

For the sake of simplicity, let us limit ourselves to the four-city version shown in the graph in Figure 8.3.[1] An agent playing the role of an ant starts at city A, chooses (probabilistically) from the three edges available to it, and by following this edge, reaches the next city, say, B. At this point, the agent chooses (again, probabilistically), one of the remaining two edges and reaches the next city, say, C. Finally, it continues to the last remaining city, D, from which it returns to A.

The same is done by all the other ants, easily hundreds of them, in a realistic application. Each ant follows its own trajectory and, while doing so, it leaves behind pheromone trail. Frequently traveled edges receive more pheromone than rarely visited ones. Along neglected edges, pheromone evaporates, and its low density makes these edges less attractive, and they gradually fall into disuse. If the mechanism of adding and evaporating pheromone is well-tuned, ACO relatively quickly converges to the shortest—or almost-shortest—path along the graph (similarly as in the case of other AI techniques, this one does not guarantee the best solution, only a "reasonably good" one).

8.2.3 Initialization

At the beginning, all edges are given the same initial pheromone density. In nature, this of course means no pheromone at all, $\tau_i = 0$. In a computer implementation, this would not

[1] Realistic applications work with at least dozens of cities, often much more than that.

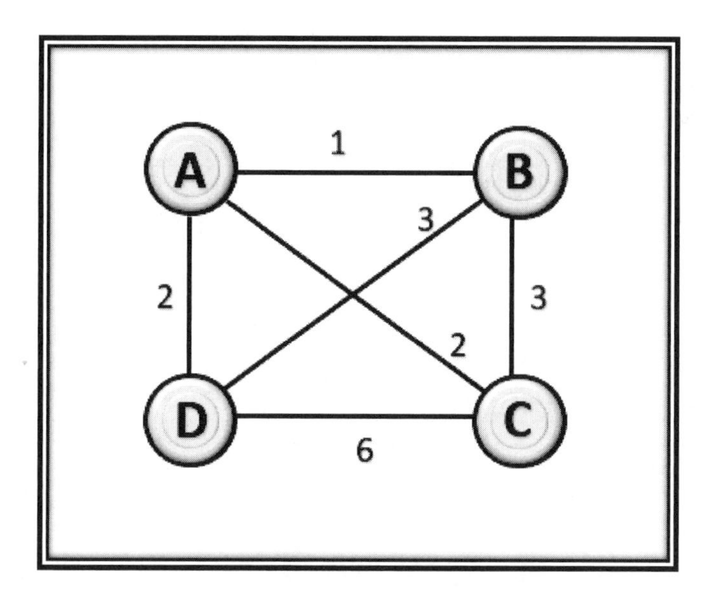

FIGURE 8.3 A simple four-city version of the traveling-salesman problem.

be practical. Remember Formula 8.1 that calculates the probabilities with which edges are selected? In this formula, the denominator would initially be zero because $\delta_i = \tau_i \cdot \frac{1}{d_i} = 0 \cdot \frac{1}{d_i} = 0$ for each i, which means that also $\Sigma_i \delta_i = 0$. A zero in a denominator is of course unacceptable.

To avoid this inconvenience, it is customary to initialize pheromone densities to a non-zero value, usually $\tau_i = 1$.

8.2.4 Establishing the Probabilistic Decisions

The four-city system from Figure 8.3 gives each ant the choice of six alternative routes: ABCDA, ABDCA, ACBDA, ACDBA, ADBCA, and ADCBA. The length of any of these routes is the sum of the distances of the four edges that the route involves. The concrete values are calculated in Table 8.2 where the first column specifies the concrete trajectories, and the second column calculates their lengths.

During an ant's trip through the graph, each edge is chosen with a certain probability. Here is how the probabilities are established. At the beginning, in city A, the ant has three choices. Since the initial pheromone densities are everywhere the same, the ant's decision where to go will only depend on the next city's proximity. The mechanism is the same as in the simplified case from the previous section, the probabilities being calculated using Equation 8.1.

Having calculated the probabilities, the ant makes its probabilistic choice and proceeds to the city at the end of the selected edge. At this new location, it goes through the same decision-making, and keeps repeating the same procedure each time it faces a choice, until it finally returns to the starting point, A.

8.2.5 Numeric Example

Let us take a look at how all this is carried out in the concrete four-city world from Figure 8.3. Let us repeat that, at the beginning, all edges have pheromone density $\tau_i = 1$,

The distance covered along each route is the sum of the lengths of its edges.
The last column gives the amount of pheromone (per distance unit) added by a single ant to each edge in the route.

Route	Distance	Pheromone per Ant
ABCDA	$1+3+6+2=12$	1/12
ABDCA	$1+3+6+2=12$	1/12
ACBDA	$2+3+3+2=10$	1/10
ACDBA	$2+6+3+1=12$	1/12
ADBCA	$2+3+3+2=10$	1/10
ADCBA	$2+6+3+1=12$	1/12

TABLE 8.2 The Distances Covered by an Ant Along the Different Routes, and the Amount of Pheromone (per Distance Unit) Added by the Ant Along Each Route

which means that each edge's desirability at this stage is equal to the next city's proximity, $\delta_i = \tau \cdot \eta = 1 \cdot \eta = \eta = \frac{1}{d_i}$. In city A, the distances of B, C, and D are 1, 2, and 2, respectively; the proximities of these three cities are therefore 1, 0.5, and 0.5. Here is how the probabilities of the three available edges are calculated:

$$P_{AB} = \frac{\delta_{AB}}{\Sigma_z \delta_{Az}} = \frac{1.0}{1.0 + 0.5 + 0.5} = \frac{1}{2}$$
$$P_{AC} = \frac{\delta_{AC}}{\Sigma_z \delta_{Az}} = \frac{0.5}{1.0 + 0.5 + 0.5} = \frac{1}{4}$$
$$P_{AD} = \frac{\delta_{AD}}{\Sigma_z \delta_{Az}} = \frac{0.5}{1.0 + 0.5 + 0.5} = \frac{1}{4}$$

Each ant's decision is made using the *softmax* function from Equation 8.1. If 100 ants start at A, about 50 of them will choose the edge leading to B, about 25 will choose the edge leading to C, and about 25 will choose the edge leading to D.

At the location that the ant has thus reached, there are only two options left. For instance, in city B, the ant can go either to C or D, with the distances being in both cases the same, $d_{BC} = 3 = d_{BD} = 3$. This means that, once the ant has reached B from A, the probabilities of choosing either of the remaining two edges are the same, $P_{BC} = \frac{1/3}{1/3+1/3} = 0.5$ and, likewise, $P_{BD} = 0.5$.

The procedure is then repeated for each new edge. In our very small domain, with only four cities involved, the situation becomes trivial. At the third city, there is no choice left. The agent simply proceeds to the last remaining place, and then returns to A. In other words, these last probabilities are 100%. For the reader's convenience, all the numbers thus calculated are shown in Figure 8.4.

8.2.6 How Much Pheromone Is Deposited by a Single Ant?

In the ACO's computer model, each ant drops the same fixed amount of pheromone uniformly along the entire route it has taken, whether the route was long or short. If the distance is 10, then 0.1 of the ant's pheromone is left on the trail per distance unit.

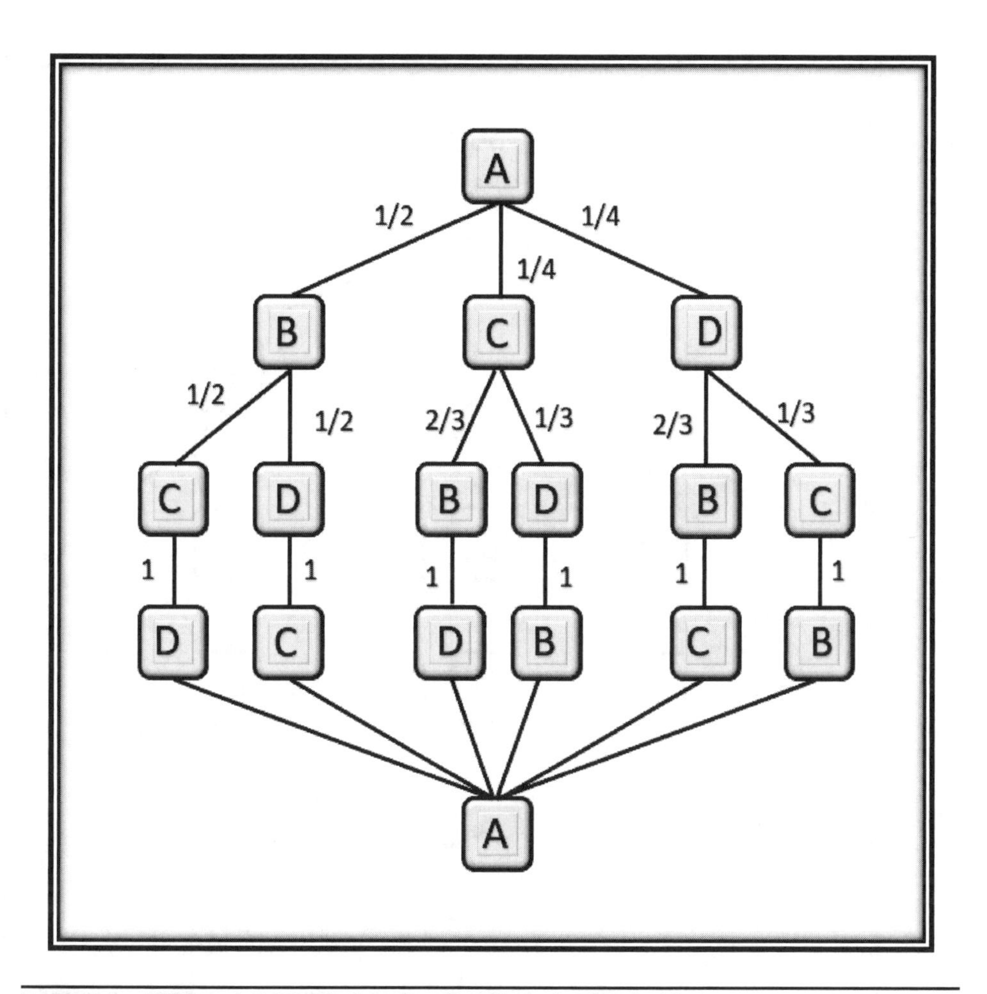

Figure 8.4 ACO's probabilities in the given traveling-salesman problem.

If the distance is 5, then 0.2 is left per distance unit. Again, we assume that each ant has at its disposal the same amount of pheromone for each single trip.

The last column in Table 8.2 gives for each route the amount of pheromone deposited by a single ant per distance unit. The reader can see that this number is obtained by dividing 1 (the ant's unit store of pheromone) by the distance.

8.2.7 Number of Ants Along Each Route

Whereas the previous paragraph focused on the amount of pheromone added by a single ant, we must not forget that the ACO's basic philosophy is that *many* ants wander around the graph. The total amount of pheromone added by all of them is then obtained as sum of the deposits of all individuals along the edges.

In the previous numeric example, the probabilities of choosing, in city A, from among the three edges were established as $P_{AB} = 1/2, P_{AC} = 1/4$, and $P_{AD} = 1/4$. These numbers are indicated in Figure 8.4, and the same graph also shows the probabilities for

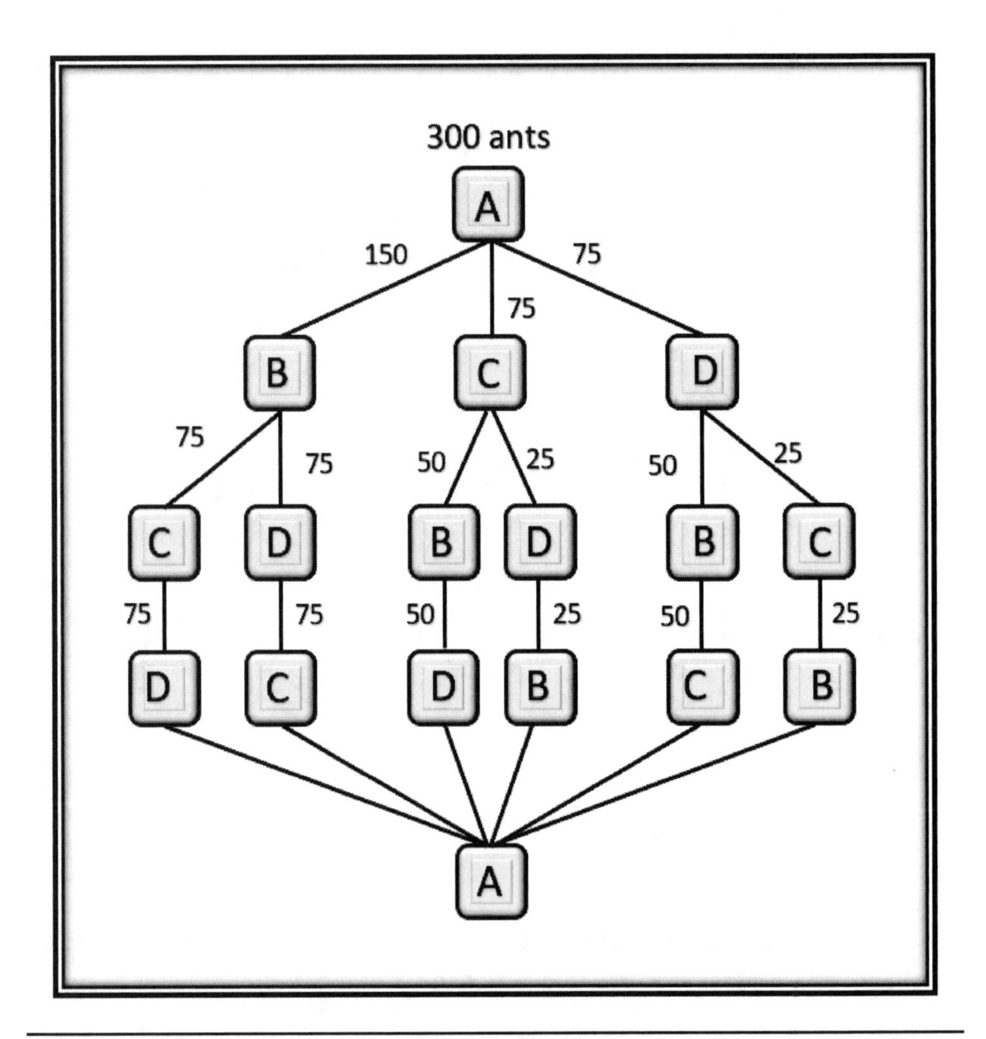

FIGURE 8.5 Average numbers of ants following the individual edges, assuming that the entire colony consists of 300 ants.

the choices made at the following cities. Suppose the program works with many ants, say, $N = 300$. In this event, it makes sense to expect that the numbers of ants choosing the individual edges are (with the given probabilities) approximately those given along the edges of Figure 8.5. For instance, $P_{AC} = 1/4$ means that about one-quarter of the 300 ants (i.e., 75) will begin their journey by moving from A to C; of these, two-thirds (i.e., 50) will continue to B, after which there is no other choice but to proceed to D (i.e., all 50 will proceed to D), and then return to A. This means that 50 ants follow the route, ACBDA.

8.2.8 Pheromone Added to Each Edge

Let us focus, for instance, on edge AB. From the different routes listed in Table 8.1, the following four contain this edge: ABCDA, ABDCA, ACDBA, and ADCBA. In the first two routes, the ants move from A to B, and in the second two, they are following the same edge in the opposite direction, from B to A. We assume that the direction does not affect the amount of pheromone deposited on the edge.

The last column in Table 8.1 gives the amount of pheromone per distance unit per one trip. For instance, each ant following ABCDA adds 1/12 of the unit amount per unit distance. Figure 8.5 is telling us that the number of ants following this route, ABCDA, is 75. Consequently, the amount of pheromone added by this cohort is $75 \times 1/12 = 75/12$.

Following the same line of reasoning, we can obtain the amounts of pheromone that have been deposited by ants that followed the remaining three trajectories: ABDCA, ACDBA, and ADCBA. In the end, the total amount of pheromone added to each unit of distance along edge AB is calculated as follows:

$$\Delta\tau_{AB}^{unit} = \frac{75}{12} + \frac{75}{12} + \frac{25}{12} + \frac{25}{12} = \frac{200}{12} = 16.7$$

The length of the edge happens to be just one distance unit, so the total amount of pheromone added to this edge is $\Delta\tau_{AB}^{unit} = 16.7 \times 1 = 16.7$. Once this calculation has been completed for all edges (always multiplying $\Delta\tau_{xy}^{unit}$ by the length of the given edge), we have the amount of pheromone added to each edge by all $N = 300$ ants in one round.

8.2.9 Updating the Values

Now that we have obtained all $\Delta\tau_{xy}^{unit}$ values, we are ready to update the pheromone densities. This is done in two steps. First, we add the sum total of what has been deposited by the ants in the single round:

$$\tau_{xy} = \tau_{xy} + \Delta\tau_{xy}^{unit}$$

In the second step, we reduce the pheromone along each edge by the loss caused by evaporation:

$$\tau_{xy} = \tau_{xy}(1 - \rho) \tag{8.3}$$

With this, we have completed one round.[2] The process is repeated for a predefined number of rounds, or until a user-specified termination criterion has been satisfied. For instance, the process may stop when no meaningful improvement has been observed over a predefined number of rounds.

8.2.10 Full-Fledged Probabilistic Formula

So far, this section has illustrated the ACO procedure only using the first iteration where the pheromone densities along all edges have been initialized to 1. This made it possible to base the previous calculations only on distances/proximities—note that Equation 8.1 calculates the probabilities from edge desirabilities, δ_i, but does not explicitly mention pheromone densities τ_i which are implicit, in the formula.

In reality, a more detailed version of the formula is used. Let τ_{xy} be the pheromone density along the edge connecting x and y, and let $\eta_{xy} = 1/d_{xy}$ be the proximity of y as

[2] This one round is sometimes referred to as one *iteration*, less commonly as one *generation*.

Input: List of the cities with all city-to-city distances
Number of ants, N
Parameters: ρ, α, β

1. All ants are placed in the city that represents the start. Pheromone densities along all edges are initialized to 1.
2. Each ant follows its own random route using the probabilities established by Equation 8.4.
3. The amount of pheromone to be added to each edge is established.
4. The amount of pheromone thus calculated is added to the edges, and the result is then subjected to evaporation by Equation 8.3.
5. Unless a termination criterion has been satisfied, return to step 2.

Table 8.3 Pseudo-Code of the ACO Technique Addressing the Traveling Salesman

observed from x. The probability with which the ant selects this edge is calculated as follows:

$$P_{xy} = \frac{\tau_{xy}^{\alpha} \cdot \eta_{xy}^{\beta}}{\Sigma_z \tau_{xz}^{\alpha} \cdot \eta_{xz}^{\beta}} \tag{8.4}$$

Importantly, the sum in the denominator is taken over all possible edge choices available at x. Parameters α a β allow the user to control the relative importance of proximity versus pheromone density. Usually, $\alpha > 0$ and $\beta \geq 1$ are used (previously, we always assumed $\alpha = 1$ and $\beta = 1$).

8.2.11 Outline of ACO's Handling of the TSP

The entire procedure is divided into fixed time steps that we have called iterations or rounds. At each round, each food-searching ant makes its probabilistic edge selections and adds the above-calculated amount of pheromone to each edge along its entire walk. Once all these additions have been made, the density along each edge is reduced by evaporation. In summary: first add some pheromone, then evaporate. The procedure is summarized by the pseudo-code in Table 8.3.

8.2.12 Closing Comments

ACO is a powerful tool capable of finding a very good route through a graph, and of doing so at computational costs much lower than those of many previously used methods. For instance, classical search techniques are here much less efficient.

For convenience, this section discussed only the basic version of the TSP, which allowed us to introduce all sorts of simplifying assumptions. For instance, our ants deposited pheromone wherever they moved, whereas in reality, they do so only if they find food. However, any reader who has grasped the basic principles will find it relatively easy to employ them in more difficult domains such as, say, job-shop scheduling. Note also that ACO is flexible enough to deal with non-stationary domains where concrete circumstances (e.g., distances and food locations) are subject to changes in time.

8.2.13 Main Limitation

Just like the basic versions of the heuristic search techniques, ACO primarily targets domains with discrete decision-making. An engineer facing a continuous-valued domain will consider some other swarm-intelligence tools, perhaps those introduced in the rest of this chapter.

Control Questions

If you are unable to answer the following questions, return to the appropriate place in the preceding text.

- How does an ant decide which direction to take at each junction? Write down the probabilistic formula.
- How does ACO determine the amount of pheromone to be deposited along each edge once all ants have completed their run through the graph? What is pheromone evaporation and how is it modeled?
- Outline ACO's general algorithm and explain the individual steps.

8.3 Particle-Swarm Optimization

Unlike the ants' colony, the technique known as particle-swarm optimization, PSO, is meant for applications where the agent is to identify the best combination of values from *continuous* domains. The attentive reader will recall that such domains were mentioned in Section 3.6 in the context of heuristic search. That section also said that heuristic search is possible even here, the best-known approaches being those relying on neural networks—which belong rather to a textbook of machine learning. Nevertheless, AI does have at its disposal a few techniques capable of dealing with continuous domains. This chapter will discuss two of the most popular: the PSO introduced in this section, and the artificial-bees colony relegated to the next.

8.3.1 Particles or Birds?

Somewhat counter-intuitively, PSO's mechanism is often explained using the metaphor of a flock of birds. When adjusting the direction of its flight, each bird balances two pieces of information: (1) its own memory of the best location it has seen so far, and (2) the knowledge of what was the best location ever experienced by any member of the whole flock.

An experienced programmer does not care much about whether the author's preferred metaphor relies on birds or particles. The behavior of birds, however, seems to convey the idea more graphically.

8.3.2 Find the Maximum of a Multivariate Function

Figure 8.6 shows a function of two variables. The function has three peaks; one of them is the *global* maximum and the other two are *local* maxima. If the function's exact mathematical formula is known, the global maximum can be found by calculus.

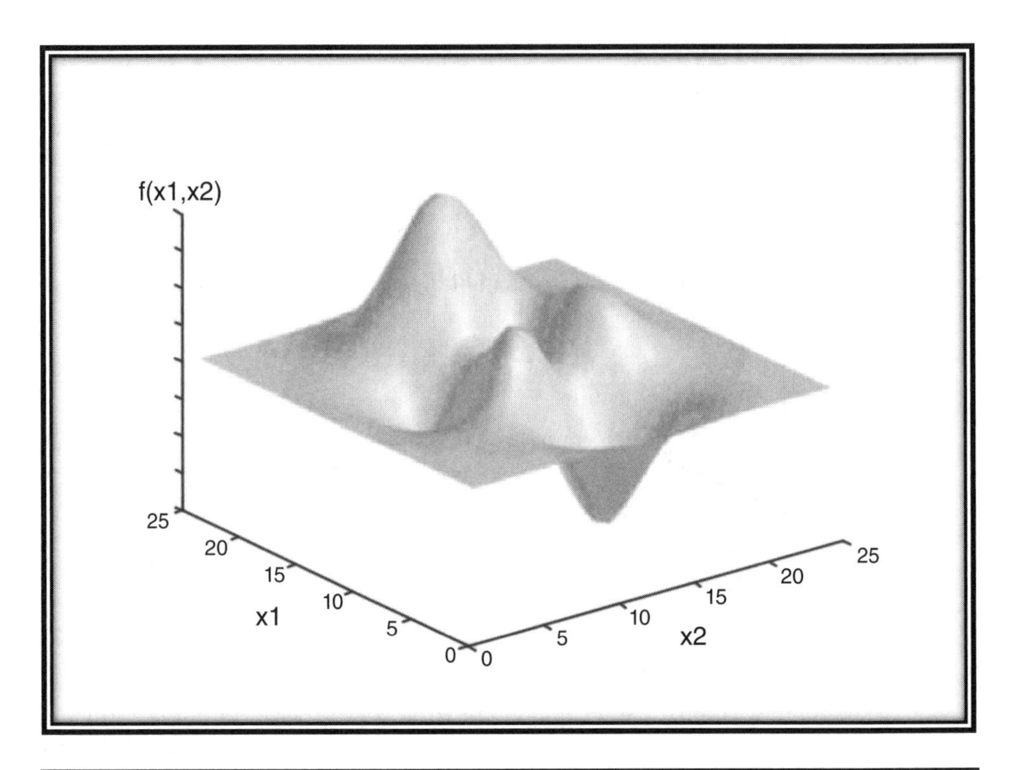

FIGURE 8.6 PSO seeks the global maximum of a complex multivariate function where calculus-based analysis is either too difficult, or even impossible.

The function from Figure 8.6 is relatively simple and its values depend on only two input variables, x_1 and x_2. In some engineering applications, however, the function is so complicated, and relies on so many variables, that finding the extreme by differentiation is well nigh impossible. Worse still, the exact formula may not even be known, and the function's values can only be established experimentally or by the *tournament* approach that was explained in Chapter 6.

For problems of this kind, swarm intelligence can offer some very efficient solutions. This section will focus on *PSO*.

8.3.3 Terminology

Swarm intelligence operates with sets of agents. In this section, each agent emulates the behavior of a bird. As already mentioned, some experts prefer to call them particles. To help the reader get used to both terms, this section will use them interchangeably, as if they were perfect synonyms.

For the sake of consistency, this text will always assume that the goal is to find a function's maximum. However, the technique's general principles remain the same whether we want maximum or minimum.

In Figure 8.6, the search space is defined by the horizontal plane that represents two variables, x_1 and x_2. In reality, of course, the search space can have any number of dimensions. At any moment, the agent/bird/particle finds itself in one concrete spot of the search space. This spot will be referred to as *location*. In some contexts, calling it *position* will appear more fitting, which is why these two terms, too, will be used interchangeably.

8.3.4 Three Assumptions

First of all, PSO assumes that none of the participating agents knows the location of the function's global maximum, let alone its value. What each agent does know is its own current location and this location's value returned by an evaluation function (this function's role is here essentially the same as in heuristic search or in the genetic algorithm).[3]

Second, PSO assumes that, during the flock's flight, each of its individuals has passed through many locations. The bird does not remember these locations. What it does remember is the highest value it has experienced so far, and the coordinates of the location where this highest value was experienced. This location is denoted by $\mathbf{P}_{best}$. The letter **P** reminds us that we are dealing with a position. It is boldfaced because the position (location) is a vector. By contrast, the best value, F_{best}, is a scalar. The pair $[\mathbf{P}_{best}, F_{best}]$ is sometimes called the agent's *local* information.

Third, PSO assumes that the agent knows the location of the function's highest value observed so far by any member of the flock—essentially, the maximum of the F_{best} values of all individuals. This is the current version of the *global* maximum and its location is denoted by $\mathbf{G}_{best}$. Note that this, too, is boldfaced because the location is a vector.

8.3.5 Agent's Goal

Figure 8.6 shows a function of two variables, x_1 and x_2. In the search for the function's maximum, the individual agents move around in this two-dimensional space in a way reminiscent of a flock of birds. For the programmer implementing the algorithm, "moving around" means changing an agent's location. How much the location is modified along each dimension depends on the agent's *velocity*. This velocity keeps changing in time, too.

To calculate an individual's velocity changes, PSO uses a formula that combines the knowledge of this individual's own best location so far, $\mathbf{P}_{best}$, with the knowledge of the best location, $\mathbf{G}_{best}$, experienced so far by any other member of the flock.

In the case of real birds, an individual's location and velocity are defined by two variables, latitude and longitude (if we ignore altitude). Technical implementations of PSO allow for *any* number of variables, perhaps a great many of them. The principle, however, remains the same.

8.3.6 Updating Velocity and Position: Simple Formula

All movements are sampled at unit time intervals. At each moment, each agent has a certain location and velocity within a system of coordinates whose number of dimensions

[3]Note that PSO thus relies on information different from what is available to birds. Scavengers, such as vultures or condors, infer from the intensity of a dead mouse's smell its distance. This is rather like knowing the distance from the maximum, which is not the same as the bird's current location's value. Again, biology offers only general inspiration.

equals the number of variables. Here is how the agent's location (position) along the i-th axis changes from one time step to the next:

$$P_i = P_i + v_i \cdot \Delta t$$

If we identify Δt with the unit-length time interval, we can put $\Delta t = 1$, and the formula simplifies as follows:

$$P_i = P_i + v_i \tag{8.5}$$

This means that to update the location along each axis, we only need to determine v_i, the agent's momentary velocity along this axis. Let P_i be the agent's current location along the i-th axis, let $P_{i,best}$ be the i-th dimension of this agent's best location so far, and let by $G_{i,best}$ be the i-th dimension of the best location that has been experienced by any bird in the flock. Here is a simple way of establishing the agent's velocity by combining its distance from the group's best location with its distance from its own currently best location. The formula is applied separately to each dimension.

$$v_i = v_i + (P_{i,best} - P_i) + (G_{i,best} - P_i) \tag{8.6}$$

Note that the term $(P_{i,best} - P_i)$ "pulls" the value toward the individual's best location. The greater the difference between the terms in the parentheses, the greater the change in velocity (along the i-th dimension). Likewise, $(G_{i,best} - P_i)$ "pulls" the velocity toward the global best.

8.3.7 Full-Scale Version of Velocity Updates

The shortcoming of Equation 8.6 is that it does not allow the engineer to adjust the relative importance of $P_{i,best}$ and $G_{i,best}$. Suppose we introduce two *learning factors*, c_1 and c_2, whose task is to weigh the relative importance of the two terms. Here is how they are used in an improved version of the formula:

$$v_i = v_i + c_1(P_{i,best} - P_i) + c_2(G_{i,best} - P_i) \tag{8.7}$$

The reader can see that $c_1 > c_2$ indicates that $P_{i,best}$ is more important than $G_{i,best}$. Conversely, $c_1 < c_2$ indicates that the flock as a whole has more influence than the individual's own past experience.

This is still not the end of the story. Realistic applications employ another pair of coefficients, r_1 and r_2, whose values are obtained from a random-number generator that chooses them with uniform distribution from the unit interval, $(0, 1)$. This is what the ultimate formula looks like:

$$v_i = v_i + c_1 r_1(P_{i,best} - P_i) + c_2 r_2(G_{i,best} - P_i) \tag{8.8}$$

Each time Equation 8.8 is invoked, a different pair of random values is used, the motivation being to make the whole process as probabilistic as possible, to increase the role of chance.

We can see that the engineer does have some control, while a certain degree of randomness remains. Regardless of how the programmer adjusts the values of c_1 and c_2, we may still observe that $P_{i,best}$ has more influence than $G_{i,best}$ on one occasion, while on another occasion it is the other way round. The random coefficients have been introduced to make the process probabilistic, which makes search process more exploratory.

8.3.8 What Values for c_1 and c_2?

Let us now take a closer look at the parameters that the engineer *can* control. Here, $c_2 > c_1$ means that the global maximum is valued more than the individual agent's maximum. In this event, experience shows that a solution is found relatively soon, thanks to the power of the flock's collective knowledge. The efficiency comes at a price, though. Very often, the quickly discovered solution turns out to be only a local extreme, perhaps very inferior to the global maximum. The higher the dominance of c_2 over c_1, the higher the danger of getting trapped by a local maximum.

In the face of local extremes, the opposite case, $c_2 < c_1$, is safer. However, the convergence may then be painfully slow. This is because the experience of individual agents plays only minor role in the flock's behavior, and an early sign of an individual's success is easily ignored. Obviously, the engineer has to balance computational efficiency with the dangers of getting trapped at local extremes. For this reason, some practitioners prefer to make c_1 and c_2 time-dependent. For instance, one may begin with a high c_1 to find a rich choice of promising locations, and then gradually reduce c_1 to accelerate the convergence toward the best of them.

The issue is complicated by the two additional parameters, r_1 and r_2, that are randomly generated. If the values of c_1 and c_2 are much smaller than r_1 and r_2, randomness will dominate. On the surface, this may be hoped to mitigate the two aforementioned dangers (slowness and the danger of local minima). However, we must not forget that excessive randomness will render the search chaotic, non-systematic, and unpredictable.

8.3.9 PSO's Overall Algorithm

Let us assume the existence of an evaluation function that for each location returns its value. We want to find the location with the maximum value. The user specifies the number of birds, N, and the values of the parameters, c_1 and c_2. At the beginning, PSO randomly distributes the birds over the entire search space, denoting the location of the k-th bird by $\mathbf{P}^k$. This initial location becomes $\mathbf{P}^k_{best}$ and its value is $F^k_{best} = F^k$. The location with the highest value in the flock becomes $\mathbf{G}_{best}$.

What happens next is summarized by the pseudo-code in Table 8.4.[4] First, velocities and locations of all agents are updated by Formulas 8.5 and 8.8. Then evaluation function for each agent returns the value of its new location. If this value is higher than this agent's previous best, the current location becomes $\mathbf{P}^k_{best}$. In the next step, the best locations of all agents are compared, and the location with the highest value becomes the global best, $\mathbf{G}_{best}$.

8.3.10 Possible Complications

The baseline algorithm works well in simple domains where it is enough to identify one global maximum. Sometimes, however, we need more. For instance, the concrete application may necessitate that *all* global maxima be identified. In that event, the procedure has to be modified accordingly. Perhaps the easiest way to do so is by adjusting the evaluation function. Once the first global maximum has been found, the evaluation function

[4]To accustom the user to both terminologies, the pseudo-code uses the more traditional term, *particle* instead of *bird*.

Input: Evaluation function that for each location returns its value.
Number of particles, N, and the values of parameters c_1 and c_2.
For each particle, randomly initialized location, $\mathbf{P}^k$, and velocity.
For each particle, the initial location is its $\mathbf{P}^k_{best}$, with value, F^k_{best}.

Until a termination criterion has been satisfied, run the following loop:

1. For each particle, update its velocity and location. The new location of the kth particle is $\mathbf{P}^k$.
2. For each particle, indexed by k:
   i. Evaluation function returns the value, F^k, of the current location.
   ii. If $F^k > F^k_{best}$, set $F^k_{best} = F^k$ and $\mathbf{P}^k_{best} = \mathbf{P}^k$.
3. Let $\mathbf{G}_{best}$ be the highest value of all $\mathbf{P}^k_{best}$.
4. Return to step 1.

Table 8.4 Pseudo-Code of the PSO Technique

may be instructed to return much lower values for the neighborhood of this maximum's location. After this, the birds' locations are re-initialized, and the search starts anew with the previously discovered maximum now "disqualified."

Another complication is a non-stationary domain where the locations of the maxima vary in time. The nature of PSO is such that, in due time, all birds concentrate at one spot, hopefully in the vicinity of a maximum (whether local or global). If the maximum shifts, the flock may lack the flexibility to respond to the new circumstance because all birds are now so close together that they do not cover the search space adequately. In domains of this kind, the engineer may want to write a sub-program that recognizes the changed circumstances and reacts to them by the creation of additional birds with randomly initialized locations. This renews the system's flexibility. Another possibility is a sharp (though temporary) reduction of the values of parameters c_1 and c_2, which increases the influence of the randomly generated r_1 and r_2.

8.3.11 Dangers of Local Extremes

PSO is easy to implement, and the flock usually reaches the function's maximum very quickly—but the discovered maximum may not be the ultimate solution. The technique is prone to get trapped in local extremes from which PSO's basic version cannot easily escape when the birds get close to each other. Once each $\mathbf{P}^k_{best}$ approaches the same local maximum, so does $\mathbf{G}_{best}$, and the flock is stuck.

To mitigate the danger of getting trapped in a local minimum (or, once there, to get out of it), all sorts of advanced techniques have been suggested. However, the resulting computer program then tends to be dominated by these extra "tricks," with only a minor part being the pure PSO.

8.3.12 Multiple Flocks

A relatively simple way of mitigating sensitivity to local extremes is to work with multiple flocks, where a flock is defined as a group of individuals sharing this group's $\mathbf{G}_{best}$. Each flock consists of a different set of randomly initialized birds and relies on its own specific parameter values. Moreover, the information (e.g., the evaluation function) available to each flock can be intentionally corrupted in an attempt to make the flock avoid the nearest local extreme. Finally, the individual flocks do not have to be created at the same time. Rather, they can be introduced one by one at predefined time intervals.

In due time, each flock reaches some solution, possibly in a different location in the search space. The engineer then chooses the best.

Control Questions

If you are unable to answer the following questions, return to the appropriate place in the preceding text.

- In what sense do we say that the agents simulate the behaviors of birds in search of prey? Summarize the three basic assumptions underlying the technique.
- Explain the basic principles of each agent's behavior; then write down the formulas for velocity update and location update. Discuss the impact of PSO's various parameters.
- Discuss the possible complications facing PSO and suggest some techniques to address these complications. What do you think is PSO's main shortcoming?

8.4 Artificial-Bees Colony, ABC

A powerful alternative to PSO is the *artificial bees colony*, ABC, a technique that seeks to mimic food-searching behavior of honey bees. Primary field of application is the same as in the case of PSO: search for the maxima or minima of continuous functions that cannot be found by calculus. ABC is more difficult to learn than PSO, and it also seems to take more effort to implement. On the positive side, it is more robust in the presence of local minima.

For the reader's convenience, we will present its simplified version that is easier to explain and easier to understand than the fully loaded original. Even so, the technique is still capable of dealing with the given class of tasks.

8.4.1 Original Inspiration

How do honey bees search for food? Initially, many foraging individuals just wander around at random, independently of each other. Sooner or later, some of them succeed, then return home and inform the rest of the colony about the food's locations and quantities. After this, other bees set out in the indicated directions, and explore not only these precise locations just discovered, but also their neighborhoods. The successes or failures of these follow-up explorations then instruct further searches, and so on.

In parallel with this systematic type of exploration, random foraging continues, too, so that other—as yet unknown—resources may thus be discovered.

8.4.2 What the Metaphor Offers to AI

The previous description, while a grave simplification of the real process, offers inspiration. Again, let us suppose that the task is to identify the maximum of a complex multivariate function.[5] There are at least two situations in which mathematical analysis cannot be used. First, the function's defining formula is unknown, and its values in

[5] The principle is the same if we decide to look for the function's minimum. To prevent unnecessary confusion, we will always assume we need the maximum.

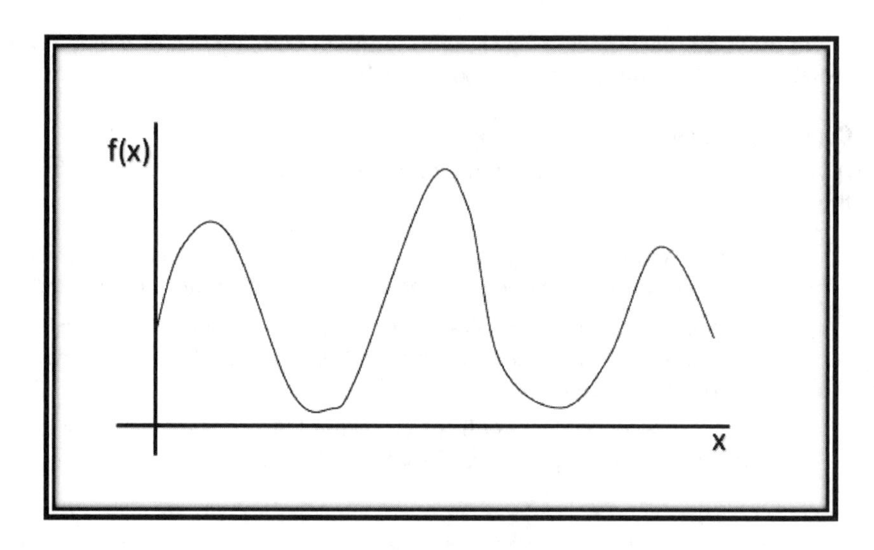

FIGURE 8.7 Typical task for ABC: find the maximum of a function. Unlike the simple example depicted here, the targeted function is usually multivariate.

concrete locations can only be established experimentally. Second, the formula *is* known, but it is so complicated as to discourage any attempt to use calculus.

As already mentioned, the bees want to find rich food resources, whereas the engineer wants to find a function's maximum. In both cases, however, the procedure remains essentially the same. A location is defined by its coordinates in a multivariate space. We assume the existence of an evaluation function (or an experimental procedure) that for any location returns its value. Suppose a foraging bee has reached a certain location. If its value is attractive, other bees explore the location's neighborhood. Chances are that even higher values will thus be discovered.

In this sense, one can say that ABC, too, combines *local* search with *global* search. Global search means foraging in the entire search space. Local search explores the vicinity of promising locations.

8.4.3 Task

Figure 8.7 shows a function, $f(x)$, that has one global maximum and two local maxima. ABC's goal is to find the global maximum's x-coordinate. In applications where this appears excessively ambitious, the practically minded engineer may be content with a location that is at least "reasonably close" to the maximum.

Again, the graph in Figure 8.7 is an extreme simplification, introduced here only for educational purposes. The reader already knows that, in reality, the function's value will be determined not only by x, but by a whole set of other variables, sometimes a great many of them. The number of variables is N_{dim} (the *dim* in the subscript standing for dimensions). No explicit formula may exist, in which case the values have to be obtained by some kind of experimentation that may be very expensive.

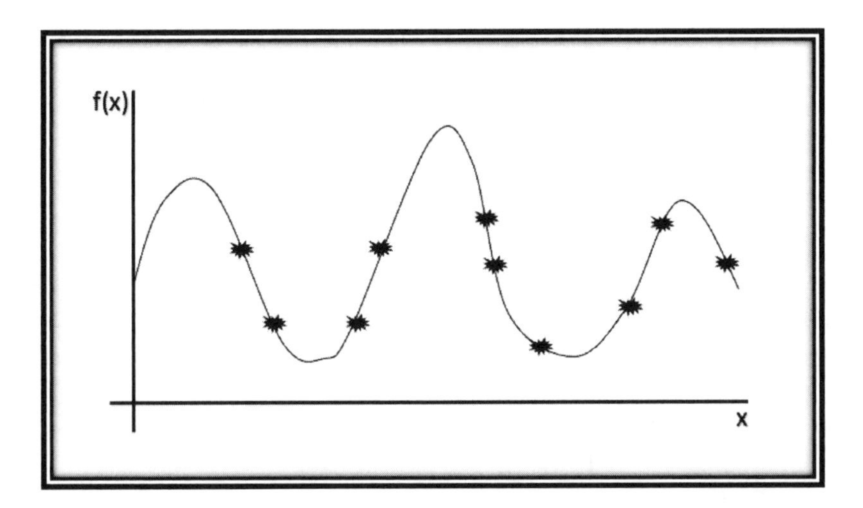

Figure 8.8 Ten bees randomly pick their targets (locations) along the *x*-axis. An evaluation function returns for each location its value along the *y*-axis.

8.4.4 First Step

As illustrated by Figure 8.8, the process begins by a set of *foraging bees* being dispatched to random locations along the x-axis. Again, a realistic application will rely on N_{dim} variables, not just x.

For each of the random locations reached by these bees, the evaluation function returns its value (along the vertical axis). Inevitably, some bees will do better than the others in the sense that the locations they have reached received higher values from the evaluation function. Figure 8.8 shows example locations identified by ten foraging bees.

8.4.5 How to Select Promising Targets

Some of the areas reached by foraging bees are going to be explored by *following bees* (see below). Here is how these destinations are selected. The primary criterion used by ABC is the evaluation function. The system identifies a few of the best locations discovered by foraging bees. In Figure 8.9, these are pointed to by the arrows labeled as *best bees*. These locations are deemed more attractive than others.

To mitigate the dangers posed by local maxima, ABC gives a chance also to some other locations. To this end, the system randomly selects a few of the non-best bees from the remaining eight. In Figure 8.9, three such instances are pointed to by the arrows labeled as *random additions*. They are added just in case they may prove useful in the future.

The number of *best bees* and the number of *random additions* are two user-specified parameters that control ABC's behavior.

8.4.6 *Following Bees*

Once the *best bees* and *random additions* have been identified as the locations of primary interest, ABC defines around each of them a neighborhood in the form of an N_{dim}-dimensional hyper-sphere. In Figure 8.10, the neighborhoods are indicated by the

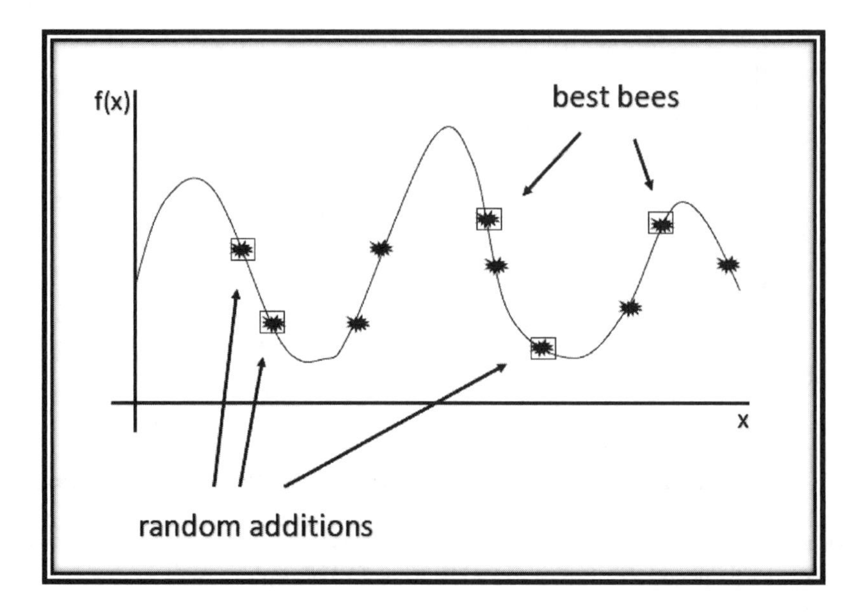

Figure 8.9 Two locations with the best function values are identified. In addition to these, three other locations are selected randomly.

circles that are, perhaps somewhat misleadingly, drawn around the "stars." More accurately, the neighborhood in this one-dimensional space should be represented by a line segment along the x-axis. In a two-dimensional space, the neighborhood would be a circle, and if there are more dimensions, a sphere or a hyper-sphere.

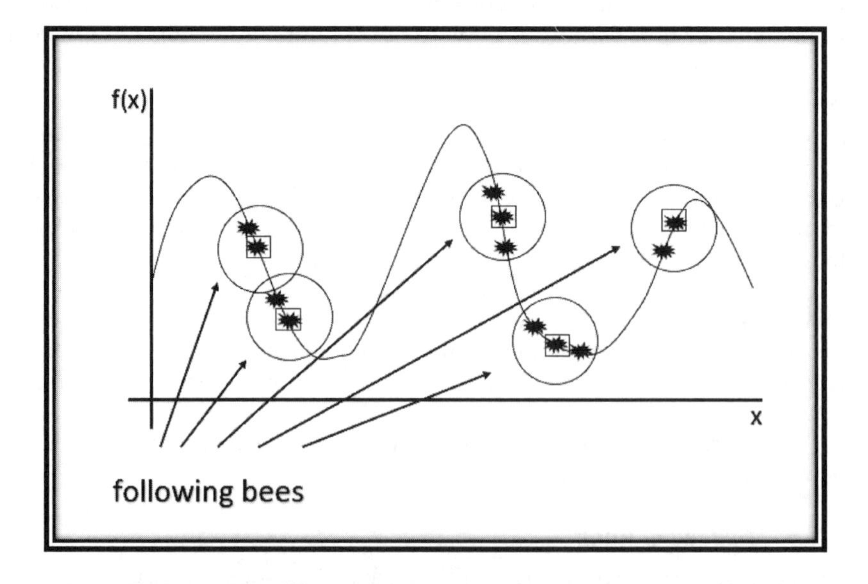

Figure 8.10 *Following bees* visit randomly chosen locations in the neighborhoods of the best bees and the locations of a few randomly chosen other bees.

Each of the *following bees* chooses a random location inside one of these neighborhoods. The engineer may decide that, say, five following bees should target each of the *best-bees* neighborhoods, and two following bees should target each of the *random-additions* neighborhoods. The idea is illustrated in Figure 8.10 where each of these neighborhoods is visited by two *following bees*.

8.4.7 Updating the Best Locations

Wherever the concrete *following bee* may land, the evaluation function returns the value of this new location. Of course, there is a chance that some following bees will reach locations that have higher values than those of the previous bees that defined the neighborhoods. This, too, is illustrated by Figure 8.10. For instance, in the leftmost circle, one of the following bees has scored a higher value than the original *random-addition* bee (the one marked in the picture by the square). In another circle, we will find *following bees* that have lower values than their predecessors. At any rate, ABC identifies in each neighborhood the bee that landed in the highest-valued location.

From now on, the locations of these highest-valued bees will become the new centers of the neighborhoods that will thus shift slightly toward the local maxima. Among these neighborhoods, the new set of *best bees* and *random additions* will be identified, and their neighborhoods will be explored by the next set of *following bees*—and the same story is repeated all over.

8.4.8 *Supporting Bees*

Section 8.3 specifies one of the weaknesses of PSO's baseline version: it gets easily trapped in local minima. This is due to the tendency of the birds in the flock to converge to the same area of the search space.

To avoid this pitfall, ABC sends a set of *supporting bees* to random locations, unrelated to those highlighted in Figure 8.10. Most *supporting bees* will reach locations with inferior values, and as such will hardly expedite the search. Occasionally, however, a lucky *supporting bee will* discover a previously unexplored peak in the function. When this happens, some *following bees* will target this location's neighborhood as if it were discovered by a *best bee*. This, then, is ABC's main "trick" for avoiding local maxima.

8.4.9 Parameters

Even in this simple version of ABC, the number of parameters affecting the technique's behavior is high. For instance, the engineer has to tell the program how many *foraging* and *following bees* should be involved; which of the *following bees* should be dispatched to the "best" locations and which to the "added" locations, how large should be the neighborhoods explored by the *following bees* along each dimension; and some other details. Some of the most important parameters are listed in Table 8.5 for the reader's convenience.

Of course, the program's efficiency, and its ability to find a good solution to the problem at hand, depends on the choice of these parameters' values. As a rule of thumb, the more bees are employed, in each category, the higher the chances that a good solution will be found—though at higher computational costs.

The areas explored by the *following bees* should perhaps be relatively large at the beginning, gradually shrinking as the program gravitates to better and better solutions.

N_{dim}	... number of dimensions (variables)
$N_{foraging}$	... number of foraging bees
$N_{following}$	... number of following bees
$N_{supporting}$	... number of supporting bees
N_{best}	... number of best locations
N_{add}	... number of random additions
S_i	... the size of each neighborhood along the i-th dimension
Stopping criteria (e.g., a predefined number of generations or the lack of improvement over a series of generations)	

TABLE 8.5 Some of the Most Important Parameters Affecting ABC's Behavior

The engineer should also keep in mind that the neighborhoods' sizes may have to be different along each dimension.

8.4.10 Algorithm

Previous paragraphs outlined the technique's individual steps. It is time to tie them all together in a coherent algorithm. This is done by the pseudo-code in Table 8.6.

The process begins by random initial exploration conducted by *foraging bees*. To each of these bees, the random-number generator assigns a location in the given system of coordinates where each variable is represented by one axis. For each of these locations, evaluation function returns a value.

ABC then chooses N_{best} locations with the highest values, and then N_{add} additional locations selected at random, regardless of their values. Around each of these locations, a neighborhood area of a pre-defined size is defined.

Following bees are then dispatched to random locations within these neighborhoods; most *following bees* target the neighborhoods of the best locations, but some target the neighborhoods of the random additions. Evaluation function returns the value of each of these *following bees'* locations. Within each neighborhood, the best site reached by any

Input: Sets of bees of different categories (*foraging*, *following*, etc.)
Evaluation function and user-specified parameters
Termination criterion

1. For each *foraging bee*, the random-number generator chooses a location in the multi-dimensional search space.
2. Evaluation function returns the value for each location.
3. The following steps are repeated until a termination criterion has been satisfied.
   i. Among the locations thus identified, select N_{best} locations; add to them N_{add} additional locations chosen at random (regardless of their values).
   ii. Define the neighborhoods of these selected locations. Send *following bees* to random places within these neighborhoods.
   iii. Send *supporting bees* to randomly chosen sites in the whole search space.
   iv. Evaluation function returns the value for each location visited by a bee.

TABLE 8.6 Pseudo-Code of the ABC Technique

bee is identified. Independently of all this, *supporting bees* are dispatched to randomly generated locations that are unrelated to any of these neighborhoods.

The process is repeated until a user-specified termination criterion has been satisfied.

Control Questions

If you are unable to answer the following questions, return to the appropriate place in the preceding text.

- What is the main goal of the ABC technique? What main weakness of PSO does it eliminate?
- What does ABC do with the locations reached by *foraging bees*? Where does ABC send *following bees*? Why is it not enough to use only the neighborhoods of the best bees? What is the job of *supporting bees*?
- Summarize the entire ABC algorithm in a few sentences. Comment on ABC's sensitivity to local extremes.

8.5 Practice Makes Perfect

To improve your understanding, take a chance with the following exercises, thought experiments, and computer assignments.

- Return to Figure 8.3, change the distance along edge AB from the original 1 unit to 3 units, and then hand-simulate one step of the traveling-salesman problem as addressed by the ACO technique from Section 8.2.
- Write a computer program that implements the ACO algorithm. The input consists of the values of ACO's parameters and of a matrix with all city-to-city distances. Run the program on some simple experimental data and measure the time the program needs to find the best (or almost best) route.
- Write a one-page essay explaining the reasons why PSO easily gets trapped in a local extreme of the evaluation function. Suggest measures to mitigate this problem. Apart from the simple idea mentioned in this chapter, and apart from ideas suggested in relevant literature, can you invent your own?
- Write a program implementing PSO for a user-specified number of agents. Create a multi-peak multi-variate function and test your PSO program's ability to find its global maximum, or at least a very good local maximum. Run extensive experiments to evaluate the program's computational costs under diverse circumstances. Discuss the trade-off between these costs and the program's chances to reach the global maximum.
- Write a program implementing the artificial-bees colony, ABC. Make sure the program allows the user to experiment with diverse values of the system's many parameters.
- Using the programs developed in the previous two exercises, run a comparative study that evaluates the performance of PSO and ABC using two or three different test-beds. Consider the following criteria: computational costs, ability to avoid local maxima, sensitivity to user-specified parameters.

- Write a one-page essay explaining what makes the ABC technique to be so robust in the face of local extremes.
- Search the web for other techniques from the field of swarm intelligence and summarize their principles in a two-page essay.

8.6 Concluding Remarks

A solution to the traveling-salesman problem by ACO was introduced in Gambardella and Dorigo (1995), a paper that largely relied on ideas from the PhD dissertation of one of its authors, Dorigo (1992). The principles of PSO appeared at about the same time, being presented by Kennedy, and Eberhart (1995). The technique of artificial-bees colony is younger, having been introduced in a journal paper by Karaboga and Basturk (2007).

The efficiency with which the swarm-intelligence techniques often dispose of arguably difficult problems has shaken the worldview of many AI specialists. The potential of this new paradigm is staggering, and its current popularity makes classical search look well-nigh obsolete.

The excitement is justified—and yet the reader should not forget that this is not the first major breakthrough in the history of AI. One has to be open-minded, but also clear-headed. Experience has taught us that despising the labors of earlier generations may be unwise. Some ideas that have been shrugged off as old-fashioned may experience resurrection, even if under a different guise. Still, this author has to confess that he himself is fascinated by these novel tools. Their distributed multi-agent philosophy has opened brand new vistas.

This chapter could introduce only some of the most popular techniques, and even these had to be presented in their most elementary versions. Other swarm-intelligence ideas have recently been proposed, seeking to emulate the behavior of cats, wasps, termites, and even some donkeys-and-smugglers strategies. Each of them has its advantages and shortcomings. Each fares brilliantly in some domains—and less brilliantly in others. It is only to be regretted that an introductory text cannot devote more space to their countless aspects.

CHAPTER 9
Elements of Automated Reasoning

Problem solving is not enough; intelligent behavior has other aspects, too. It should be capable of reasoning, of drawing conclusions from known facts, of supporting these conclusions by convincing arguments. This is what people are good at, and this is what we want to implement in artificial intelligence (AI) programs.

To prepare the soil for a later elaboration of these notions, this chapter introduces the rudiments of automated reasoning. When doing so, the text relies on a formalism borrowed from *Prolog*, a programming language that about a generation ago was all the rage in AI circles, and not only there. The idea is to describe a given problem in terms of logic so that a *reasoning mechanism* can answer non-trivial queries such as those typical of, say, medical diagnosis. After this introduction, later chapters will explain the nature of automated reasoning, outline the difficulties that impede its practical implementations, and describe the techniques that scientists have developed to overcome these difficulties.

9.1 Facts and Queries

A critical aspect of intelligence is the ability to answer questions that are non-trivial in the sense that their processing requires background knowledge. Let us illustrate the idea using a simple toy domain.

9.1.1 List of Facts

Textbooks of AI are fond of examples based on family relations because they are simple, and just about everybody has a good grasp of the underlying concepts. Honoring this tradition, we will begin with a few facts that Table 9.1 presents in the form of a graph as well as in a way reminiscent of a *Prolog* program. The latter mechanism is intuitively clear and easy to read—which is why the book will adhere to its principles in most of the following chapters.

The formalism relies on *predicates* and *arguments*. The specific knowledge base from Table 9.1 knows only one predicate, `parent`, with two arguments in the parentheses. We will interpret this as telling us that the first argument is a parent of the second. For instance, the first fact states that `bill` is parent of `eve`. Note that the order of the

```
parent(bill,eve).
parent(jane,eve).
parent(eve,pete).
parent(eve,ann).
parent(fred,pete).
parent(fred,ann).
```

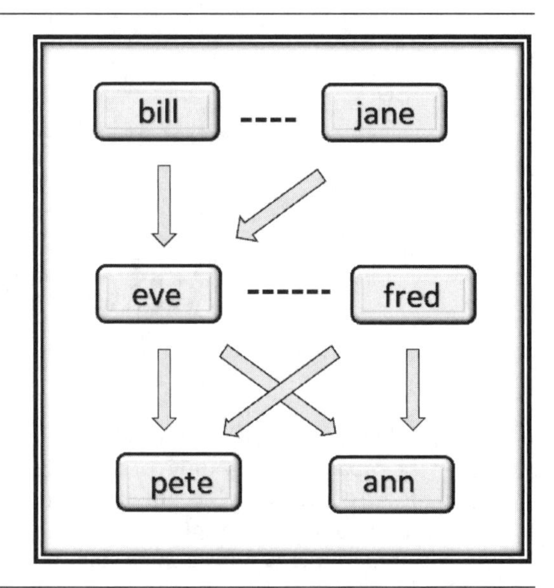

Table 9.1 A Simple Family-Relations Knowledge Base Consisting of a set of Facts. Only One Predicate, `parent`, Is Considered Here

arguments is not arbitrary. In our case, we decided that the first argument is the parent of the second, but it could just as well be the other way round, the decision being up to the programmer. Once the interpretation has been established, however, one has to be consistent throughout the entire program.

The `parent` predicate has two arguments. Other predicates may require three, four, or even more arguments, and yet others may have only one argument or none at all.

The *knowledge base* from Table 9.1 is extremely simple, containing as it does nothing but six *facts*. Each fact is a predicate followed by a period. The reader can see the equivalence between the knowledge base on the left, and its graphical representation on the right.

9.1.2 Answering Users' Queries

The primary goal of a *Prolog* program is to answer queries expressed in a concrete format. For example, a user who wants to know whether `bill` is `eve`'s parent will present the following query:

? :- parent(bill,eve).

The symbol ":-" plays here the role of an arrow pointing toward the question mark. The query is followed by a period. In response to the query, the system scans its list of facts and, finding the right predicate, generates the following answer:

YES

The same answer will be generated also when the user wants to know if eve is parent of ann (see the fourth predicate in our list of facts). However, on receiving the following query,

? :- parent(bill,pete).

the system fails to discover this fact in the knowledge base, and therefore generates the following answer:

NO

9.1.3 Queries with Variables

A user who wants to know the name of **eve**'s parent will formulate the query in the following way:

? :- parent(X,eve).

Here, X represents a variable whose *instantiation* (a concrete value) the user wants to know. The system recognizes X as a variable because it is capitalized. Conversely, the names of constants (e.g., the persons' names) always begin with lowercase letters.

Having received this query, the system scans its knowledge base, looking for a predicate that has eve as the second argument. In our case, two such facts can be found. The first is parent(bill,eve), which is why the system initially answers the query with bill, an instantiation of X. If the user reacts to this by typing a semicolon, the system continues the search, realizing that the very next fact in the knowledge base, parent(jane,eve), also has eve as the second argument. The system therefore generates another answer, jane. The output then looks something like this:

```
X = bill;
X = jane
```

If the user enters another semicolon, the system will once again resume its search through the knowledge base, seeking yet another instance of parent whose second argument is eve. Failing to find one, the system generates the negative answer:

NO

9.1.4 More Than One Variable

In the query from the last paragraph, only one variable appeared. However, another query can just as well contain two or more variables. Suppose the user enters the following:

? :- parent(X,Y).

In this event, the system will respond with pairs of answers, each new one triggered by the user's entering the semicolon:

```
X = bill
Y = eve;

X = jane
Y = eve;
   ⋮
```

These answers will continue until the end of the knowledge base has been reached. When this happens, the final answer is NO.

9.1.5 Compound Queries

More interesting are queries consisting of more than just one predicate. Here is a typical example:

? :- parent(X,Y), parent(Y,pete).

The comma separating the two predicates is interpreted as logical AND. The user wants the system to find whether some instantiations of X and Y can lead to the positive answer. In response, the system will try to find such X (the parent of Y in the first predicate) that satisfies the requirement that the child, Y, is parent of `pete` in the second predicate. In plain English, the user wants to learn who is `pete`'s grandparent. There are two possible answers (followed by a NO):

```
X = bill
Y = eve;

X = jane
Y = eve;
```

9.1.6 Exercise

By way of a simple exercise, the reader is encouraged to write a query that identifies `jane`'s grandchildren. Depending on the concrete formulation, one possible version of the system's response is the following sequence (followed by a NO):

```
X = eve
Y = pete;

X = eve
Y = ann;
```

Here, the interim variable, X, represents the grandchild's parent.

9.1.7 Binding Variables to Concrete Values

When receiving a multi-variable query, such as those from the last paragraphs, the system attempts to find instantiations that result in the query being answered in the affirmative. We say that it tries to *bind* the variables to concrete instantiations.

In a query consisting of two or more predicates, it is customary that the search for potential bindings starts with the leftmost predicate. Thus in the case of the previous example, parent(X,Y), parent(Y,pete), AI programs usually begin by looking for the bindings of X and Y in parent(X,Y).

9.1.8 How to Process Compound Queries

Let us once more return to the last query:

? :- parent(X,Y), parent(Y,pete).

The task is to find instantiations (bindings) that would satisfy the requirement that the second argument in the query's second predicate be pete. Let us hand-simulate the procedure.

The first fact in the knowledge base is parent(bill,eve). Consequently, the first pair of bindings for the first predicate, parent(X,Y), is X = bill and Y = eve. This means that the instantiation of the second predicate, parent(Y,pete), would have to be parent(eve,pete). Since this fact is indeed present, in the knowledge base, the search stops with success and returns the just-discovered bindings of X and Y.

On the user's entering a semicolon, the system applies the same procedure to the second fact in the knowledge base, parent(jane,eve). Here the instantiations are X = jane and Y = eve, which means that the instantiation of the second predicate, parent(Y,pete), would have to be parent(eve,pete). Again, this fact is present in the knowledge base, and the bindings of X and Y can be returned to the user.

If the user again enters a semicolon, the system will attempt to apply the same procedure to the remaining facts in the knowledge base, but will fail to find bindings resulting in the second argument of the second predicate being pete. This is when the system generates answer NO.

The reader will benefit from trying to repeat the hand-simulation of this procedure without looking at the text. Appreciating the details of this query answering is necessary for a better grasp of the automated-reasoning mechanisms to be presented later in this chapter—and in the following chapters.

9.1.9 Ordering the Predicates

To obtain the desired information, the user can usually choose from two or more alternative formulations of the query. For instance, wishing to learn the names of pete's grandparents, the user can present either of the following two queries:

? :- parent(X,Y), parent(Y,pete).

? :- parent(Y,pete), parent(X,Y).

Note that the second query consists of the same two predicates as the first, only in reversed order. Both queries are equivalent because conjunction (AND, represented here by the comma) is commutative. However, this does not mean that the choice of the query is arbitrary. The two versions differ in the number of bindings each of them permits.

In the first formulation, the first predicate contains two variables, X and Y. The number of possible instantiations is the number of persons represented by variable X, multiplied by the number of persons represented by variable Y. This amounts to $6 \times 5 = 30$ because any of the 6 persons can be represented by X, and any of the remaining 5 can be represented by Y. For each instantiation, the program has to determine whether the knowledge base contains `parent(Y,pete)`. In our specific example, we were lucky: The very first instantiation of X and Y led to `parent(eve,pete)`. Generally speaking, however, the correct instantiation could have been only the last out of many possibilities, in which case the query-processing procedure would be computationally expensive.

In the second formulation of the query, we can find five possible instantiations of Y in the first predicate, `parent(Y,pete)`. Of these, however, only two lead to predicates found among the facts in the knowledge base: `Y = eve` and `Y = fred`. For each of the two, 5 possible instantiations of X in the second predicate have to be checked. We can see that there are fewer instantiations to check than in the first query. The example has convinced us that the order of the predicates matters.

9.1.10 Query Answering and Search

It is now clear that attempts to answer the user's query involve *search* for correct bindings of the variables to constants in the knowledge base. The simplest solution will employ the blind-search techniques from Chapter 2. The reader is encouraged to try and formulate an algorithm that will answer the queries with depth-first search or breadth-first search. Since the previous paragraph pointed out the importance of the number of bindings, one may even suggest a simple evaluation function to guide some heuristic search such as hill climbing.

9.1.11 Nested Arguments

An introductory text has to keep things simple and easy to follow. One thing that this section has ignored is the fact that an argument can itself be a predicate. Here is an example:

father(fred,brother(X,ann)).

This is interpreted as telling us that `fred` is the father of `ann`'s brother (assuming that the knowledge base contains the information about brothers). In predicates of this kind, we say that the arguments are *nested*. The depth of this nesting can be greater than here. For instance, `X` can itself be a predicate, and so on.

In the 1990s, it was fashionable, in the circles of *Prolog* programmers, to demonstrate one's skills by employing deep nesting. Such programs were deemed elegant. Today, such practice is less popular because nesting tends to reduce the expression's clarity. For this reason, this textbook will avoid examples with nested arguments wherever possible.

Control Questions

If you are unable to answer the following questions, return to the appropriate place in the preceding text.

- What is a *predicate*? What can you say about the number of arguments in a predicate and about their order?

- How do *Prolog*-like systems express facts, and how do they express queries? What do we mean by an *instantiation* (or binding) of a variable?
- How do such systems process the queries, including compound ones? How does the number of possible bindings affect the efficiency of query answering?

9.2 Rules and Knowledge-Based Systems

The basic principles of query answering were explained with the help of a trivial knowledge base that contained nothing but facts. Such simplicity, however, is unrealistic. Automated reasoning starts being useful only with the introduction of *rules*.

9.2.1 Simple Rules

Let us return to the knowledge base from Table 9.1. Here is an example of a rule that can possibly be added to it:

offspring(Y,X) :- parent(X,Y).

As before, the symbol ":-" stands for an arrow. Here is how we read the rule. If X is Y's parent, then Y is X's offspring. In this sense, the rule presents a logical implication: If the predicate to the right of ":-" is *true*, then also the predicate to the left of it is *true*.

9.2.2 Longer Rules

The last rule had only one predicate on the right-hand side. Other rules have two or more such predicates, connected by commas. As before, the comma represents conjunction, logical AND. Here is an example:

grandparent(X,Y) :- parent(X,Z), parent(Z,Y).

Let us begin with the first predicate on the right-hand side, `parent(X,Z)`. For the bindings {`X = bill, Z = eve`}, the knowledge base contains the corresponding fact, `parent(bill,eve)`. If we find an instantiation of Y for which the knowledge base contains the second predicate, `parent(eve,Y)`, then both predicates on the right-hand side are *true*.

The rule is telling us that, in this particular case, X is Y's grandparent. A quick look at Table 9.1 confirms that this is indeed the case for `Y = pete` because the knowledge base contains the predicate `parent(eve,pete)`. We can thus conclude that `bill` is `pete`'s grandparent. Note that it is possible to find another successful instantiation that informs us that `bill` is `ann`'s grandparent.

9.2.3 Formal View of Rules

Rules take the form of `H :- B` where `H` is the rule's *head* and `B` is the rule's *body*. The body, also called *antecedent*, is either a single predicate or a conjunction of two or more predicates that are here usually called *conditions*.

The head, sometimes called *consequent* or *conclusion*, will for our needs be represented by a single predicate. The consequent is deemed *true* if each of the conjuncted conditions in the body has been evaluated as *true*. If at least one condition in the body has been shown to be *false*, then the conclusion is evaluated as *false*, too.

9.2.4 Closed-World Assumption

When processing a rule, the system first of all has to establish whether some instantiation of the variables will result in the body being *true*, in which case also the head is *true*. If such instantiation cannot be found, the system's response is NO.

Generally speaking, the system assumes that all facts in the knowledge base are *true*. Besides, all conclusions that can be derived from the available facts and rules by a sound inference mechanism (such as the one described above) are regarded as being *true*. Conversely, anything that is *not* explicitly listed in the knowledge base, and cannot be derived from it, is assumed to be *false*.

The principle is known as the *closed-world assumption*. In reality, `jack` can be `fred`'s parent. But since the knowledge base does not contain `parent(jack,fred)`, any automated-reasoning program will say NO. Remember: Only such facts are deemed *true* that exist in the knowledge base. All other facts are deemed *false*.

More generally, the closed-world assumption says that anything that *cannot be inferred* from the given knowledge is *false*, too.

9.2.5 Knowledge-Based Systems

The reader now understands that a mechanism for automated reasoning consists of two things: the *knowledge base* and the *inference mechanism* operating over it.

The knowledge base in Table 9.2 is a slightly expanded version of the previous one. Note how the definitions of later concepts rely on earlier concepts. For instance, `father` is defined with the help of predicates `parent` and `male`. Somewhat later, the definition of `uncle` uses the concept of `father`. By way of a simple exercise, the reader is encouraged to formulate additional family-related concepts such as `niece`, `son_in_law`, and so on.

```
parent(bill,eve).            female(ann).
parent(jane,eve).            female(eve).
parent(eve,pete,).           female(jane).
parent(eve,ann).             male(bill).
parent(fred,pete).           male(pete).
parent(fred,ann).            male(fred).

offspring(X,Y) :- parent(Y,X).
grandparent(X,Y) :- parent(X,Z), parent(Z,Y).
father(X,Y) :- parent(X,Y), male(X).
mother(X,Y) :- parent(X,Y), female(X).
sibling(X,Y) :- parent(Z,X), parent(Z,Y).
uncle(X,Y) :- parent(Z,Y), sibling(X,Z), male(X).
...
```

TABLE 9.2 An Extension of the Family-Relations Knowledge Base. What Has Been Added Here Are Some Facts About Gender, and a Few Rules Defining Slightly More Advanced Concepts

This knowledge base is still very small, and can thus be used only for queries that seem almost trivial. Real-world implementations work with hundreds, or even thousands of rules that cover the knowledge from the given field of expertise. Besides the rules, such knowledge bases contain many facts—to which the system may usually add further facts that have been obtained in the course of the machine's dialog with the user.

A *knowledge-based system* thus consists of a knowledge base, an inference mechanism, and a communication module to interact with the user.

Control Questions

If you are unable to answer the following questions, return to the appropriate place in the preceding text.

- How will you formally define a rule? What do we mean by its *body* and what do we mean by its *head*?
- Explain the principle of the *closed-world assumption*.
- What do you know about *knowledge-based systems*?

9.3 Simple Reasoning with Rules

The simple query-answering mechanism described in Section 9.1 was limited to knowledge bases that contained nothing but facts. It is now time to turn our attention to a more advanced query-answering technique that is capable of operating with rules, too.

9.3.1 Answering a Query

The best way to explain the technique's essence is to illustrate it by a simple example. Suppose the knowledge base is the one in Table 9.2, and suppose the user has submitted the following query:

? :- offspring(ann,eve).

One possible series of actions that might lead to an answer is shown in Table 9.3. The first step considers a trivial solution: Can this particular predicate, `offspring`, with these two constant arguments, ann and eve, be found among the list of facts? In the case

---

Suppose the user submitted query `? :- offspring(ann,eve)`. One possibility to process the query is to take the following steps:

1. Try to find the fact `offspring(ann,eve)` in the knowledge base.
2. If the fact is not found, try to find a rule with `offspring` at its head. This results in discovering `offspring(X,Y) :- parent(Y,X)`.
3. Investigate `parent(Y,X)`.
   i. Instantiate X to `ann` and Y to `eve`, thus instantiating `parent(Y,X)` to `parent(eve,ann)`.
   ii. Search the knowledge base for fact `parent(eve,ann)`.
4. Having found the fact, return YES.

Table 9.3 Example of the Actions Taken in Response to a Simple Query

of a positive answer, the program would return YES and stop. Failing this, the program searches through the list of rules: Is there one headed by this particular predicate? Yes, the knowledge base indeed does contain a rule that has `offspring(X,Y)` as a consequent; its antecedent is `parent(Y,X)`. Suppose the variables X and Y are instantiated to `ann` and `eve` (as in the user's query). Does the knowledge base contain the corresponding fact, `parent(eve,ann)`? A brief scan of the database shows that this is indeed the case, and the query is thus answered in the affirmative.

9.3.2 Beyond the Basics

The example from Table 9.3 was still very simple: The discovered rule had a single condition in the antecedent, `parent(X,Y)`, and this predicate could be immediately instantiated to a fact that could be found in the knowledge base.

More typically, the process has to be applied recursively. For illustration, suppose the user wants to know whether `fred` is `ann`'s uncle. The concrete fact `uncle(fred,ann)` not being listed in the knowledge base, the program finds a rule whose head is `uncle(X,Y)`:

uncle(X,Y) :- parent(Z,Y), sibling(X,Z), male(X).

The variables in the head are instantiated to `X=fred` and `Y=ann`. In the first step, the program has to establish whether it is possible to find an instantiation for Z such that `fred` is Z's sibling. This calls for the evaluation of another rule whose consequent is `sibling` and whose antecedent contains two `parent` predicates. Instantiations of the two predicates have to be found. In the event of success, one has to proceed to the third condition in the query, namely, `male(X)`.

The last example has convinced us that the query-answering process may involve a whole series of steps, and this may be computationally intensive in the case of very large knowledge bases. Again, it is a good exercise to try to figure out how blind search would carry out these steps, and perhaps even to write a computer program to handle query answering in this way.

9.3.3 Concepts Defined by More Than One Rule

Quite often, a concept cannot be defined by a single rule. This is typically the case with definitions in the form of logical disjunctions. For instance, we may want to establish that X is Y's sibling either if X is Y's brother or if X is Y's sister. In the formalism we have been using so far, this concept will be defined by the following two rules.

sibling(X,Y) :- brother(X,Y).

sibling(X,Y) :- sister(X,Y).

When evaluating query `?:-sibling(ann,pete)`, the knowledge-based system first scans the knowledge base in search of this concrete fact. Failing this, it focuses on the first of the two rules; it instantiates the variables X and Y to `ann` and `pete`, respectively, and then tries to find out whether the corresponding fact, `brother(ann,pete)`, exists in the knowledge base. Failing this, the program proceeds to the second rule and examines the instantiation `sister(ann,pete)`. Only if this fails, too, will the system return NO.

In this example, the concept `sibling` was defined by two rules. However, it is not uncommon that three, or even more rules are employed in the case of some complicated concepts.

9.3.4 Disjunctive Normal Form

In the last paragraph, the concept was defined by a disjunction (logical OR) of two rules; either the first rule was *true* or the second. Within each rule, however, the individual predicates are conjuncted (logical AND). For the consequent to be *true*, all conditions in the body have to be satisfied (*true*). We say that this formalism relies on the *disjunctive normal form*, DNF. Here is an example of an expression in DNF:

$$(X_1 \wedge X_2) \vee (Y_1 \wedge Y_2) \vee (Z_1 \wedge Z_2 \wedge Z_3)$$

Informally, the terms inside the parentheses are "ANDed" and the parenthesized expressions are "ORed." Besides, any term inside the parentheses can be negated. Logicians have been able to prove that any reasonable logical expression can be converted into this format.

9.3.5 Recursive Concept Definitions

Some concepts are best defined with the use of recursion. The following example illustrates the idea:

ancestor(X,Y) :- parent(X,Y).
ancestor(X,Y) :- parent(X,Z), ancestor(Z,Y).

For X to be Y's ancestor, the body of at least one of the two rules has to be evaluated as *true*. The reason why we say that `ancestor` is here defined *recursively* is that one of the conditions in the second rule has the same predicate name as the head of the rule (even though the arguments are different).

9.3.6 Evaluating Recursive Concepts

Suppose we have added to the knowledge base from Table 9.1 the two `ancestor` rules from the previous paragraph, and suppose the user has presented the following query:

? :- ancestor(bill,ann).

Let us simulate the behavior of a program that decides whether to respond to the query with YES or NO. The first rule says that X is Y's ancestor if X is Y's parent. The program thus instantiates the two variables accordingly, `X=bill` and `Y=ann`, and then tries to figure out whether the knowledge base contains the fact `parent(bill, ann)`. Unable to find such fact, the program has to conclude that the first rule has failed.

The next try is to consider the second rule, the one that says that X is Y's ancestor if it is possible to find such Z that X is Z's parent, and Z is Y's ancestor. It is easy to see that `X = bill` permits only one instantiation of Z for which the corresponding fact can be found in the knowledge base: `Z = eve`. With this, the truth of `parent(bill,eve)` has been established—which, of course, is not enough; it still remains to be seen is whether `bill` is `eve`'s ancestor.

The same procedure is thus invoked recursively. The first rule says that `eve` is `ann`'s ancestor if `eve` is `ann`'s parent. A brief scan of the knowledge base confirms that indeed it does contain `parent(eve,ann)`. This means that the second condition, `ancestor(eve,ann)`, has been confirmed to be *true*. With both conditions of the second rule thus being shown *true*, the program concludes that `bill` is `ann`'s ancestor.

9.3.7 Comments on Recursion

In the above example, the recursion had to be invoked only once.[1] In practical applications, however, it is not uncommon to see the recursion invoked dozens of times. In the family-relations domain, this would correspond to confirming an ancestor several centuries ago.

Experienced programmers know that successful implementation of recursion necessitates well-thought-out stopping criteria. One has to be careful not to permit never-ending recursive calls. For this reason, recursive definitions require at least two rules, the first of which represents the stopping criterion.

In our simple example, the recursion has been stopped by the confirmation that `bill` is `eve`'s parent. By way of a simple exercise, the reader is encouraged to give some thought to how to stop the recursion in the case where the query is to be answered with NO. Another useful exercise will hand simulate the above procedure for a query that contains variables.

9.3.8 Summary

A knowledge base consists of facts and rules. Each rule has a head (a consequent) and a body (an antecedent); the body consisting of either a single condition, or a conjunction of two or more conditions, separated by commas. The definition of some concepts requires two or more rules; of these, the conditions of at least one rule have to be satisfied if the concept is to be deemed *true*. Since all conditions in the rule's body are conjuncted, and the rules disjuncted, a rule-based definition takes the form of a DNF. Finally, we have seen that some concepts are best defined recursively.

Once the knowledge base has been created and encoded, we need a mechanism to carry out the automated reasoning over this knowledge base. The technique described in this chapter was simple, but perhaps not ideally suited for implementation in a computer program. The next two chapters will describe a more robust approach that is based on first-order logic and on the inference procedures of *modus ponens* and *resolution*.

Control Questions

If you are unable to answer the following questions, return to the appropriate place in the preceding text.

- Explain the simple mechanism for answering a user's queries. Illustrate its behavior on a hand-simulated example.

[1]The knowledge base knowing only three generations, recursion cannot go beyond grandparents.

- Provide examples of concepts whose definitions require multiple rules and/or recursive rules.
- Explain the mechanism that can be used by a program that is to answer user's queries based on the available knowledge base.

9.4 Practice Makes Perfect

To improve your understanding, take a chance with the following exercises, thought experiments, and computer assignments.

- Using the knowledge base from Table 9.2, formulate a query that will find out if `bill` has a nephew; if yes, we want to know this nephew's name. Try to do the same for such concepts as niece, uncle, and aunt (of course, assuming that none of these concepts has been defined by some rule(s) in the knowledge base). How would you formulate a query that will tell you whether `bill` and `ann` are at all related?
- Can you think of a family-relations concept that cannot be defined by a single rule, so that two or even three rules may be needed?
- Create your own knowledge base from your own field (different from the family relations of this chapter). One possibility is number theory or some other mathematical domain where it is possible to formulate rules that do not allow exceptions. Suggest concepts whose definition requires two or more rules.
- Offer examples of concepts whose definitions require recursion. You may consider ideas from graph theory or operations with advanced programming data structures such as oriented lists or binary trees.
- Using a general-purpose language such as C++ or Python, write a program that will answer a user's query using the mechanism explained in this chapter. Make the program sufficiently general. For instance, the program should begin with reading the concrete knowledge base needed for the given domain. Choose a data structure that you believe is best-suited for the representation of the rules and facts.

9.5 Concluding Remarks

The chapter has introduced a simple mechanism capable of conveying to the computer some specific knowledge that can later be used in answering users' queries and in automated reasoning.

The concrete mechanism discussed here relies on certain principles borrowed from the once-famous programming language *Prolog* (programming in logic). This language was developed in 1972 by Alain Colmerauer and Philippe Roussel with the intention to facilitate computer implementations and exploitation of the Horn clauses that had been studied by Robert Kowalski. Students interested in the history of these efforts are referred to Colmerauer and Roussel (1996), and to Kowalski (1988).

Prior to being able to start with *automated reasoning*, the engineer has to gain some proficiency in converting expert knowledge to rules such as those considered in the preceding text. The most popular way of doing so relies on first-order logic. Then, once the facts and rules are available, we need algorithms to employ them in automatic inferences. These issues will be addressed by Chapters 10 and 11.

Practical attempts to implement automated-reasoning programs have led to an important piece of experience: Outside very simple domains, such as the family relations from this chapter, it is all but impossible to come up with rules that are totally reliable. Even in the case of family relations, the task can be daunting. For instance, one rule may specify that a woman's husband is the father of her children. Of course, this is not always the case, the children may come from her previous marriage. At any rate, pure logic is not enough; practical implementations call for mechanisms to quantify the uncertainty in the rules, and to propagate this uncertainty throughout the reasoning process. Techniques to deal with imperfect knowledge will be addressed in Chapters 13 through 15. Alternative ways of knowledge representation will be introduced in Chapter 12.

In the 1970s, the ideas outlined in this chapter gave rise to so-called *expert systems* that were intended for quasi-intelligent analyses of problems that could only be solved with the help of non-trivial background knowledge. These will be briefly discussed in Chapter 16.

CHAPTER 10

Logic and Reasoning, Simplified

Chapter 9 outlined the essence of automated reasoning based on logic—and indeed, this is the most common approach. The scenario is simple. Someone creates a knowledge base consisting of rules and facts. The user presents a query, and the AI software seeks to answer this query by subjecting the knowledge base to an *inference procedure*, essentially a sequence of logical operations. Here is the novelty: Whereas classical programming relied on data and algorithms, automated reasoning deals with data, algorithms, and knowledge bases.

Since logic seems to be this paradigm's crucial ingredient, this chapter will revise some of its most relevant aspects. To begin with, it explains the elements of first-order logic and shows how to convert its expressions into *Prolog*-like rules and facts when creating the knowledge base. With the fundamental principles of knowledge encoding thus established, the chapter moves on to the basic reasoning techniques that are based on the classical *modus ponens* and the modern *resolution principle*.

In this chapter, the ideas are explained using a highly simplified formalism that ignores predicates and variables. A more general exposition of these reasoning mechanisms is relegated to Chapter 11.

10.1 Entailment, Inference, Theorem Proving

Chapter 9 convinced us that knowledge can be expressed in terms of rules and facts. Before we turn our attention to automated reasoning, we need to get acquainted with some terminology and basic ideas.

10.1.1 Entailment

Suppose that John lives in Miami. The statement, "all people living in Miami live in Florida" then *entails* that John, too, lives in Florida, or else logic would never make any sense. Likewise, a typical knowledge base entails things that are not explicitly specified in it. The symbol we use to indicate entailment is "$\models$." For instance, the circumstance that the knowledge base consisting of statements $p_1, \ldots, p_n$ entails q is expressed as follows:

$$p_1 \wedge \ldots \wedge p_n \models q$$

10.1.2 Inference Procedure

It is not enough to suspect that a knowledge base entails something. The engineer wants a program that automatically *infers* what is entailed. This is a job for an *inference procedure*. The symbol "$\vdash$" indicates that such procedure infers from the set of statements $p_1, \ldots, p_n$ another statement, q:

$$p_1 \wedge \ldots \wedge p_n \vdash q$$

Whereas entailment means that q inevitably follows from the knowledge base, inference means that a procedure has derived q. The circumstance that q has been inferred does not guarantee that q is actually entailed. After all, the inference procedure may be unsound, incorrect.

10.1.3 Modus Ponens in Its Simplest Form

Perhaps the best known inference mechanism is *modus ponens*. If we know that p implies q and if we learn that p is *true*, then we conclude that q is *true*, too. A logician will express this law formally as follows:

$$p \wedge (p \rightarrow q) \vdash q$$

Admittedly, a non-logician finds this expression abstract and difficult to read. Some textbooks therefore prefer to express *modus ponens* in the following way:

$$\begin{array}{c} p \rightarrow q \\ p \\ \hline q \end{array}$$

10.1.4 Example

Let us denote by p the fact that x lives in Miami, and let us denote by q the fact that x lives in Florida. The implication $p \rightarrow q$ then indicates that whoever lives in Miami, lives also in Florida. This is what the first line in the above definition of *modus ponens* says.

Suppose we are now told that John lives in Miami, which means p is *true* for John. This is the second line in the above table. Combining this with the previous implication, we conclude that John lives in Florida.

10.1.5 Other Inference Mechanisms

Apart from *modus ponens*, logicians know some other useful laws. Among these, perhaps the best-known is *modus tollens*, formally specified as follows:

$$\neg q \wedge (p \rightarrow q) \vdash \neg p$$

Let us illustrate this law using the same example. Suppose that p states that x lives in Miami, q states that x lives in Florida, and $p \rightarrow q$ states that anybody living in Miami also lives in Florida. If x does *not* live in Florida, then q is *false*, which means that its opposite, $\neg q$, is *true*. *Modus tollens* then guarantees $\neg p$, which means that x does not live in Miami, either. Such conclusion is obviously correct.

Closely related to *modus tollens* is the law of *contraposition* that is formally specified as follows:

$$(p \rightarrow q) \vdash (\neg q \rightarrow \neg p)$$

Returning to our Miami-Florida example, the law of *contraposition* is telling us that, from the fact that "living in Miami implies living in Florida," we can infer that "not living in Florida implies not living in Miami."

Note the subtle difference between the last two laws. Whereas *modus tollens* infers the fact $\neg p$, the law of *contraposition* infers the implication $\neg q \rightarrow \neg p$.

10.1.6 Soundness of an Inference Procedure

We say that an inference procedure is *sound* if every consequence inferred from a knowledge base by this procedure is also entailed by it. Let $p = p_1, \ldots, p_n$ be the knowledge base. An inference procedure denoted by "$\vdash$" is sound if it satisfies the following criterion:

$$\text{for any } q \text{, if } p \vdash q \text{, then } p \models q$$

For instance, *modus ponens* is a sound inference procedure because any conclusion inferred from a knowledge base by *modus ponens* is indeed entailed by this knowledge base.

10.1.7 Completeness of an Inference Procedure

Conversely, we say that an inference procedure is *complete* if it is capable of discovering any consequence entailed by the knowledge base. Let $p = p_1, \ldots, p_n$ be the knowledge base. An inference procedure denoted by "$\vdash$" is complete if it satisfies the following criterion:

$$\text{for any } q \text{, if } p \models q \text{, then } p \vdash q$$

For instance, *modus ponens* is *not* complete unless special limitations have been imposed on the knowledge base. In a knowledge base where the rules and facts are formulated in a general form, some consequences entailed by the knowledge base cannot be discovered by the application of *modus ponens*. We will return to the question of *modus ponens*'s completeness in Section 10.2.

10.1.8 Theorem Proving

One popular application of inferences based on *modus ponens* is automated theorem proving. This is actually a whole sub-field of artificial intelligence. The task is defined as follows: Given a knowledge base, $p = p_1, \ldots, p_n$, and a set of inference rules (such as *modus ponens*), prove that some new statement, q, which is not listed in the knowledge base, is a direct consequence of p.

The task can be cast as a *search problem*. The initial state is the knowledge base; the final state is a knowledge that contains the original facts and rules, and also q that has been gradually derived from the original knowledge base. The search operators depend on the employed inference rules. For instance, one such operator may identify a pair of statements to which *modus ponens* can be applied, and then draw the logical conclusion. In reality, however, theorem proving is more likely to rely on the advanced version of *modus ponens* to be introduced in Section 10.2.

In principle, other search operators can be used such as those based on *modus tollens* or the law of *contraposition*. However, *modus ponens* is the most common among them. Even more popular is the *resolution principle* to be introduced in Section 10.3

10.1.9 Semidecidability

Proving that q is entailed by knowledge base p is computationally less demanding than proving that q is *not* entailed by p. If q does not follow from p, then a procedure that proves its opposite, $\neg q$, suffers from computational costs that grow exponentially with the size of the knowledge base. Generally speaking, computer science considers all exponentially growing costs to be prohibitive. In this sense, proving $\neg q$ is unrealistic.

The circumstance that proving *truth* is possible while proving *falsity* is not is sometimes referred to by the term, *semidecidability*.

Control Questions

If you are unable to answer the following questions, return to the appropriate place in the preceding text.

- Explain the difference between the two critical terms introduced in this section: entailment and inference.
- When do we say that an inference procedure is sound, and when do we say that it is complete? What do we mean by *semidecidability*?
- Explain the principle of *modus ponens*. Is it sound and complete? Define also *modus tollens* and the law of *contraposition*.

10.2 Reasoning with Modus Ponens

Now that we have grasped the basic concepts, we can turn our attention to this chapter's main goal: the study of the mechanisms permitting automated reasoning. First, we will introduce a technique based on *modus ponens*. Then, in Section 10.3, we will focus on the more popular *resolution principle*.

10.2.1 General Form of *Modus Ponens*

The basic version of *modus ponens* that was introduced in the previous section is rather simplistic and its practical use in automated reasoning is limited. Logicians prefer the more general formulation that is presented in Table 10.1.

Two questions have to be answered: first, *when* the law can be applied; second, *how* to apply it. As for the first, the reader will have noticed, in Table 10.1, that one of the terms in the antecedent of the first rule, a_i, is identical to the consequent of the second rule. This defines the situation where *modus ponens* can be applied. Second, the resulting rule, logically consistent with the first two (entailed by them), is created as follows: The consequent, b, is inherited from the first rule, and the antecedent is the concatenation of the antecedents of the first two rules, with the exception of a_i which has been eliminated.

$$
\begin{array}{ll}
a_1 \wedge \ldots \wedge a_m & \rightarrow b \\
c_1 \wedge \ldots \wedge c_n & \rightarrow a_i \\
\hline
a_1 \wedge \ldots \wedge a_{i-1} \wedge a_{i+1} \wedge \ldots \wedge a_m \wedge c_1 \wedge \ldots \wedge c_n & \rightarrow b
\end{array}
$$

Note that the consequent of the second rule, a_i, is found among the terms in the antecedent of the first rule. Whenever this is the case, the third rule follows from the first two.

Table 10.1 *Modus Ponens* in Its General Formulation

It is easy to see that the formulation of *modus ponens* from Section 10.1 is a special case of the general formulation in Table 10.1. In the version from Section 10.1, the antecedent of the second rule is empty, and this consequent, p, is identical to the antecedent of the first rule.

10.2.2 Horn Clauses

Automated reasoning is much easier to implement if we simplify the knowledge base in the sense that we allow only rules in the form of *Horn clauses* that are defined by the following expression:

$$a_1 \wedge a_2 \wedge \ldots \wedge a_n \rightarrow b \tag{10.1}$$

Note that a Horn clause does not contain any negations, the consequent consists of only one term, and the only logical operator in the antecedent is conjunction. A knowledge base where all rules take the form of Horn clauses is called a *Horn-clause knowledge base*. This formalism has been adopted in *Prolog*.

The main advantage gained by constraining the knowledge base to Horn clauses is that *modus ponens* then becomes not only *sound*, but also *complete*. Practical experience shows that limitation to Horn clauses does not present any major obstacle in our attempts to encode our knowledge.

10.2.3 Truth and Falsity of Facts

To make it easier to implement an inference mechanism that applies *modus ponens* to Horn-clause knowledge bases, we will express all facts in the form of rules. To be specific, the circumstance that fact p is *true* will be formalized as follows (here, T stands for *true*):

$$T \rightarrow p \tag{10.2}$$

Conversely, the circumstance that p is known to be *false* is formalized as follows (here, F stands for *false*):

$$p \rightarrow F \tag{10.3}$$

The reason this formalism makes sense can be found in the truth table for implication (see Table 10.2). In the first case, $T \rightarrow p$, the only possibility for the rule to be *true* if the

The rule $p \rightarrow q$ is *false* if it leads to a situation where truth would imply falsity. In all the other cases it is *true*.

p	q	$p \rightarrow q$
true	*true*	*true*
true	*false*	*false*
false	*true*	*true*
false	*false*	*true*

TABLE 10.2 Truth Values of Implication

antecedent is *true* is that p must be *true*, too. In the second case, $p \rightarrow F$, the only possibility for the rule to be *true* when the consequent is *false* is that p must be *false*, too.

10.2.4 Concrete Example

The mechanism that uses *modus ponens* in automated reasoning is illustrated by the simple example shown in Table 10.3. The input of the reasoning process consists of the first five rules that are labeled with integers 1 through 5. The goal is to prove that e is *true*.

At each step of the reasoning process, the program first has to identify a pair of rules that can be subjected to *modus ponens*: The consequent of one rule has to be found among the conditions in another rule's antecedent. One such pair consists of rules 1 and 4 where the term satisfying the requirement is a. Subjecting the two rules to *modus ponens* generates rule 6 that is then added to the knowledge base. Similarly rules 2 and 5 result in rule 7 being added to the knowledge base. The process then continues in like manner until the moment when the application of *modus ponens* to some pair of rules results in $T \rightarrow e$ (rule 10). Since this is just another way of stating that e is *true*, the goal has been reached, and the procedure can stop.

Using the five-rule knowledge base below, prove that e is *true*:

1	$a \wedge b \rightarrow c$	
2	$d \rightarrow b$	
3	$c \wedge a \rightarrow e$	
4	$T \rightarrow a$	
5	$T \rightarrow d$	
6	$b \rightarrow c$	from 1 and 4
7	$T \rightarrow b$	from 2 and 5
8	$c \rightarrow e$	from 3 and 4
9	$T \rightarrow c$	from 6 and 7
10	$T \rightarrow e$	from 8 and 9

TABLE 10.3 Illustration of Reasoning by *Modus Ponens*

10.2.5 Practical Considerations

Note how easy it is to implement the process with the blind search algorithms from Chapter 2. In the previous example, we could see that, very often, there is a choice of two or more pairs of rules that can enter *modus ponens*. This determines the search's branching factor. The engineer can also consider addressing the problem with some version of heuristic search. This, however, requires an evaluation function; a beginner who lacks experience with automated reasoning may find the design of this function a challenging task.

When implementing the reasoning process by search, it is advisable to represent each rule's antecedent by a list of conditions, L. *Modus ponens* can be employed if we find a rule whose consequent, s, satisfies the requirement $s \in L$.

The reader now appreciates the practicality of expressing the *truth* of p as $T \rightarrow p$. This formalism guarantees that each rule and fact has an explicit antecedent as well as a consequent, and can thus be treated in a unified way, with the lists L.

10.2.6 Inference in Horn-Clause Knowledge Bases

It is time to formalize the reasoning process that the example from Table 10.3 illustrated. Let us begin by noticing that longer rules are gradually shortened by the removal of some conditions from their antecedents. For instance, when rules 1 and 4 were combined, the resulting rule 6 could be regarded as a shorter version of rule 1 (condition a in the antecedent has been eliminated).

The principle is captured by the algorithm whose pseudo-code is summarized in Table 10.4. Since conjunction is commutative, the order of the conditions listed in the antecedent is arbitrary. The sequence of the gradually shortened rules in the table can be therefore accepted without loss of generality.

Of course, the table illustrates only the essence. Reality can be more complicated than that. Thus in the example from Table 10.3, rule 1 was shortened a bit, resulting in rule 6. After this, other rules were shortened, and only in the penultimate step the procedure returned to rule 6, shortening it again by the removal of yet another condition, which resulted in rule 9.

---

1. Identify in the knowledge base a term a that is known to be *true*, $T \rightarrow a$.
2. If there is a rule $p_1 \wedge \ldots p_n \rightarrow q$ such that it is possible to prove that $p_1 = a$, add to the knowledge base the following rule:

$$p_2 \wedge \ldots p_n \rightarrow q$$

Repeated application of the last two steps generates the following sequence:

$$\begin{aligned} p_1 \wedge p_2 \wedge \ldots \wedge p_n &\rightarrow q \\ p_2 \wedge \ldots \wedge p_n &\rightarrow q \\ \ldots & \\ p_n &\rightarrow q \\ T &\rightarrow q \end{aligned}$$

Table 10.4 Inference by *Modus Ponens* in a Horn-Clause Knowledge Base (Principle)

The final comment relates to the stopping criterion. In the previous simple example, the process stopped when $T \rightarrow e$ was reached. In reality, such stopping criterion (the proof of e) may be not be satisfied in a reasonable number of steps. Besides, one has to be prepared for the circumstance that the truth of e cannot be proved at all, simply because it is *false*. Either way, the programmer has to instruct the program what to do when the reasoning process does not seem to be getting anywhere.

Control Questions

If you are unable to answer the following questions, return to the appropriate place in the preceding text.

- Explain the principle of *modus ponens* in its general form and discuss the relation of this general form to the simple version from Section 10.1.
- What is a Horn clause? Why is it useful to formulate the knowledge base in the form of Horn clauses?
- Explain how to employ blind search for making inferences in Horn-clause knowledge bases.

10.3 Reasoning Using the Resolution Principle

When applied to Horn clauses, *modus ponens* is both sound and complete. Even so, most implementations of automated reasoning (including *Prolog*) prefer to employ the *resolution principle* which is sound and complete for a broader range of rules: for those in the so-called *normal form*.

10.3.1 Normal Form

A knowledge base satisfies the normal-form requirements if each of its rules takes the form of the following expression:

$$a_1 \land a_2 \land \ldots a_n \rightarrow b_1 \lor \ldots \lor b_m \qquad (10.4)$$

This means that a rule in the normal form has to satisfy the following requirements. First, it does not contain any negations; second, the only logical operator permitted in the antecedent is conjunction; third, the only logical operator permitted in the consequent is disjunction; and finally, no parentheses are used (neither in the antecedent nor in the consequent).

Note that the Horn clause from Section 10.2 is a special case of the normal-form rule where the consequent consists of a single term.

10.3.2 Resolution Principle

For more than two millennia, logical inferences were dominated by Aristotle's *modus ponens*. It was only in the twentieth century that the more general *resolution*

Suppose it can somehow be proved that $d_j = a_i$ where a_i is one of the conditions of the first rule's antecedent and d_j is one of the conditions of the second rule's consequent. In this event, the third rule follows from the first two.

$$a_1 \wedge \ldots \wedge a_m \rightarrow b_1 \vee \ldots \vee b_k$$
$$c_1 \wedge \ldots \wedge c_n \rightarrow d_1 \vee \ldots \vee d_l$$

$$a_1 \wedge \ldots a_{i-1} \wedge a_{i+1} \ldots \wedge c_1 \wedge \ldots \wedge c_n \rightarrow b_1 \vee \ldots b_k \vee d_1 \ldots \vee d_{j-1} \vee d_{j+1} \vee \ldots d_l$$

The antecedent of the third rule contains all conditions in the antecedents of the first two rules, except for a_i. The consequent of the third rule contains all the conditions in the consequents of the first two rules, except for d_j.

TABLE 10.5 *Resolution Principle* in Its General Formulation

principle was discovered. The principle is summarized in Table 10.5.[1] Here is how it works:

1. The third rule follows from the first two if we can prove that one of the conditions, d_j, in the consequent of the second rule is equivalent to one of the conditions, a_i, in the antecedent of the first rule: $d_j = a_i$.
2. The antecedent of the third rule consists of all the conditions in the antecedents of the first two rules except for a_i which has been dropped.
3. The consequent of the third rule consists of all the conditions in the consequents of the first two rules except for d_j which has been dropped.

Importantly, *modus ponens* can be shown to be a special case of the resolution principle for the case where any consequent consists of a single term.

10.3.3 Theoretical Advantage

Regardless of whether the knowledge base has taken the form of Horn clauses or normal-form rules, theoreticians have been able to prove that the resolution principle is sound and complete in the sense of the terminology introduced in Section 10.1. By contrast, *modus ponens* is complete only in Horn-clause knowledge bases.

10.3.4 Concrete Example

The mechanism that applies resolution to automated reasoning is illustrated by the simple example in Table 10.6 where the (unrealistically small) knowledge base consists of just three rules. Note that the first rule is in the normal form; it is not a Horn clause because the consequent is a disjunction of two conditions. For this reason, *modus ponens* cannot be used here (recall that it can be used only for Horn clauses).

[1] Recall that the rules' consequents are here disjunctions of conditions, whereas in Horn clauses the consequent was only allowed to contain a single condition.

Using the three-rule knowledge base below, prove that c is *true*:

1	$T \rightarrow a \vee b$	
2	$a \rightarrow c$	
3	$b \rightarrow c$	
4	$T \rightarrow a \vee c$	obtained by resolving 1 and 3
5	$T \rightarrow c$	obtained by resolving 2 and 4

Table 10.6 Illustration of Reasoning with the *Resolution Principle*

The example is simple and easy to follow. The reasoning process is here far from deterministic. For instance, the very first step resolves the first rule with the third, but the agent could just as well have begun by resolving the first rule with the second. The solution would have been reached in the same number of steps.

10.3.5 Practical Considerations

The reader will agree that automated reasoning based on the resolution principle can be implemented using the search techniques from Chapters 2 and 3. The method is similar to the one used in the previous section for reasoning with *modus ponens*.

Blind search seems a natural choice here unless the engineer knows how to design the evaluation function for heuristic search. This evaluation function may not be easy to suggest. Here is one general idea. When choosing between two pairs of rules, give preference to the pair that promises faster arrival at the solution. However, what is faster is not necessarily obvious unless a very-deep look-ahead strategy (see Section 3.2) is employed.

10.3.6 Computational Costs

Applying the resolution principle in the simple form just described can be computationally expensive. In knowledge bases with hundreds, or even thousands of rules, the branching factor of the underlying search can be very high, and the depth of the search can be considerable, too.

As a rule of thumb, the involved computational costs grow exponentially with the size of the knowledge base. They can easily become prohibitive unless the engineer has found ways to reduce the branching factor. One possibility is the technique known as *backward chaining*.

10.3.7 Backward Chaining

One way to keep the costs of the search in check is to reduce the size of the search tree by applying the technique of *backward chaining*. Here is the principle.

Suppose we want to establish the *truth* of a term denoted by p. Backward chaining first adds to the knowledge base the opposite statement, $p \rightarrow F$; in other words, it assumes that p is *false*. Then it searches the knowledge base for a rule that has the term p as a consequent. Such rule is then resolved with $p \rightarrow F$, and the result is again added to the knowledge base. The procedure then continues, with F always on the right-hand side of the resulting rule, until the search fails (no pair of resolvable rules can be found), or until the following contradiction has been reached: $T \rightarrow F$.

Since the contradiction has been reached after a sound inference mechanism, it is interpreted as a proof that the original assumption, $p \rightarrow F$, is incorrect. Of course, if this assumption is incorrect, then the opposite has to be correct. This means that p is *true*.

10.3.8 Concrete Example

Let us illustrate the idea of backward chaining on the simple example from Table 10.7. The task is to prove that e is *true*. The technique adds to the knowledge base the opposite of what is to be proved (i.e., $e \rightarrow F$) and then applies a series of resolutions in a way that always maintains falsity, F, on the right-hand side of the resulting rules. If the sequence of resolutions reaches a contradiction, $T \rightarrow F$, then $e \rightarrow F$ cannot be correct, which means that its opposite, $T \rightarrow e$, must be correct.

To be able to maintain falsity, F, on the right-hand side of the resolution's result, the process must always search for the right pairs of rules to be resolved. In Table 10.7, the first resolution is applied to rule 6 and rule 3. The reason for this choice is that rule 3 has e on the right-hand side, and rule 6 has e on the left-hand side. Since the consequent of rule 6 is F, the consequent of the resulting rule 7 is F, too.

The task is to prove that e is *true* by applying resolution with backward chaining to the five-rule knowledge base below.

The first step is to add to the knowledge base the opposite of what is to be proved: $e \rightarrow F$.

Then, resolution is applied to such pairs that make sure the consequent of the resolution's result is always F.

1	$b \rightarrow a$	
2	$c \wedge a \rightarrow d$	
3	$d \wedge c \rightarrow e$	
4	$T \rightarrow c$	
5	$T \rightarrow b$	
6	$e \rightarrow F$	added as opposite of what is to be proved
7	$d \wedge c \rightarrow F$	from 6 and 3
8	$d \rightarrow F$	from 7 and 4
9	$c \wedge a \rightarrow F$	from 8 and 2
10	$a \rightarrow F$	from 9 and 4
11	$b \rightarrow F$	from 10 and 1
12	$T \rightarrow F$	from 11 and 5

After a series of resolutions, a contradiction has been reached: $T \rightarrow F$.

Conclusion: Based on the given knowledge base, the assumption that e is *false* resulted in a contradiction. Therefore, the opposite statement must be correct: the one that says that e is *true*.

TABLE 10.7 Illustration of Reasoning by Resolution with Backward Chaining

In the next step, rule 7 is resolved with rule 4 whose consequent, c, can be found among the conditions in the antecedent of rule 7. Alternatively, rule 7 could have been resolved with rule 2 whose consequent is d—the other term in the seventh rule's antecedent's conditions. The reader can see that resolution with backward chaining can easily be cast as search whose branching factor is determined by the number of rule-pairs that satisfy the requirement of maintaining F on the right-hand side. If heuristic search is to be used, an appropriate evaluation function has to be defined.

10.3.9 Resolution as Search

Table 10.8 contains the pseudo-code of a procedure that uses blind search when inferring the truth or falsity of a statement by the use of the resolution principle. Note that this is *not* the pure backward chaining from the previous paragraphs. The intention of presenting here this algorithm is just to illustrate the possibilities of applying search to automated reasoning.

Starting with a list L that contains nothing but the opposite of what is to be proved, the procedure always seeks to resolve a rule $p \in L$ with either a rule from the knowledge base or with another rule from L. If this succeeds, the new rule obtained by resolution is added to L; if it fails, p is removed from L.

We can obtain the backward-chaining version of this algorithm by a minor modification. To wit, the programmer may decide to add the requirement that the rule p from step 3 should not be selected at random, but that preference should always be given to rules whose consequent is F.

Control Questions

If you are unable to answer the following questions, return to the appropriate place in the preceding text.

- What are the main aspects of the resolution principle? What is the relation between the resolution principle and *modus ponens*? How will you implement the procedure using the lists of conditions in the antecedent and consequent?
- How would you characterize the principle of *backward chaining*? What is its main advantage when compared to plain resolution?
- Summarize the algorithm that implements resolution with backward chaining using AI search.

Input: a knowledge base
Task: prove that q is *true*.

1. Create an empty list L and place in it the negation of what is to be proved: $L = \{q \rightarrow F\}$.
2. If L contains $T \rightarrow F$, then stop with success: q has been proved.
3. Select from L some rule, p, and try to resolve it either with a rule from the knowledge base or with some $p_1 \in L$. Add the result to the knowledge base.
4. If p cannot be resolved with any rule from the knowledge base or from L, remove p from L.
5. If $L = \emptyset$, stop with failure. Otherwise, go to step 2.

TABLE 10.8 Using a Search Technique to Implement Reasoning by Resolution

10.4 Expressing Knowledge in Normal Form

Section 10.3 mentioned that the resolution principle is sound and complete when applied to a knowledge base whose all rules are in normal form. This may look like a serious limitation—but it is not! It turns out that any knowledge base can be converted to normal form.

10.4.1 Normal Form (Revision)

Equation 10.4 characterized the definition of a rule in normal form: all conditions in its antecedent are conjuncted, all conditions in its consequent disjuncted, no term is negated, and no parentheses are used.

10.4.2 Conversion to Normal Form

Theoreticians have been able to prove a theorem that we present here without proof:

> Any set, S, of rules can be converted into another set, S', such that S and S' are logically equivalent, and all rules in S' are in normal form.

The conversion can be accomplished by a systematic application of a relatively simple set of logical operations. Let us first illustrate the procedure by an example.

10.4.3 Concrete Example

Suppose the knowledge base contains the following rule:

$$\neg a \wedge c \rightarrow b \vee (\neg d \wedge e)$$

This rule is clearly not in the normal form because of the negations and the parentheses. Table 10.9 shows how a sequence of logical operations gradually converts this rule into two rules that are logically equivalent to it, and *are* in the normal form. Note that the conversion increased the number of rules (there are now two instead of one).

Practical experience shows that a very similar sequence of steps can successfully convert the vast majority of non-normal-form rules into the right format. By way of a simple summary, Table 10.10 lists the steps that have been taken in our concrete example.

Control Questions

If you are unable to answer the following questions, return to the appropriate place in the preceding text.

- What are the defining features of rules satisfying the requirements of the normal form? Can any knowledge base be converted to its equivalent in the normal form?
- What steps will you take when converting a general-form rule to its equivalent in normal form?

Let us consider the following rule:

$$\neg a \wedge c \rightarrow b \vee (\neg d \wedge e)$$

The implication, $\rightarrow$, is removed by the use of the equivalence $p \rightarrow q \equiv \neg p \vee q$:

$$\neg(\neg a \wedge c) \vee b \vee (\neg d \wedge e)$$

De Morgan's law is used to remove the first pair of parentheses:

$$a \vee \neg c \vee b \vee (\neg d \wedge e)$$

Distribution is used to remove the second pair of parentheses:

$$(a \vee \neg c \vee b \vee \neg d) \wedge (a \vee \neg c \vee b \vee e)$$

The last expression can be re-written as two expressions:

1. $a \vee \neg c \vee b \vee \neg d$ 2. $a \vee \neg c \vee b \vee e$

Rearranging the terms, we obtain the following:

1. $\neg c \vee \neg d \vee a \vee b$ 2. $\neg x \vee a \vee b \vee e$

De Morgan's law is used to re-write the first expression (the second is unchanged):

1. $\neg(c \wedge d) \vee a \vee b$ 2. $\neg c \vee a \vee b \vee e$

Applying to these expressions the equivalence $p \rightarrow q \equiv \neg p \vee q$ in reversed order results in the following rules:

$c \wedge d \rightarrow a \vee b$

$c \rightarrow a \vee b \vee e$

Both of the resulting rules are in normal form.

Table 10.9 Illustration of the Process That Converts a Rule to Its Equivalent That Satisfies the Requirements of Normal Form

1. Eliminate the implication, $\rightarrow$, by using the equivalence, $p \rightarrow q \equiv \neg p \vee q$.
2. Using de Morgan's laws, eliminate negations of parenthesized terms.
3. Distribute $\wedge$ and $\vee$, obtaining a conjunction of dijunctions.
4. Each of the disjunctions in this new expression will be treated as a separate logical expression.
5. Combine the negated terms in each of these expressions.
6. Using equivalence $p \rightarrow q \equiv \neg p \vee q$, convert the expressions back to rules.

Table 10.10 Sequence of Operations That Convert a Logical Expression to Its Normal-Form Equivalent

Hand-simulate the procedure with which *modus ponens* infers from this four-rule knowledge base the conclusion that c is *true*.

1	$a \wedge b \rightarrow c$
2	$d \rightarrow b$
3	$T \rightarrow d$
4	$T \rightarrow a$

Table 10.11 An Example for Practicing *Modus Ponens*

10.5 Practice Makes Perfect

To improve your understanding, take a chance with the following exercises, thought experiments, and computer assignments.

- Suggest real-world examples illustrating the use of *modus ponens* and *modus tollens*. The examples have to be different from the Miami-Florida example from Section 10.1.
- Using the four-rule knowledge base from Table 10.11, show the individual steps that employ *modus ponens* when proving that *c* is *true*.
- Using the five-rule knowledge base from Table 10.12, show the individual steps that employ the resolution principle with backward chaining when proving that *e* is *true*.
- Convert the following logical statement into one or more rules in normal form.

$$a \wedge b \rightarrow c \vee (d \wedge \neg e)$$

- Write a two-page essay discussing the possible stopping criteria to be employed in search-based approaches to reasoning by *modus ponens*, and reasoning by resolution with backward chaining. How expensive are these automated-reasoning procedures computationally?

Hand-simulate the procedure with which the resolution principle with backward chaining infers from the following knowledge base the conclusion that e is *true*.

1	$b \rightarrow a$
2	$T \rightarrow c$
3	$c \wedge a \rightarrow d$
4	$d \wedge c \rightarrow e$
5	$T \rightarrow b$

Table 10.12 An Example for Practicing Resolution with Backward Chaining

10.6 Concluding Remarks

The history of *modus ponens* can be traced all the way back to ancient Athens and Aristotle. In the 1800s, George Boole found a way to express logic in terms of two-valued algebra, representing *true* by 1 and *false* by 0, and introducing the idea of truth tables. The more general reasoning technique of the resolution principle is relatively new, having been proposed by Robinson (1965). In the 1970s, the resolution principle was chosen as the inference mechanism underlying the programming language *Prolog*.

The goal of this chapter was to develop some basic understanding of reasoning over a knowledge base expressed in the form of Horn clauses or normal-form rules. For educational purposes, the presentation was simplified in the sense that it ignored the need to handle predicates and variables as we know them from Chapter 9. This will be rectified in Chapter 11.

Even so, the reader now has a good idea of the essence. To implement a program capable of reasoning in a certain domain, the engineer first has to identify relevant knowledge. This knowledge then has to be expressed in some formal paradigm. The most commonly used paradigm, in this context, is first-order logic which makes it easy to implement the reasoning processes. A program based on the principles presented in this chapter then answers users' queries by a search for the consequences that can be derived from the knowledge base.

CHAPTER 11
Logic and Reasoning Using Variables

Chapter 10 explained the use of *modus ponens* and resolution by way of examples that were simplified to the extreme. For instance, these examples never considered predicates and variables. Now that the basic principles are clear, it is time to move on and complete the presentation.

To begin with, we need to pay more attention to the mechanisms to encode knowledge and to manipulate it. This will be accomplished within the framework of first-order logic. This chapter will explain the idea of existential and universal quantification of variables, and the role they play in knowledge encoding. Later, the text will explain such critical notions as unification and bindings, and their consequences for automated reasoning. Thus armed, the reader will then be ready to write his or her own knowledge bases and automated-reasoning programs.

11.1 Rules and Quantifiers

Suppose we possess the knowledge needed to solve problems within a certain domain. How to express this knowledge in a formalized way that allows us to use it by an AI program? The most popular tool is *first-order logic*, a paradigm that is beneficial also in automated reasoning.

11.1.1 Objects and Functions

At the lowest level of our formalism, we find *objects* such as `bill`, `dog`, `tree`, or `house`. Above these are *functions* of these objects. Typically, a function will accept as input one or more arguments and return a set of objects. For instance, the engineer may decide that the function `car_of(bill)` should return the list of all cars belonging to `bill`.

11.1.2 Relations

Besides objects and functions, first-order logic operates also with *relations*. Informally, a relation is a function that returns either *true* or *false*. We encountered some examples in Chapter 9 where, for instance, `parent(X,Y)` was *true* for some instantiations of X and Y (e.g., $X =$ `bill` and $Y =$ `eve`), and *false* for other instantiations, depending on whether or not the specific fact could be found in the knowledge base.

The relation acquires the form of a name (e.g., `parent`) that is followed by a list of arguments in parentheses, `(X, Y)`. This list can be empty. Throughout the rest of this book, the relation's name will be called *predicate*.

11.1.3 Constants and Variables

When it comes to constants and variables, it is somewhat awkward that the tradition in first-order logic is different from the formalism preferred in *Prolog*. In first-order logic, the names of constants begin with upper-case letters, and the names of variables with lower-case letters, as in `parent(Bill,x)`, where, of course, `Bill` is a constant and `x` is a variable.

The circumstance that *Prolog* uses a convention that is the exact opposite is inconvenient, but usually does not cause any major confusion once the student has gotten accustomed to it. Starting with the next paragraph, this chapter will rely on the first-order logic convention.

11.1.4 Order of Arguments

The engineer putting together the knowledge base may decide to interpret `teaches(Bill, Fred)` as meaning that the first argument teaches the second; to wit, that `Bill` teaches `Fred`. Once this decision has been made, the order of the arguments becomes mandatory; one must keep in mind that to write `teaches(Fred, Bill)` means to reverse the roles of the teacher and the pupil. Inconsistency in the order of arguments can result in difficult-to-discover errors in the AI program.

Put another way, the engineer has a complete freedom how to conceive the predicates and their arguments; but once the decision has been made, care must be taken not to deviate.

11.1.5 Atoms and Expressions

An atom is a relation applied to a set of objects. A logical expression is a set of atoms and/or other expressions, combined by the logical operations of negation (NOT), conjunction (AND), disjunction (OR), and implication (*if-then*). We will represent these operations by the usual symbols, $\neg$, $\wedge$, $\vee$, and $\rightarrow$, respectively.[1]

Whether a given logical expression is *true* or *false* depends on the truth and falsity of its component atoms and expressions, and on the properties of the employed logical operations. Table 11.1 reminds the reader of the truth values that result from these operations. As common, in Boolean algebra, *true* is represented by 1 and *false* is represented by 0. The only operation that beginners sometimes find confusing is implication: The rule is *false* only in the case where truth appears to imply falsity; in all other cases it is *true*. Why this is the case may be intuitively grasped by considering the following rule:

```
lives(x,Miami) → lives(x,Florida).
```

In plain English, if x lives in Miami then x lives in Florida.

---

[1] Of course, logic knows also other operations, such as equivalence or eXclusive-OR. These, however, are in automated reasoning rarely used.

p	q	$\neg p$	$p \wedge q$	$p \vee q$	$p \rightarrow q$
1	1	0	1	1	1
1	0	0	0	1	0
0	1	1	0	1	1
0	0	1	0	0	1

Table 11.1 Truths and Falsities Resulting from Basic Logical Operations

Suppose that `Bill` lives in Paris so that the antecedent is *false*. In this case, also the consequent is *false* because `Bill` lives in France, and not in Florida. The situation represents the last row of the truth table for implication. The reader will agree that the rule is still correct, regardless of `Bill`'s domicile. Similar examples can be invented for the table's third row (e.g., `Bill` may live in Orlando). The only way to prove the implication wrong would be to show that x lives in Miami, and yet not in Florida.

11.1.6 Logical Expressions in Automated Reasoning

Let us consider the following three expressions where atoms are connected by logical operators.

```
offspring(Pete,Fred) ∧ ¬offspring(Fred,Pete).
parent(x,Pete) ∨ parent(x,Fred).
parent(x,y) ∧ parent(y,z) → grandparent(x,z).
```

The first expression says that `Pete` is `Fred`'s offspring but not the other way round. The second says that x is a parent of either `Pete` or `Fred`. And the third defines the meaning of `grandparent`.

Generally speaking, any of the three examples is a valid logical expression. For the needs of automated reasoning, however, it is much better to express knowledge in terms of *rules*, which is the case of the last expression (the one that defines a `grandparent`). The reader already knows that, informally, a rule is a logical expression containing the operator of implication, "$\rightarrow$."

11.1.7 Universal Quantifier

Rules in which all arguments of all predicates are constants are rarely useful in automated reasoning. Instead, realistic implementations typically operate with predicates whose arguments are variables. These variables are then to be instantiated in the course of the reasoning process in ways we saw in a earlier chapter.

If we want to emphasize that some rule is valid for any possible instantiation of a certain variable, we make it clear by the *universal quantifier*, $\forall$. For instance, the concept `grandparent` will then be defined as follows.

$$\forall x \forall y \forall x.[\mathbf{parent(x,y)} \wedge \mathbf{parent(y,z)} \rightarrow \mathbf{grandparent(x,z)}].$$

We say that variables `x`, `y`, and `z` are here *universally quantified*. The rule is read as follows: "for any instantiation of `x`, `y`, and `z`, if ... then ..."

11.1.8 Existential Quantifier

Some rules inform us about relations that are *true* only for some specific instantiations of the variables. This is the case of the English sentence "some students like AI." In other words, not *all* but only *some* students are fond of this course. To convey this meaning is the task for the *existential quantifier*, ∃. The given English sentence then takes the following form:

$\exists$x.[**student(x)** $\rightarrow$ **likes(x,AI)].**

We say that variable x is here *existentially quantified*. The rule is read as follows: "there exists at least one x such that if ... then ..."

11.1.9 Order of Quantifiers

If more than one variable in the sentence is quantified, the list of these quantifications is processed "from left to right." To see why this matters, consider the following expression:

$\forall$**p**$\exists$**h.house(p,h).**

The term informs us that, "for every person, there exists at least one house that belongs to the person." In other words, everybody has a house.[2]

Suppose, however, that we reverse the order of the quantifications as follows:

$\exists$**h**$\forall$**p.house(p,h).**

Reading the quantifiers from left to right, we realize that the expression informs us that, "there exists at least one house that belongs to every person." This may be the case of a city hall.

The reader will remember that changing the order in the list of quantifications may seriously modify the meaning of the expression.

11.1.10 Additional Examples

To illustrate the concepts thus described, Table 11.2 offers a few translations of simple English statements to their first-order logic equivalents. Note that a lot depends on the engineer's own choice of the predicates and their definitions. For instance, when translating the sentence, "all mammals have large brains," the engineer has decided that the relation `brain(y,x)` is *true* if and only if y is x's brain.

Apart from the rules suggested in Table 11.2, other formulations are possible. For instance, the second rule uses relation `teaches(x,AI)`, assuming that the truth or falsity of the predicate can somehow be inferred from the available knowledge base. However, it is also conceivable that the knowledge base lists all courses, among them `course(AI)`, in which case the engineer may prefer to replace (in that rule) relation `teaches(x,AI)` with the following three:

teaches(x,y) $\wedge$ **course(y)** $\wedge$ **equals(y,AI).**

---

[2]Note that this interpretation is only implicit because the expression does not specify that p is a person and h is a house.

"All mammals have large brains."
$\forall x\ \forall y.[\mathbf{mammal(x)} \wedge \mathbf{brain(y,x)} \rightarrow \mathbf{large(y)}].$ Here, relation `brain(y,x)` is meant to be *true* if `y` is **x**'s brain.
"No professor of AI likes spiders."
$\forall x \forall y.[\mathbf{professor(x)} \wedge \mathbf{teaches(x,AI)} \wedge \mathbf{spider(y)} \rightarrow \neg\ \mathbf{likes(x,y)}].$ Here, relation `likes(x,y)` is *true* if `x` likes `y`.
"Some professors are married."
$\exists x.[\mathbf{professor(x)} \rightarrow \mathbf{married(x)}].$
"People are either biologists or they dislike at least some animals."
$\forall x \exists y.[\mathbf{person(x)} \wedge \mathbf{animal(y)} \rightarrow \mathbf{biologist(x)} \vee \neg\ \mathbf{likes(x,y)}].$

TABLE 11.2 Examples of Logical Expressions for Simple English Statements

The reader may ask if the order of these relations is here arbitrary. In principle, it is. However, later we will see that the concrete order may seriously affect the computational efficiency of the program that uses the rule in the process of automated reasoning.

Control Questions

If you are unable to answer the following questions, return to the appropriate place in the preceding text.

- What is the difference (in first-order logic) between a function and a relation? Illustrate the two concepts by examples.
- What is an atom and what is a logical expression?
- What do you know about universal and existential quantifiers? How are they specified in logical expressions? Does their order matter?

11.2 Removing Quantifiers

Beginners find quantifiers unnatural, bothersome, and hard to get used to. Here is the good news: Quantifiers can often be avoided. A logician will argue that their removal will modify, even if ever so slightly, the meaning of some logical expressions. However, the practical-minded engineer will counter that minor loss in the accuracy of the knowledge base can be tolerated as long as the inaccuracy does not affect the reasoning process.

11.2.1 Removing Some Existential Quantifiers

Suppose we want to enter into the knowledge base the information that John has a house, employing the relation `house(h,p)` that is *true* if house `h` belongs to person `p`. Here is the corresponding logical expression with an existentially quantified variable:

$$\exists \mathbf{h}.\mathbf{house(h,John)}.$$

To avoid existential quantification, the engineer may decide to simplify the statement as follows:

house(SK,John).

Here, `SK` is the name assigned to `John`'s house. The name is a constant, and as such has to be capitalized. Are the two statements logically equivalent? Not really. The demise of variable `x` has eliminated the need to quantify it. However, the price for the simplification is that the statement's meaning is now slightly different. The modified rule works with one concrete house that is here denoted by `SK`; the original rule allows for the possibility that `John` has two or more houses.

This minor loss in accuracy may be acceptable, though. More likely than not, `John` *does* have only one house. And even if he has more than one, the number of his houses will be irrelevant if this number never enters any reasoning process. In our daily discussions and arguments, such minor inaccuracies rarely play a serious role. In the same spirit, they are unlikely to be critical in AI programming.

11.2.2 Existentially Quantified Vectors

Suppose we decide to enter in the knowledge base the information that, in our domain of interest, everybody has a house. The reader will recall that a few pages ago, this circumstance was conveyed by the following statement:

∀p∃h.house(h,p).

Again, the existential quantifier can be eliminated by making the assumption that each person has one and only one house. These houses can then be represented by a vector whose each element is a house belonging to a concrete person:

∀p.house(SK(p),p).

Here, `SK(p)` is the vector of houses, each referred by the pointer to the concrete person (in the parentheses) who owns it. Again, we have to concede that this modified statement is not exactly equivalent to the original one. Reduced accuracy, however, may be an acceptable price to pay for the elimination of the existential quantifier.

11.2.3 Frequently Overlooked Case

In the last paragraph, a vector of constants was used. This little trick, however, would be inappropriate when dealing with the following statement that differs from the previous in the order with which the quantifiers are declared:

∃h∀p.house(h,p).

The reader will recall that the statement was mentioned in Section 11.1 as a way of informing the program about the existence of a house that belongs to everybody; for instance, this may be the city hall. Since this is a single house, only a scalar constant is needed:

∀p.house(SK,p)

11.2.4 Skolemization[3]

The previous examples help us appreciate the essence of the mechanism to remove some existential quantifiers. We are now ready to introduce the general form of a technique known as *skolemization*. Consider the following relation:

$$\forall \mathbf{v}_1 \ldots \forall \mathbf{v}_n \exists \mathbf{x}.[\mathbf{v}_1,\ldots \mathbf{v}_n,\mathbf{x}]$$

Here, the existentially quantified variable, x, is preceded by n universally quantified variables, v_i. In the spirit of the previous examples, we need a constant for every combination of the values of these variables. Here is what the previous expression looks like after skolemization:

$$\forall \mathbf{v}_1 \ldots \forall \mathbf{v}_n.[\mathbf{v}_1,\ldots \mathbf{v}_n,\mathbf{SK}(\mathbf{v}_1,\ldots \mathbf{v}_n)]$$

We can see that the expression remains almost the same, except that the existentially quantified variable x has been replaced with an n-dimensional constant array where each entry corresponds to one combination of the values of v_i for $i \in [1, n]$.

The reader will have noticed that the universally quantified variables v_i precede the existentially quantified x. The number of dimensions in the skolemized array of constants is the same as the number of universal quantifiers in the original expression.

Also, note that the simple examples from the beginning of this section are special cases of this general formula.

11.2.5 Removing the Remaining Existential Quantifiers

Skolemization eliminates most of the existential quantifiers, but not all of them. Fortunately, practical experience indicates that virtually all of the existential quantifiers that have survived skolemization can easily be replaced with the universal quantifiers. Again, this *does* modify the strict logical meaning of the given expression. As before, however, the change is barely perceptible, and rarely affects the subsequent reasoning process.

For illustration, consider the following rule which, in plain English, tells us that in each difficult course taken by at least one student, an attempt to succeed involves a lot of work:

$$\forall \mathbf{c}.[\mathbf{difficult(c)} \land \exists \mathbf{s}.[\mathbf{takes(s,c)} \rightarrow \mathbf{requires\_work(c)}]$$

On some thought, the reader will agree that the replacement of the existential quantifier with the universal quantifier does not represent here any major change in the rule's meaning. Here is the result of the replacement:

$$\forall \mathbf{c} \forall \mathbf{s}.[\mathbf{difficult(c)} \land \mathbf{takes(s,c)} \rightarrow \mathbf{requires\_work(c)}]$$

This new rule says that the course will require a lot of work for any student that has taken it. From the perspective of strict logic, this new formulation *does* constitute a minor

[3] This approach was suggested by the Norwegian logician Thoralf Skolem (1887–1963). It is in his memory that textbooks (including this one) choose to denote the resulting constants with SK.

modification. However, the difference is unlikely to lead to different conclusions when the rule is employed in automated reasoning.

11.2.6 Consequence of the Disappeared ∃'s

We have seen that most existential quantifiers in the knowledge base can be eliminated by skolemization. Those existential quantifiers that survive skolemization can almost always be replaced with universal quantifiers without any serious consequences. This is good. We have reached a situation where all remaining quantifiers are universal.

This has one serious consequence. If we know that all variables in the knowledge base are quantified universally, we do not need to specify the quantifiers at all! Indeed, it is now enough to write the rules without any quantification, and simply assume they are valid for each instantiation, that is, universally. This, by the way, is why the *Prolog*-like rules in Chapter 9 never needed quantifiers.

Control Questions

If you are unable to answer the following questions, return to the appropriate place in the preceding text.

- Explain how Skolemization eliminates the majority of existential quantifiers.
- Why can the existential quantifiers that have survived Skolemization be eliminated, too? What does this circumstance mean for variable quantification in general?

11.3 Binding, Unification, and Reasoning

The principles of *modus ponens* and resolution in Chapter 10 were explained in simplified circumstances where no variables were needed. Let us now take a look at how to handle more realistic domains. To begin with, the introduction of variables necessitates that we explore two critical concepts: *binding* and *unification*.

11.3.1 Binding Variables

Suppose we want to subject to resolution the following two rules. The first is telling us that anybody teaching the AI course is professor. The second informs us that all professors (addressed in the given knowledge base) receive their salary from the University.

$$\mathbf{teaches(x, AI) \rightarrow professor(x)}$$
$$\mathbf{professor(y) \rightarrow paid\_by(y, University)}$$

The first rule has `professor(x)` in the consequent, the other has something similar in the antecedent. When we say, "something similar," we mean that there *is* a difference. The variable in the parentheses is in one case denoted by `x`, and in the other by `y`. But if we assume that `x = y`, the two relations become identical, and the two rules can thus be subjected to resolution.

We say that the assumption `x = y` *binds* the two arguments.

11.3.2 Binding List

In the process of reasoning, the AI program always looks for pairs of rules that can be subjected to *modus ponens* or resolution. Very often, the two rules only become eligible if we bind their variables accordingly (as in the previous paragraph).

Throughout the reasoning process, many bindings usually have to be established. All these bindings are then collected in a *binding list*. Suppose, for instance, some earlier reasoning process established that `teaches(John,AI)`—that John teaches artificial intelligence. This binds `x` from the previous rule's antecedent to `John`. Then we saw that when the next step bound `x` to `y`, the reasoning process could prove that `John` is being paid by the University, which is indicated by the term `paid_by(John,University)` in the second of the rules mentioned above. Adding this to the binding list, σ, we obtain the following:

$$\sigma = \{\texttt{x} = \texttt{John},\ \texttt{y} = \texttt{x}\}.$$

In realistic applications, the binding lists can be very long.

11.3.3 Bindings of Nested Relations

The end of Section 9.1 briefly mentioned the existence of nested arguments: An argument can itself be a predicate with its own arguments. This circumstance has to be considered, too. Suppose we want to find the binding for the following two expressions:

father(Fred,brother(x,Ann)).
father(Fred,y).

Whereas the first argument, `Fred`, has the same value in both relations, the second differs in the two relations. The binding relies on the assumption that the actual values should be the same. This is the case if the binding looks like this:

$$\{\texttt{y} = \texttt{brother(x,Ann)}\}.$$

11.3.4 Unification

The procedure that establishes all necessary bindings is called *unification*. The bindings may be much more complicated than those in the previous paragraphs. The general algorithm of *unification* is fairly complicated because of the need to accommodate the most diverse types of nesting. Fortunately, deep nesting is today much less common than it was a generation or so ago. Mastering the entire unification technique is therefore not as urgent as it used to be.

Perhaps it will be enough to convey the basic principle only by means of the three simple examples in Table 11.3. Encoding the mechanism in a computer program should not pose difficulties.

11.3.5 Modus Ponens and Resolution Using Variables

In a knowledge base that contains variables, the inference mechanisms based on *modus ponens* and resolution are essentially the same as in the simplified context from Chapter 10. The only novelty is that here the program has to provide all the bindings

The following examples always contain two expressions, p and q, for which the binding list is to be established.

$p =$ `professor(Bill)`, $q =$ `professor(x)`
binding list: {`x = Bill`}

$p =$ `teaches(Bill, x)`, $q =$ `teaches(y,AI)`
binding list: {`y = Bill, AI = x`}

$p =$ `f(x,y)`, `q = g(y,a)`
binding list: {`x = y, y = a`}

Note that a minor post-processing of the last example may result in an alternative binding: {`x = a, y = a`}.

TABLE 11.3 Example Results of Unification

under which the reasoning procedure has been conducted. The algorithm from Table 10.8 is thus modified in the sense that the two rules can only be resolved if they have been unified by a specific binding list, σ.

The complete versions of inference rules are summarized in Table 11.4. The reader will have noticed the symbols "$|_\sigma$" at the end of the third rule in each of the two procedures.

Control Questions

If you are unable to answer the following questions, return to the appropriate place in the preceding text.

- What is a binding list? What is unification? What roles do they play in our reasoning techniques?
- In what way do the full-fledged versions of *modus ponens* and resolution differ from their simplified versions discussed in Chapter 10?

Modus ponens: Suppose that d in the second rule and a_i in the first rule unify under the binding list σ. Here is the behavior of *modus ponens*:

$$a_1 \wedge \ldots \wedge a_m \rightarrow b$$
$$c_1 \wedge \ldots \wedge c_n \rightarrow d$$

$$a_1 \wedge \ldots \wedge a_{i-1} \wedge a_{i+1} \wedge \ldots \wedge a_m \wedge c_1 \wedge \ldots \wedge c_n \rightarrow b \;|_\sigma$$

Resolution: Suppose that d_j in the second rule and a_i in the first rule unify under the binding list σ. Here is the behavior of resolution:

$$a_1 \wedge \ldots \wedge a_m \rightarrow b_1 \vee \ldots \vee b_k$$
$$c_1 \wedge \ldots \wedge c_n \rightarrow d_1 \vee \ldots \vee d_l$$

$$a_1 \wedge \ldots a_{i-1} \wedge a_{i+1} \ldots \wedge c_1 \wedge \ldots \wedge c_n \rightarrow b_1 \vee \ldots b_k \vee d_1 \ldots \vee d_{j-1} \vee d_{j+1} \vee \ldots d_l \;|_\sigma$$

TABLE 11.4 Reasoning Procedures Using Variables

11.4 Practical Inference Procedures

The information from the previous section should be sufficient for anyone to implement a simple automated-reasoning program. To make it easier, however, let us illustrate the reasoning process on a simple example. After this, we will discuss a few issues related to the reasoning process's computational costs.

11.4.1 Concrete Example

Table 11.5 presents a trivial knowledge base that consists of just three rules. The question we want the program to answer is whether some course is difficult. When comparing the reasoning process with the algorithms from Chapter 10, we can see that now the query contains a variable, `y`. The table shows how the query is answered by means of resolution with backward chaining, the approach that always keeps "F" on the right-hand side.

The table also shows the concrete bindings under which the resolution occurs. These bindings are then used by the program when returning its answer to the query. In this particular example, the program will respond with `y` = `AI`, `p` = `Bill`. We encountered answers of a similar nature in Chapter 9.

Knowledge Base:

1. $T \rightarrow$ `new_course(AI)`.
2. $T \rightarrow$ `tough_prof(Bill)`.
3. `new_course(x)` $\wedge$ `tough_prof(p)` $\rightarrow$ `difficult(x)`.

Query:

Informally: "Is course `y` difficult?"
Formally: **? :- difficult(y)**

The sequence of steps followed by the resolution principle with backward chaining.

First, add to the knowledge the opposite of what is to be proved.

Note that the last column lists the bindings.

	expression	source	bindings
a	`difficult(y)` $\rightarrow$ `F`	query	
b	`new_course(x)` $\wedge$ `tough_prof(p)` $\rightarrow$ `F`	resolved: a,3	`x=y`
c	`new_course(x)` $\rightarrow$ `F`	resolved: b,2	`p=Bill`
d	$T \rightarrow F$	resolved: c,1	`x=AI`

Contradiction has been reached, which means that the original query was correct under the given bindings.

The list of bindings: σ = {`x` = `y`, `p` = `Bill`, `x` = `AI`}.

The list of bindings after post-processing: σ = {`y` = `AI`, `p` = `Bill`}.

Interpretation: The course is difficult if it is `AI` and if the professor teaching the course is `Bill`.

TABLE 11.5 An Example of a Reasoning Procedure That Uses a Knowledge Base That Contains Variables

Task: Find the bindings for the following rule antecedent:

`professor(x), teaches(x,AI),father(x,y) →...`

predicate	# bindings
`professor(x)`	1,000
`teaches(x,AI)`	1
`father(x,y)`	6,000
`father(const,y)`	2

Processing the predicates from left to right results in 2,000 bindings.

If we change the order to `teaches(x,AI), professor(x), father(x,y)`, the number of bindings drops to 2.

Table 11.6 The Number of Variable Bindings Depends on the Order in Which the Predicates Are Processed

As before, one column lists the pairs of rules that have been subjected to resolution (e.g., the second line contains, `resolved: a,3`). This information can be used in a more advanced system capable of informing the user about how exactly the concrete solution has been reached. Such explanations are commonly used in so-called *expert systems* that were popular in the 1980s and 1990s. They will be briefly discussed in Chapter 16.

11.4.2 Multiple Solutions

At certain steps of automated reasoning, alternative pairs of rules (sometimes *many* different pairs of rules) can possibly be subjected to *modus ponens* or resolution.[4] Which pair should be chosen, at the given step, which pair leads to fastest query processing? This question should be considered by any programmer who plans to implement the reasoning procedure by heuristic search. In the case of purely random choices, the process is likely to be expensive.

A number of techniques have been invented and successfully implemented in working programs. Many of them rely on suggestions how to formulate the rules, how to formulate queries, and what heuristics the reasoning program should use when deciding how to proceed. Descriptions and analyses of all these techniques and "tricks" might easily fill a whole book; besides, most of them are rather advanced. Let us therefore mention just some of the most basic ones, just for the sake of illustration.

11.4.3 Number of Bindings

At the top of Table 11.6, we can see a rule antecedent that consists of three predicates. The rule deals with a professor of AI, (this person is denoted by x), and the professor's child, y. When trying to resolve this rule with another rule in the knowledge base, the system has to identify many bindings of the two variables, x and y. How many such bindings are there? Let us use for the calculation the numbers from Table 11.6.

[4] For instance, in the case of the knowledge base from Table 10.6, the process could start either by resolving rules 1 and 2, or by resolving rules 1 and 3.

We start with the first predicate, `professor(x)`. The table is telling us that the knowledge base contains 1,000 instances of professors. For each, we have to establish whether x teaches AI. Since this is a yes-or-no decision, only one binding needs to be considered here. This identifies a concrete x. For each x, the third predicate, `father(x,y)` establishes the child, y. The table is telling us that each concrete father has on average two children, which means two bindings. The total number of bindings is therefore $1,000 \times 1 \times 2 = 2,000$.

Suppose we change the order of the predicates as indicated at the bottom of Table 11.6. Now we start with `teaches(x,AI)`. The second row of the table is telling us that there is only one professor teaching this course. The next predicate says that this person, x, is a professor. Again, this is a yes-or-no decision because the binding was established with the evaluation of the previous predicate, `teaches(x,AI)`. Finally, all children of x have to be identified. On average, there are two. The total number of binding is thus $1 \times 1 \times 2 = 2$.

We have convinced ourselves that the total number of bindings depends on the order in which the predicates are processed.

11.4.4 Starting from the Left

The observation from the last paragraph suggests what a cost-efficient strategy of predicate evaluation might possibly look like. An automated-reasoning program that seeks to minimize costs might begin by establishing the numbers of variable bindings in the individual predicates encountered in the rule's antecedent. Once these numbers are known, the program can determine which order minimizes the number of bindings.

Practical implementation of this approach may not be easy, though. Much more common is therefore the approach that makes the program always process the predicates from left to right. The responsibility for suggesting the predicates' most efficient order is then relegated to the programmer who is likely to know more about the numbers of bindings than the machine.

The pseudo-code in Table 11.7 summarizes the search algorithm built around the idea just indicated.

11.4.5 Accelerating the Reasoning Process

Automated reasoning based on the resolution principle can be computationally expensive even when its backward-chaining version is used. Knowing this, AI experts have

Input: a knowledge base
Task: prove that q is *true*.

1. Create an empty list L and put in it the negation of what is to be proved: $L = \{q \rightarrow F\}$.
2. If L contains $T \rightarrow F$, then stop with success: q has been proved.
3. If the first atom in p can be resolved with some rule from the knowledge base or with some $p_1 \in L$, do so, and add the result to the knowledge base.
4. If p cannot be resolved with any rule from the knowledge base or from L, remove p from L.
5. If $L = \emptyset$, stop with failure. Otherwise, go to step 2.

Table 11.7 Pseudo-Code of Ordered Resolution

developed auxiliary techniques to speed up the process. Let us briefly mention at least two of them, just to give the reader an idea of the employed principles.

11.4.6 Look-Ahead Strategy

When binding a variable, check whether each other condition (in the antecedent) that involves the same variable has at least one binding. For illustration, consider the following rule antecedent that concerns the cases where a professor's child is taking his or her father's course:

$$\textbf{professor}(\mathrm{x}), \textbf{father}(\mathrm{x}, \mathrm{y}), \textbf{teaches}(\mathrm{x}, \mathrm{z}), \textbf{studies}(\mathrm{y}), \textbf{takes}(\mathrm{y}, \mathrm{z}) \rightarrow \ldots$$

The number of bindings of these terms is high. There are many professors, each of them having two children on average, the professor may teach a few courses, and so on. Here, all these bindings are not of immediate interest. For instance, when investigating the second predicate, `father(x,y)`, one can just as well ignore those children, `y`, for which `studies(y)` is *false*. To avoid these unnecessary bindings, the program has to "look ahead" and consider the bindings of the same variable for the antecedent's remaining predicates.

11.4.7 Back-Jumping

To appreciate the nature of this second "trick," take another look at the rule antecedent from the previous paragraph:

$$\textbf{professor}(\mathrm{x}), \textbf{father}(\mathrm{x}, \mathrm{y}), \textbf{teaches}(\mathrm{x}, \mathrm{z}), \textbf{studies}(\mathrm{y}), \textbf{takes}(\mathrm{y}, \mathrm{z}) \rightarrow \ldots$$

Suppose the program investigates the predicates from left to right, as recommended earlier. Suppose that the evaluation of the first three predicates has currently established the following partial binding list:

$$\sigma = \{\texttt{x = Bill, y = Eve, z = AI}\}$$

At this moment, the program realizes that `studies(y)` is *false* because `Eve` does not study. The classical hill-climbing approach will backtrack to the previous predicate, and try another biding for `z` in `teaches(Bill,z)`. This, however, will not help because `studies(Eve)` will remain *false* regardless of `z`'s bindings (regardless of the courses her father, `Bill`, teaches). In a situation of this kind, it thus makes sense to backtrack not just to the previous predicate, but to "jump" over this predicate all the way to `father(x,y)` and consider another instantiation of `y`, which means another of `Bill`'s children. This technique is called *back-jumping*.

True, the situation was to some degree caused by the programmer who ordered the predicates in an awkward manner, but a well-written reasoning program that uses back-jumping can deal with it.

Control Questions

If you are unable to answer the following questions, return to the appropriate place in the preceding text.

- Why does the reasoning procedure illustrated by Table 11.5 contain also one column containing the variable bindings?
- Explain why the number of the bindings to be explored depends on the order in which the predicates in the antecedent are listed.
- Outline the strategies that this section suggested for an improvement of the reasoning process's efficiency.

11.5 Practice Makes Perfect

To improve your understanding, take a chance with the following exercises, thought experiments, and computer assignments.

- Consider the logical meaning of the following English statements, and then express them by rules formulated in first-order logic (you may use two rules for a single English sentence). Make sure you use appropriate quantifiers.
  "Some courses are difficult."
  "Some students do not choose popular courses."
  "Professors like to teach advanced classes."
  "Some professors of computer science are married."
  "Some reptiles are poisonous, in which case they are not long."
  "People are either biologists or they dislike at least some animals."
- Write a two-page essay discussing the role and meaning of quantifiers in first-order logic, and the procedures we can use to eliminate them in our knowledge bases.
- Unify the following pairs of expressions:

  $p =$ `likes(x,y)`, $q =$ `likes(Bill,AI)`
  $p =$ `difficult(course(AI))`, $q =$ `x`
  $p =$ `uncle(x,y)`, $q =$ `uncle(brother(x,y),z)`

- Using the numbers from Table 11.6, calculate the total number of bindings involved in the evaluation of the query in the following order:

  `father(x,y), teaches(x,AI), professor(x)`

- Invent your own example that will illustrate the look-ahead strategy introduced in Section 11.4. Explain why the look-ahead strategy really accelerates the reasoning process.
- Invent your own example that will illustrate the back-jumping strategy introduced in Section 11.4. Explain why back-jumping helps here.
- Using an internet browser, find another approach to increase the efficiency of rule evaluation. Describe it in a one-page essay.

11.6 Concluding Remarks

The reader now has an idea how to implement simple programs capable of automated reasoning, even if still only in a rather rudimentary form. What is new, in this chapter, is the introduction of variables, and the need to bind them when conducting reasoning by *modus ponens* or by resolution. The great number of possible bindings and instantiations leads in realistic knowledge bases to considerable computational costs. A lot of research has gone into ways of reducing these costs by "anticipating" the numbers of bindings.

An automated-reasoning program may return a list of the employed variable bindings. The list is then used in an answer to the user's query—the reader knows how to generate responses to queries such as those from Chapter 9. Accomplishing the task in a professional manner is of course more complicated than what this chapter indicated, and it requires a lot of creativity on the programmer's part. The underlying principles, however, should by now be fairly clear.

The reasoning procedures described in this chapter are still too simple to handle the needs of realistic domains. For instance, they assume that all rules are perfectly reliable, and that the knowledge base does not suffer from any kind of uncertainty or ignorance. In realistic domains, such perfection is rare. What is needed are special measures to deal with the imperfections of human knowledge. These will be the subject of later chapters.

Will the reader who has studied the last three chapters be able to write a *Prolog* compiler? An honest answer is, "yes and no." Our exposition relied on a very simplified version of *Prolog*. For this, a dedicated student may now be prepared to write the software needed to interpret the language's statements, and to perform some kind of rudimentary reasoning over these statements. This said, full-scale *Prolog* has many other functionalities that were ignored here.

CHAPTER 12

Alternative Ways of Representing Knowledge

First-order logic is the most commonly used paradigm for knowledge representation and for automated reasoning. It is not the only one, though. Other mechanisms exist. Perhaps the best-known among them are frames and semantic networks, two approaches closely related to each other. While their current popularity is a pale shadow of what it was in the 1990s, they do possess certain advantages, and an AI specialist definitely needs to know about their existence.

This chapter offers a brief introduction to these two alternatives, and shows what kind of inferences they permit. The reader will learn about their positive and negative aspects, and will develop some understanding of what made these two approaches so popular in the past, and why they are so rarely employed nowadays.

12.1 Frames and Semantic Networks

Perhaps the most popular alternative to rules and logic is the paradigm known as *frames*. Equally important is this technique's graphical equivalent, semantic networks. Although the two paradigms are today much less popular than they used to be, it is good to know about their strengths and weaknesses.

12.1.1 Concrete Example of Frames

Table 12.1 shows a very simple knowledge base expressed in *frames* such as `mammal`, `dog`, or `fido`. Each frame contains one or more *slots* such as `moving` and `covered_by` in the cases of `mammal`. The slots are to be filled with concrete values. These values tell us, for instance, that members of the class `mammal` are covered with hair and that they walk. Filling the slots with values is not mandatory; a slot can be empty.

The simple example shown in Table 12.1 contains only a few frames, and none of these has more than three slots. In reality, there will be at least hundreds of frames, and some of them can have many slots. Moreover, some slots can be used repeatedly. For example, `fido` is an instance of `dog`, but also an instance of domestic animals, and so on. To reflect all these properties, the slot `instance_of` may be used several times within the same frame, each time with a different super-class.

Each *frame* has one or more *slots* to be filled with concrete values.

If a slot or a value is missing, then the instance inherits the corresponding value from the class to which it belongs. Sub-classes, too, inherit the values of their super-classes, unless specified otherwise.

mammal
  moving: walks
  covered_by: hair
dog
  subclass_of: mammal
fido
  instance_of: dog
dolphin
  subclass_of: mammal
  moving: swims
  covered_by: scales
bird
  moving: flies
  covered_by: feathers
hen
  subclass_of: bird
  moving: walks
pipi
  instance_of: hen

Table 12.1 An Example of a Frame-Based Knowledge Base

12.1.2 Inherited Values

In Table 12.1, one of the slots in frame `mammal` characterizes locomotion: Mammals move around by walking. Such slot is absent from frame `dog`, but another of this latter frame's slots specifies that `dog` is a subclass of `mammal`, a frame where the `moving`-slot *is* available; its value applies to all sub-classes. We say that `dog` *inherits* the slot values from the higher-level frame `mammal`, and it is thus supposed to move by walking.

A larger knowledge base might contain another class, say, `vertebrate`, and the class `mammal` may possess slot `subclass_of` specifying that `mammal` is a subclass of vertebrates. In that event, `mammal` would inherit all slot values of `vertebrate`; likewise, `dog` would inherit slot values of `mammal` and `vertebrate`.

Finally, note the slot named `instance_of`. Its meaning is similar to that of `subclass_of`, with all the consequences for inherited values.

12.1.3 Exceptions to Rules

One advantage of the inheritance mechanism is that it offers a natural way of dealing with exceptions. Let us clarify the point by a simple example. The frames in Table 12.1 contain the information about `mammal`'s way of locomotion: walks. We know, however, that dolphins do not walk in spite of being mammals. In our knowledge base, this exception is taken care of by the presence of the slot `moving` in the frame `dolphin` where its value is `swims`. True, `dolphin` is a subclass of `mammal`, but the explicit specification of the value in its `moving` slot overrides whatever value might have been inherited.

Generally speaking, the value is inherited from the super-class frame if the sub-class does not have the slot. Also, if the sub-class does have the slot, but its value is not specified, then the inheritance mechanisms still applies. In this way, the frame-based knowledge representation makes it possible to formulate general rules while allowing exceptions.

12.1.4 Semantic Networks

Closely related to frame-based knowledge representation is the idea of a *semantic network*, SN. One such SN is shown in Figure 12.1. The reader will have noticed that this SN represents exactly the same knowledge as the frames listed in Table 12.1. The good point

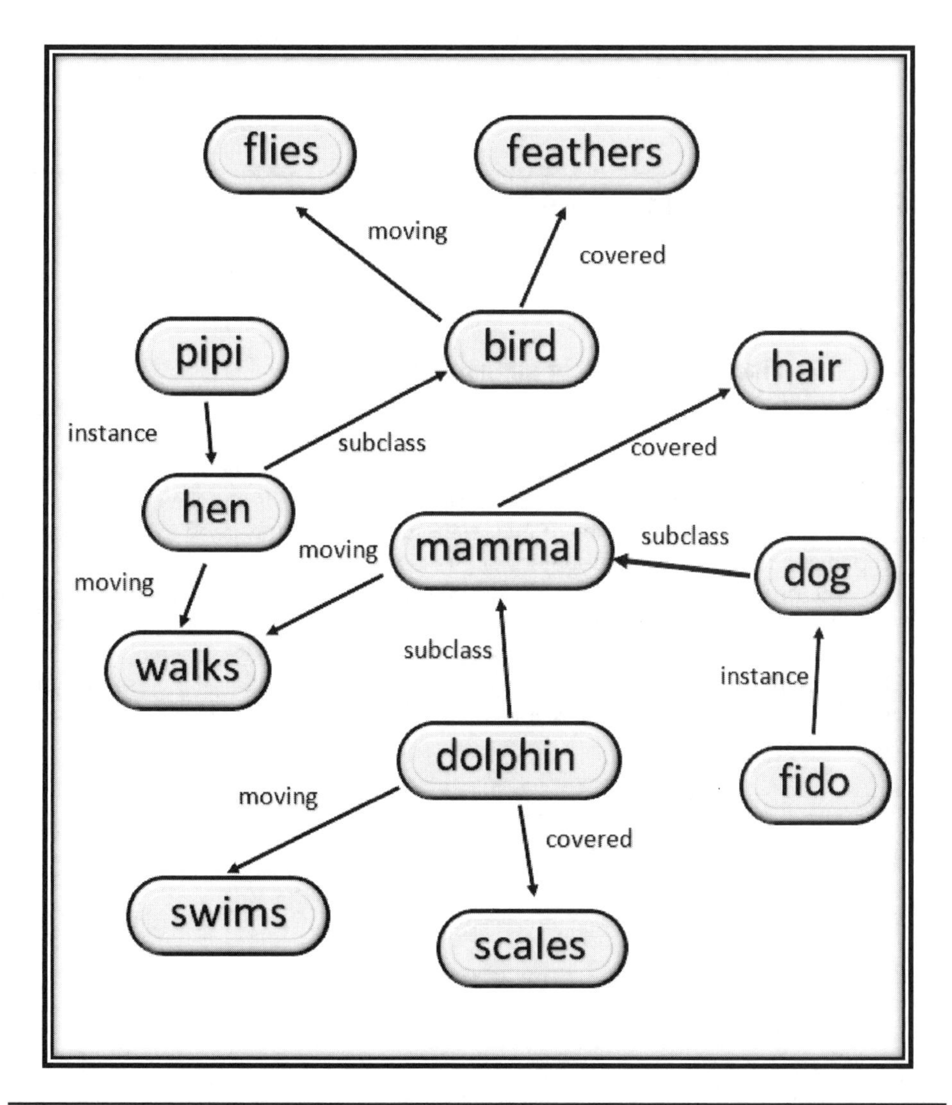

Figure 12.1 A semantic network that represents the same knowledge as the frame-based knowledge base from Table 12.1.

about a SN is that it represents the knowledge bases visually, in the form of graphs. These are often easier to interpret and navigate.

In computer programs, SNs are easy to implement. Perhaps the simplest way of doing so is by representing each edge in the graph by the triplet, `[start,edge_name, end]`. Of course, an experienced programmer is likely to employ more advanced data structures.

Control Questions

If you are unable to answer the following questions, return to the appropriate place in the preceding text.

- How do *frames* represent knowledge? Explain the meaning of *inheritance*, and comment on how it allows us to handle exceptions to rules.
- What are *semantic networks* and how are they related to frames?

12.2 Reasoning with Frame-Based Knowledge

One of the benefits of knowledge representation by frames and SNs is that, in these paradigms, inference can be very efficient. Two examples will illustrate the point.

12.2.1 Finding the Class of an Instance

Perhaps the simplest task in a frame-based knowledge base is to find the class of an object. For instance, object `fido` does not offer any direct information regarding its being covered by hair. However, the frame does have a slot informing us that it is an instance of `dog` which, in turn, is a sub-class of `mammal`. And for the latter, the `covered_by` slot is available here. This is just one of many situations where it is good to have a mechanism capable of identifying super-classes of an object.

The pseudo-code in Table 12.2 summarizes an algorithm that determines whether or not a given object belongs to a user-specified class `X`. The principle is simple. If the object contains an `instance_of`-slot or a `subclass_of`-slot, these may point directly to `X`. Alternatively, the object may be an instance or subclass of a class that may itself be a subclass of `X`, and so on, recursively. The reader will find it easy to implement the technique by blind search.

Query: `instance(Object,X).`
Interpretation of the query: Is `Object` an instance of class `X`?

1. Let S_C be the set of all classes indicated by slot `instance_of` in `Object`.
2. If $S_C = \emptyset$, stop and return NO.
3. Select some $C \in S_C$. If $C =$ `X`, return YES.
4. Add to S_C all classes indicated to by slots `subclass_of` in C.
5. Return to step 2.

Table 12.2 Pseudo-Code of a Technique to Determine Whether a Given Object Belongs to a Given Super-Class

Query: `value(Object,P,v).`
Interpretation of the query: What is the value, `v`, of slot `P` in `Object`?

1. If `Object`'s slot for property `P` contains a value, return this value.
2. Otherwise, let S_C be the set of all classes pointed to by `instance_of` in `Object`.
3. If $S_C = \emptyset$, stop and return *failure*.
4. Select some $C \in S_C$. If it specifies a value of property `P`, stop with success and return this value.
5. Otherwise, add to S_C all classes pointed to by slots `subclass_of` in C.
6. Return to step 2.

TABLE 12.3 Establishing a Slot Value in a Frame.

12.2.2 Finding the Value of a Variable

Only slightly more advanced is the question how to establish the *value* of a feature that has not been specified in a given frame. We encountered this need a few paragraphs ago when explaining the *inheritance* principle using the example of `fido`'s way of locomotion.

The pseudo-code in Table 12.3 summarizes an algorithm that determines the value of a concrete feature of a given object. The principle is similar to the one from the previous paragraph. The first choice is that the feature value is provided directly in the given object. Failing this, the algorithm attempts to find the feature in the class to which the object belongs, or in any of its super-classes.

12.2.3 Reasoning in Semantic Networks

There is a clear correspondence between SNs and frames: For any frame-based knowledge base, we can create its equivalent SN and the other way round. This means that, for reasoning in SNs, essentially the same algorithms can be used as those from Tables 12.2 and 12.3.

Concrete implementation will depend on how the computer program represents the SN. Figure 12.2 graphically illustrates the object-value establishing procedure using a simple example where the task is to determine `fido`'s way of locomotion.

12.2.4 Computational Costs of Reasoning in Frames

A quick look at the last two algorithms convinces us that their computational costs grow more or less linearly in the number of the SN edges that need to be traversed. This is much less than the costs of reasoning in rule-based systems (whether by *modus ponens* or by resolution). This observation offers yet another reason why frames and SNs used to be so popular.

Control Questions

If you are unable to answer the following questions, return to the appropriate place in the preceding text.

- For frame-based knowledge bases, summarize the technique that establishes whether an object belongs to a specific class, and the technique that establishes the value of an object's concrete property.

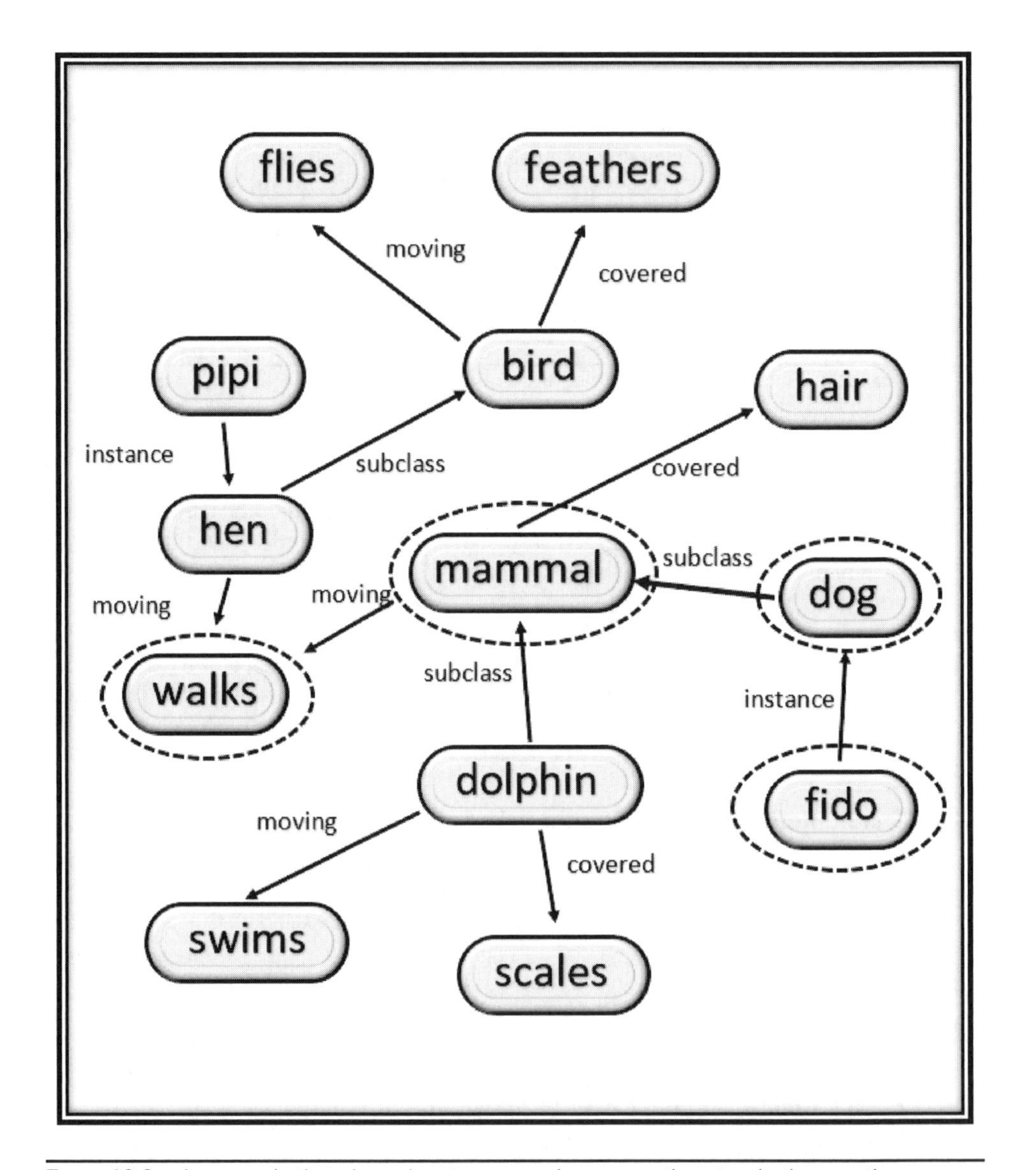

Figure 12.2 An example that shows how to answer, in a semantic network, the question concerning `fido`'s way of locomotion. Slot `moving` is missing in object `fido`, nor do we find it in the super-class, `dog`. However, the `dog`'s super-class, `mammal`, does have this slot, and its value, `walks`, is then inherited by `fido`.

- Discuss the possibilities of implementing these techniques within the paradigm of SNs.
- What are the computational costs of reasoning in frame-based knowledge bases and in SNs?

12.3 *N*-ary Relations in Frames and SNs

In the examples from the last section, frame-based representation was shown to be very intuitive, easy to deal with, and readily visualized by a SN. All this convenience, however, resulted from our tacit restriction to rely exclusively on simple properties such as the `covered_by` slot.

In many applications, this is not enough. Certain aspects of knowledge call for predicates with two or more arguments, such as the binary relations `parent(x,y)` and `bigger_than(x,y)`, or in the ternary relation `between(x,y,z)` (meaning that `y` is located between `x` and `z`). Here, frame-based representation is no longer as elegant as the previous examples seemed to promise.

12.3.1 Binary Relations and Frames

Table 12.4 shows one possibility of representing binary relations with frames and slots. The first frame represents the generic `bigger_than` relation. Note that none of this frame's two slots is filled with a concrete value.

The second frame defines the pair `dolphin` and `hen` as an instance of the relation just defined. The frame contains slots `first` and `second` whose values are meant to fill the empty slots of `bigger_than`—of which `dolphin_and_hen` is an instance. Figure 12.3 expresses the same idea visually by a SN. Importantly, the `dolphin_and_hen` frame may also contain other `instance_of` slots to represent the binary relations satisfied by `dolphin` and `hen`. For example, the frame can be an `instance_of` some `smarter_than` frame.

The reader can see that binary relations *can* be represented in this paradigm if the corresponding frames are skillfully designed. This said, the mechanism illustrated by the frame `bigger_and_then` is less-than-practical in domains with thousands of instances of a great many binary relations.

12.3.2 Frame-Based Reasoning with Binary Relations

One thing is to know how to formulate knowledge in a given paradigm; another is to know to employ this knowledge representation in query answering. To this end,

---

bigger_than
  first: ???
  second: ???

dolphin_and_hen
  instance_of: bigger-than
  first: dolphin
  second: hen

Table 12.4 An Example Showing How to Represent a Binary Relation in a Frame-Based Knowledge Base. The Idea is to Introduce a Generic Relation `bigger_than` of Which `dolphin_and_hen` Is an Instance.

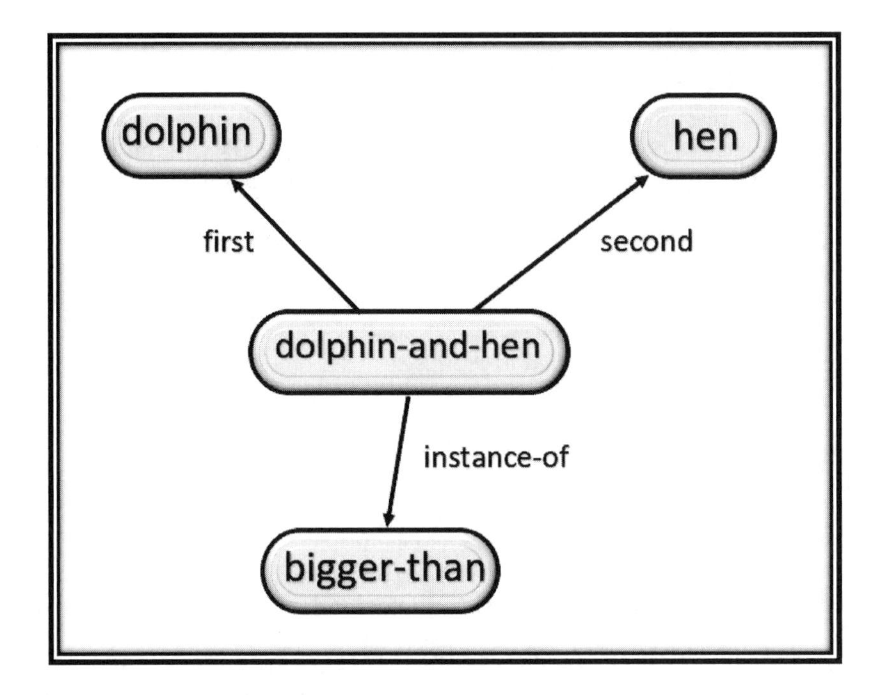

Figure 12.3 A semantic network that corresponds to the frame from Table 12.4.

proponents of frames have developed algorithms capable of reasoning over frames and SNs such as those from Tables 12.4 and 12.3. These, however, are no longer as elegant and computationally efficient as those that can be used in domains where all knowledge takes the form of unary predicates.

In the context of this textbook, perhaps the more straightforward way of implementing here automated reasoning is by a program that (somehow "behind the scene") converts the frames into rules.

12.3.3 Translating Binary Relations to Rules

The frame from Table 12.4 is easily converted to classical predicate logic. Here is one possibility of how to specify the corresponding facts.

$T \rightarrow$ **value(dolphin_and_hen, first, dolphin)**.
$T \rightarrow$ **value(dolphin_and_hen, second, hen)**.
$T \rightarrow$ **instance_of(dolphin_and_hen, bigger_than)**.

To reason with these facts, we need specific rules (see below).

12.3.4 Rules to Facilitate Reasoning with Binary Relations

If the three facts from the previous paragraph are to be used in rule-based reasoning (e.g., to be processed by resolution), we need a mechanism to interpret them in the framework of predicate logic. Here is one possibility:

value(p, first, x), value(p, second, y), instance(p, r) → holds(r, x, y).
instance(a, x), instance(b, y), holds(r, x, y) → holds(r, a, b).

The first rule converts to predicate logic the `instance_of` slot in the frame-based knowledge base. The second rule then specifies under what circumstances we can say that what holds for `x` and `y` will hold also for their respective instances, denoted here by `a` and `b` (the inheritance mechanism).

12.3.5 Difficulties Posed by *N*-ary Relations

Even greater difficulties are encountered in the case of ternary relations, those that in first-order logic are represented by predicates with three arguments such as `between(x,y,z)`. A skilled programmer is sure to find a way of encoding them in frames, but the mechanism will inevitably be even more awkward than the one from Table 12.4. The complications involved in attempts to employ these representations in automatic reasoning then outweigh the primary advantage of frames: their simplicity and intuitive appeal.

Attempts to move on to n-ary relations, for $n > 3$, are even more problematic. This is one of the reasons why this paradigm is now so rarely used, in spite of its erstwhile popularity. On the positive side, its proponents are probably right when they claim that n-ary predicates can often (though not always) be avoided, or at least re-expressed by unary or binary predicates.

Control Questions

If you are unable to answer the following questions, return to the appropriate place in the preceding text.

- Describe the simple way of representing binary relations in frames and SNs. Provide some examples, different from those in this section.
- Discuss the problems posed by n-ary relations. How do they affect the utility of these two knowledge-representation paradigms?

12.4 Practice Makes Perfect

To improve your understanding, take a chance with the following exercises, thought experiments, and computer assignments.

- Choose a domain of your own, and create for it a knowledge base in the form of frames. Make sure the domain is interesting enough to require at least 20–30 frames, each consisting of up to 10 slots. Draw a part of the SN representing this knowledge base.
- Suggest a concrete mechanism of implementing a SN in a computer program. Write a program that will draw inferences over SNs. For these, use techniques based on those from Tables 12.2 and 12.3.

- Discuss the possibilities of converting a frame-based knowledge base into rules. Write a pseudo-code of a technique that will carry out such conversion. Do not forget that the input (the frames) has to be represented in a data structure that facilitates the conversion.
- Building on the general outline from Section 12.3, suggest a mechanism to implement a program capable of inferences from a knowledge base where some frames represent binary relations.
- Write a 2-page essay discussing the possibilities of representing frames and SNs in a general-purpose programming language such as C++ or Python.
- Do some research on the web (e.g., Wikipedia) and write a two to three page essay about the software packages (from the 1980s) that were meant to support implementation of knowledge in frames.

12.5 Concluding Remarks

In the 1980s, and even some time afterwards, many AI experts expected that knowledge representation and automated reasoning would soon be dominated by frames and SNs. These expectations were encouraged by the intuitive clarity of these paradigms, by their general appeal, and by the apparent simplicity and efficiency of the related inference mechanisms.

Advocates even developed specialized programming languages to support these mechanisms. Perhaps the most famous among these languages was KL-ONE introduced by Brachman and Schmolze (1989). Their work rested on the results of several papers from earlier years. As a matter of fact, KL-ONE cannot be associated with a single scientific paper or with a few concrete authors or developers. Quite a few scientists participated in its development in a nice illustration of the benefits of cooperation.

Unfortunately, practical experience soon revealed that unary predicates offer only limited opportunities for representing knowledge. Almost inevitably, binary or ternary relations are sometimes needed. While some proponents demonstrated that it *was* possible to cast n-ary relations in frames, and even subject them to reasoning, it soon became clear that disadvantages here tend to outweigh benefits.

The popularity of both mechanisms gradually faded, until it shrank to the point where this author had to ask himself whether their current impact merited a whole chapter in his textbook. What decided, in the end, was their conceptual simplicity and clarity which—who knows—may one day motivate their resurrection, even if in another guise.

CHAPTER 13

Hurdles on the Road to Automated Reasoning

Successful theory is one thing, its practical application is another—in artificial intelligence no less than anywhere else. The potential of logic in automated reasoning once inspired no small amount of excitement. Alas, efforts to convert this potential into convincing applications left a lot to be desired. Scholars began to suspect that theoretical expectations missed an important point.

Indeed they did. The world is not as simple as the toy domains from introductory classes may lead us to believe. Human reasoning is flexible and prone to resist the straitjacket of logic. Many rules we use in our daily lives make sense most of the time, but not always. Some are so subtle as to make logic seem toothless. And yet others are so obvious that the programmer never even thinks of entering them in the knowledge base. Before we start implementing a program that aspires to be called intelligent, we need to understand the nature of the hurdles in our way.

This chapter takes a look at some of the most serious issues, thus motivating the field of uncertainty processing that will be the subject of later chapters.

13.1 Tacit Assumptions

To create a useful knowledge base is more difficult than it seems. To begin with, human reasoning relies on certain assumptions that are firmly ensconced in our subconsciousness. So obvious do they seem that it is all too easy to forget that the machine does not know about them. Let us clarify the notion by two examples.

13.1.1 The *Frame* Problem

Figure 13.1 shows two cubes, A and B, placed on a desk. Suppose we want to write a program that will instruct a robot to lift A from where it currently is, and drop it on cube B. This seems so elementary that we easily overlook that, even here, we rely on tacit expectations that we find obvious, but the machine needs to be told.

The first assumption is that B remains in its place while A is being moved. The second is that, once A has been moved, it will no longer find itself in its original location. You may find the two assumptions self-evident, but why should they be warranted? In the copy-and-paste operations of any graphic software, the copied object *does* stay where it

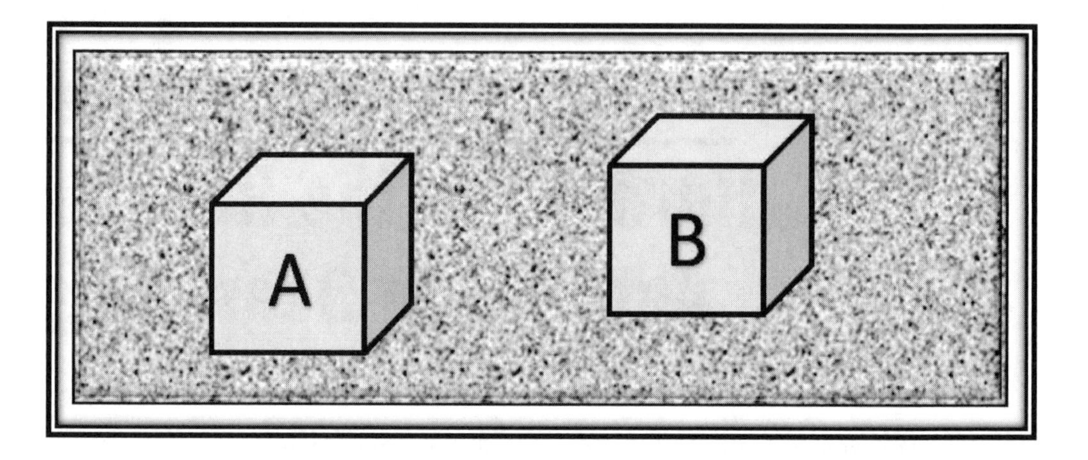

Figure 13.1 When writing a program for a robot to place cube A onto cube B, the engineer assumes that B will remain where it is now, and that A, having been moved, will no longer find itself in its original location.

was even after the completion of the operation. And as for the first assumption, why should B stay were it is? Anyone might have snatched it while A was being lifted.

Both assumptions satisfy the mechanical engineer's early intuition, but they may prove misleading in many realistic applications. AI literature refers to them as the *frame problem*.

13.1.2 Tacit Assumptions

The frame problem illustrated one major difficulty faced when developing a knowledge base. We take too much for granted, easily forgetting that the machine is the philosopher's ultimate *blank slate*, incapable of making assumptions that any child finds obvious.

As an illustration of our tendency to take things for granted, consider something truly commonplace: rules of etiquette. Even medieval society appreciated good manners, and some manuscripts dealing with the topic are older than the printing press. One old manuscript informed the readers they should not spit on the floor during an official dinner in the presence of an archbishop.

Do you see the point? I challenge you to find me a twenty-first century text on etiquette that will warn you against this very transgression. The reason they do not even mention it is that every schoolboy knows he should not spit on the floor while eating. There is no need to make this explicit in a textbook—or in a knowledge base.

Control Questions

To make sure you understand this topic, try to answer the following questions. If you have problems, return to the corresponding place in the preceding text.

- Explain the two basic aspects of what artificial intelligence calls the *frame problem*. Under what circumstances are these aspects less than obvious?

- Discuss the problem of other *tacit assumptions* that the author of a knowledge-based system may overlook. Suggest examples different from those mentioned in this section.

13.2 Non-Monotonicity

While appreciating the power of logic, we need to be aware of its limitations. One such limitation is the classical logic's inability to deal with non-monotonicity.

13.2.1 Monotonicity of Reasoning

Aristotelian logic rests on the idea of consistency. Suppose that our knowledge base contains rules covering the various aspects of family relations. If we add to this knowledge base a new fact or a new rule, we expect that the knowledge base thus modified will lead to the same conclusions about family relations as its previous version (plus perhaps some new conclusions).

This requirement is sometimes referred to as the *monotonicity* of classical reasoning: Extending an existing knowledge base should not change the results of inferences; it should only enhance them by new skills.

13.2.2 Do Hens Fly?

Outside classical logic, monotonicity is less common than a theoretician may want us to believe. Our everyday discussions make sense even when new facts appear to contradict our previous assumptions.

For instance, we know that `hens` are unable to fly. However, if we are told that some concrete hen is sitting onboard an airplane, we admit that this one *will* fly, and we do not perceive this circumstance as a refutation of our previous beliefs. From the theoretician's point of view, this newly added fact about the hen on a plane renders our reasoning non-monotonic because what we know now leads to different predictions. And yet, we do not think the new fact invalidates all our previous knowledge.

13.2.3 Do They *Not* Fly?

In view of the previous paragraph, we may decide to add to our knowledge base a rule that says that any object flies when onboard an airplane. Does this eliminate all difficulties? Far from it. While the airplane is waiting on a tarmac, it does *not* fly.

Suppose we add to the knowledge base this new admission: Non-flying objects do not fly even when in an airplane, provided that this airplane does not fly itself. Even with this improvement, we may still fail to draw the correct conclusion. For instance, the bored passengers may be tossing the hen around—in which case the bird *does* fly even when the aircraft does not.

13.2.4 Normal Circumstances

Early attempts to deal with this situation suggested that the programmer should explicitly state that normal circumstances are being assumed. For instance, we know that most birds fly. Admitting the possibility of exceptions, the engineer may specify this piece of knowledge as follows:

$$\texttt{bird(x)} \wedge \texttt{normal(x)} \rightarrow \texttt{fly(x)} \tag{13.1}$$

In other words, "normal birds fly" which is the same tacit assumption that any person would make. Of course, for this to make sense, the knowledge base has to contain the information about which birds are "normal"; for instance, the fact that states that $T \rightarrow$ `normal(eagle)`. In the case of `hen`, analogous fact is missing, and the closed-world assumption implies that `hen` is not normal, and the antecedent of Rule 13.1 is not satisfied. Hens do not fly.

13.2.5 Abnormal Circumstances

The same rule can be reformulated in the following way. This is usually more convenient:

$$\texttt{bird(x)} \wedge \neg\texttt{ab(x)} \rightarrow \texttt{fly(x)} \tag{13.2}$$

This specifies that birds fly unless abnormal. The interpretation is essentially the same as in the first rule, the difference being in what facts the engineer plans to enter in the knowledge base. In this second formulation, we expect the knowledge base to contain $T \rightarrow$ `ab(hen)`, which means that $\neg$ `ab(hen)` is *false*, and rule's consequent, `fly(hen)` is thus *false*, too. Hens do not fly. If the "abnormality" fact is not specified, then the closed-world assumption results in `ab(x)` being regarded as *false*, which means that $\neg$ `ab(x)` is deemed *true*.

13.2.6 Which Version to Prefer?

A useful guideline follows from the closed-world assumption. Suppose that, in a given application domain, most birds are normal. Then we prefer to specify only the rare case of `ab(hen)`, treating abnormality as the default value. This makes Rule 13.2 preferable to Rule 13.1.

The first rule would be preferred only in domains where a small percentage of objects were `normal`. It is in the nature of things, however, that abnormality is rare. For this reason, the first formulation is less practical than the second.

13.2.7 Theories, Assumptions, and Extensions

A more sophisticated way of handling exceptions divides the knowledge base into *theory*, its *assumptions*, and its *extension*.

Let us clarify the three concepts by an example. At the top of Table 13.1, we see a very simple theory consisting of four rules. Below the theory, two assumptions are offered. An analysis would reveal, however, that only the second assumption is consistent with the theory; the first assumption can therefore be ignored. Assumptions consistent with the theory constitute the theory's *extension*.

The principle just mentioned requires that the system be able to check the knowledge base for consistency, thus assisting the engineer in the specifications of the `ab` predicates.

The following four rules constitute a *theory*:

```
mammal(x) ∧ ¬ab(x) → walk(x).
          whale(x) → mammal(x) ∧ ¬walk(x).
                 T → mammal(Fred).
                 T → whale(Moby).
```

In principle, the theory allows the following *assumptions*:

```
T → ¬ab(Fred).
T → ¬ab(Moby).
```

An *extension* is a set of all assumptions that are consistent with the given theory. In our case, the *extension* can contain only one of the two assumptions:

```
T → ¬ ab(Moby).
```

The other assumption is inconsistent with the theory.

TABLE 13.1 Example of a *Theory*, a List of *Assumptions*, and an *Extension*

13.2.8 Multiple Extensions

When this approach was first proposed, the idea of analyzing assumptions and extensions seemed attractive. Nevertheless, early optimism did not survive simple examples such as the one from Table 13.1. Scientists realized that the same theory very often permitted two or more extensions, each of them consistent with the given theory, but contradicting the other extensions to the point of mutual exclusion. Since it was impossible to work with alternative extensions simultaneously, it became obvious that some selection would have to be made. But how to choose, what criteria to apply? All hopes that the problems of exceptions and non-monotonicity would thus go away soon faded.

13.2.9 Multi-Valued Logic

Other attempts to deal with the real world's intricacies soon surfaced. Among the most promising ones was the idea of multi-valued logic. For instance, one possibility was to allow, apart from *true* and *false*, a third value, *unknown*. A few multi-valued logic systems appeared, and their properties were systematically explored. In retrospect, the main difficulty of all these approaches was that theoreticians found them attractive—and practitioners did not.

13.2.10 Frames and Semantic Networks

Another idea regarding exceptions was discussed in Chapter 12: frame-based knowledge bases and semantic networks. The reader will recall how *inheritance* made it easy to establish properties that could be granted exceptions in sub-classes or instances. For unary predicates, the approach proved elegant and easy to implement. Unfortunately, this elegance and easiness became much less obvious in attempts to proceed to *n*-ary relations.

Control Questions

To make sure you understand this topic, try to answer the following questions. If you have problems, return to the corresponding place in the preceding text.

- Explain the notion of *monotonicity* in reasoning. Why is non-monotonicity in automated reasoning difficult to address within first-order logic?
- How is the same problem addressed by the predicates `normal(x)` and `ab(x)` discussed in this section? Outline the principle of the solution based on *extensions*. What is the main limitation of this technique?
- What other ways of handling non-monotonicity do you know?

13.3 *Mycin*'s Uncertainty Factors

The reader is beginning to understand that human knowledge is plagued with exceptions, uncertainties, inconsistencies, and even contradictions, and that critical information can easily be missing. How to deal with all of these inconveniences when creating a knowledge base? Early efforts to do so within the framework of logic (see Section 13.2) succeeded only to a limited degree.

Seeking an alternative, many experts turned their attention to numerical approaches. These turned out to be much more fruitful, and today they prevail.

13.3.1 Uncertainty Processing

Human reasoning does not experience major difficulties when facing non-monotonicity: When new information arrives, the brain finds a way to reconcile it with earlier knowledge. Classical logic finds this problematic. Although quite a few logic-based attempts have been reported, none of them (perhaps with the exception of frame-based knowledge) has resulted in realistic applications convincing enough to persuade the scientific and engineering community that this is the way to go.

Gradually, more and more experts started to suspect that the solution would have to be sought outside classical logic. Their answer came to be known as *uncertainty processing*.

13.3.2 *Mycin*'s Certainty Factors

Perhaps the earliest attempt to deal with uncertain knowledge was employed in a computer program that came down to history as *Mycin*. The solution it offered was disarmingly simple—and yet efficient! Each fact or rule (let us denote it by p) was assigned a *certainty factor*, $CP(p)$, whose task was to quantify the engineer's confidence that p was *true*.

The certainty factor is a number from the interval $[-1, 1]$, where $CF(p) = 1$ means "certainly yes," $CF(p) = -1$ means "certainly no," and $CF(p) = 0$ is interpreted as a total ignorance as to p's truth value. The other values quantify the shades of uncertainty. For instance, from $CF(p_1) = 0.8$ and $CF(p_2) = 0.6$, we infer that p_1 is more certain than p_2.

13.3.3 Truth of a Set of Facts and Rules

Suppose that $p = p_1, \ldots, p_n$ is a set of facts and rules, and suppose that for each of them the certainty factor, $CF(p_i)$, is known. The certainty that p is *true* (i.e., *all* of the p_i's are *true*) is obtained by the following formula that identifies the certainty of the whole set with the certainty of its weakest link:

$$CF(p) = \min\{CF(p_1), \ldots, CF(p_n)\} \tag{13.3}$$

Conversely, the certainty that *at least one* of the p_i's is *true* is identified with the certainty of the set's "strongest link":

$$CF(p) = \max\{CF(p_1), \ldots, CF(p_n)\} \tag{13.4}$$

13.3.4 Certainty of a Negation

Let us denote by $\neg p$ the negation of p. If $CF(p)$ is the certainty that p is *true*, then the certainty that p's negation is *true* is the opposite of $CF(p)$:

$$CF(\neg p) = -CF(p) \tag{13.5}$$

13.3.5 Numeric Example

Suppose that, for p_1, p_2, and p_3, the following certainty factors have been provided: $CF(p_1) = 0.4, CF(p_2) = -0.1$, and $CF(p_3) = 0.9$. From these, the following certainty factors are obtained:

$$CF(p_1 \wedge p_2 \wedge p_3) = \min\{0.4, -0.1, 0.9\} = -0.1$$

$$CF(p_1 \vee p_2) = \max\{0.4, -0.1\} = 0.4$$

$$CF(p_1 \wedge \neg p_3) = \min\{0.4, -0.9\} = -0.9$$

13.3.6 Certainty Factors and *Modus Ponens*

Suppose that the certainty factor for p is $CF(p) = a$, and that the certainty of rule $p \rightarrow q$ is $CF(p \rightarrow q) = b$.

If $a \leq 0$, then $CF(q) = 0$ (total ignorance), which reflects our intuition that if the rule's antecedent is deemed "rather not true," the rule does not apply.

If $a > 0$, then the certainty factor for the statement that q is *true* is given by the product of the two certainty factors:

$$CF(q) = a \cdot b \tag{13.6}$$

13.3.7 Numeric Example

Let us denote by p the statement that a student has studied hard, and let us denote by q the statement that the student will succeed at the exam. Suppose that the rule "if the student has studied hard, then he will succeed at the exam" has been assigned certainty $CF(p \rightarrow q) = 0.8$. If our certainty that John has studied hard is $CF(p) = 0.7$, then our certainty that he will succeed at the exam is obtained as follows:

$$
\begin{aligned}
CF(p) &= 0.7 \\
CF(p \rightarrow q) &= 0.8 \\
CF(q) &= 0.7 \times 0.8 = 0.56.
\end{aligned}
$$

13.3.8 Combining Evidence

In automated reasoning, a concrete conclusion can sometimes be reached by different lines of argument. For instance, John can succeed at the exam not only thanks to his hard work, but also because of being smart, and perhaps due to some other reasons. When seeking to quantify John's chances of success, therefore, the reasoning software needs a mechanism to combine the amount of evidence that has come from each of these sources.

Here is how *Mycin* handles the problem. Suppose there are two lines of argument, a_1 and a_2, and suppose that the value of the certainty factor obtained when following a_1 is x, and the value of the certainty factor obtained when following a_2 is y. *Mycin*'s certainty of the conclusion that has thus been confirmed by the two lines is established as follows:

$$
CF(a_1, a_2) = \begin{cases} x + y - xy & \texttt{if } x, y > 0 \\ x + y + xy & \texttt{if } x, y < 0 \\ \frac{x+y}{1-\min\{|x|,|y|\}} & \texttt{otherwise} \end{cases} \tag{13.7}
$$

13.3.9 Intuitive Explanation

The formulas leave a somewhat *ad hoc* impression, but they are supported by considerations that appear to make sense. For instance, the very first formula, $x + y - xy$, is motivated by an observation known from the theory of probability. Figure 14.1 (see Chapter 14) shows two intersecting sets, X and Y. Suppose we want to establish the size of their union. If we simply add the size of X to the size of Y, the area of their intersection, $X \cap Y$, will be counted twice. The correct size will thus be established only if we subtract one occurrence of this "intersection" area from the sum of the areas of X and Y. The same line of reasoning applies to probabilities—and also to certainty factors.

If both x and y are negative, then $x + y$ is negative, too, and the doubly counted intersection thus has to be added (it has appeared twice as a negative number).

The least obvious (and most *ad hoc*) is the third formula which is employed in the case where the two lines of argument conflict with each other: one has a positive certainty factor, and the other negative. Here the most convincing argument supporting this formula seems to be the rather unscientific: "It has worked so well so many times."

13.3.10 Numeric Example

Suppose that a conclusion is supported by two lines of argument. For one of them, the certainty was evaluated as $CF(a_1) = 0.5$; for the other, the certainty was $CF(a_2) = 0.8$. We see that both certainties have positive values, and so the first formula is used:

$$CF(a_1, a_2) = 0.5 + 0.8 - 0.5 \cdot 0.8 = 0.9$$

Note that the certainty of the conclusion reached by the combination of the two lines of argument is higher than the certainties of the two arguments considered independently of each other.

13.3.11 Numeric Example

Let us now consider the situation where the certainties related to the two lines of reasoning have opposite signs: $CF(a_1) = 0.5$ and $CF(a_2) = -0.6$. In this event, the third formula is used:

$$CF(a_1, a_2) = \frac{0.5 + (-0.6)}{1 - \min\{\,|0.5|, |-0.6|\}} = \frac{0.1}{0.5} = 0.2$$

In this case, we can see that the certainty of the conclusion falls somewhere between the values of the certainties of the two arguments considered independent of each other.

13.3.12 More Than Two Alternatives

The formulas that combine the certainties obtained from alternative sources consider only two possible lines, denoted here by a_1 and a_2. The simplest way of handling a situation with more than two such lines, $a_1, a_2, \ldots, a_n$, is "one step at a time." First the certainty factor is calculated for the pair, a_1, a_2. The result, $CF(a_1.a_2)$, is then combined with $CF(a_3)$, the new result with $CF(a_4)$, and so on.

13.3.13 Theoretical Foundations?

As already mentioned, the rules for how to manipulate *Mycin*'s certainty factors are rather intuitive and they lack solid mathematical foundations. True, they have been shown to succeed in a great many applications. Still, mathematically minded scholars have always suspected that something more solid should be developed. The alternatives they suggested will be discussed in the next chapters.

Control Questions

To make sure you understand this topic, try to answer the following questions. If you have problems, return to the corresponding place in the preceding text.

- What is *certainty factor*, what values can it acquire, and how are these values interpreted?
- How to calculate the certainty factors of conjunctions and disjunctions? What is the certainty factor of a negation?
- How are certainty factors treated in the course of *modus ponens* reasoning?
- How are certainty factors combined in a situation where the same conclusion can be obtained by two or more lines of reasoning?

13.4 Practice Makes Perfect

To improve your understanding, take a chance with the following exercises, thought experiments, and computer assignments.

- Suggest some examples of the human tendency to rely on various *tacit assumptions* of the kind mentioned in Section 13.1. Your examples should be different from those in this chapter. Discuss the reasons why these tacit assumptions are so often neglected, and why they are sometimes difficult to enter in the knowledge base.
- Suggest examples of the human ability to reason *non-monotonically*, as explained in Section 13.2. Again, your examples should be different from those in this chapter.
- Provide an example of a domain where it is virtually impossible to put together reliable knowledge base because of the inevitable uncertainty and incompleteness of our expert knowledge.
- Consider the following knowledge base:

$$T \rightarrow a, \text{ CF} = 0.9$$

$$T \rightarrow b, \text{ CF} = 0.8$$

$$a \rightarrow p, \text{ CF} = 0.4$$

$$b \rightarrow p, \text{ CF} = 0.3$$

  What is the certainty factor of p?
- Suppose that a certain conclusion can be reached by way of three lines of arguments, a_1, a_2, and a_3. Suppose that these lines of arguments have been evaluated as having certainties $CF(a_1), CD(a_2)$, and $CF(a_3)$. How will you combine all these values into one final certainty factor?

13.5 Concluding Remarks

The earliest discussion of the frame problem seems to be Heyes (1973). By contrast, the discovery of the consequences of the non-monotonic aspects of reasoning cannot be pinned down to a single individual. Quite a few scientists recognized these pitfalls at about the same time. Perhaps the best source of information about the early history of these thoughts is the collection of papers put together by Ginsberg (1987).

Appreciation of these difficulties came as a severe blow to the once-so-optimistic predictions of the future role of first-order logic in AI. The days of complacency were over; the resolution principle no longer looked like a panacea. By the 1990s, the scientific community had accepted that things were far from simple, and that a lot of work remained to be done if automated reasoning was ever to become a reality.

The challenge motivated many research groups to look for viable solutions. The attack was conducted along two major fronts. One relied on advanced logic. This included the assumptions/extensions mentioned in Section 13.2, and many attempts to use multi-valued logic as well as some other paradigms. Most of these suggestions, however, turned out to be too sophisticated for practical implementations, and their impact on further developments in AI remained limited.

The other front focused on numeric approaches to *uncertainty processing*. The earliest success story was the development of Edward Feigenbaum's certainty factors from the late 1960s, employed in the pioneering program *Mycin*. Studies of alternative numeric paradigms soon followed suit: probability, fuzzy-set theory, and some others. In the end, these numeric approaches proved much more fruitful than those based on logic.

The following chapters focus on those uncertainty-processing paradigms that this author believes to be most influential.

CHAPTER 14
Probabilistic Reasoning

The success of *Mycin*'s certainty factors gave rise to some serious questions. Sure enough, the power of the approach had been demonstrated by convincing experiments. On the other hand, its *ad hoc* nature was easy to criticise for its lack of theoretical foundations. Suppose a more solid approach was developed, perhaps one using formulas derived from the centuries-old theory of probability. Would the results then not be even better than those of *Mycin*? Studies encouraged by these arguments spawned a new generation of reasoning systems that culminated in the technology of Bayesian *belief networks*.

After a brief revision of some basic concepts of probability, this chapter outlines the essence of Bayesian reasoning, and then explains the principles of Bayesian networks, illustrating the possibilities of their practical use on simple numeric examples.

14.1 Theory of Probability (Revision)

To prevent unnecessary confusion, let us begin by a brief revision of this paradigm's basic terminology.

14.1.1 Sources of Probabilistic Information

The theory of probability occupies itself with ways of quantifying the odds of something happening. Sometimes the odds are obtained as an expert's estimate such as "I believe there is a 70% chance of rain, tomorrow." In other cases, the odds can be derived from known physical circumstances, such as when we say there is a 50% chance of a flipped coin coming up heads. Most typically, however, the notion is tied to relative frequency. If we repeat an experiment 100 times and observe X in 40 trials, we conclude that there is a 40% chance of X occurring. Even this, of course, is nothing but an estimate. Another set of 100 experiments may result in a different value.

14.1.2 Unit Interval

For mathematical convenience, probabilities are usually scaled into the $[0, 1]$-interval (also known as *unit interval*). This means that, say, 75% probability is specified as 0.75. This convention simplifies calculations.

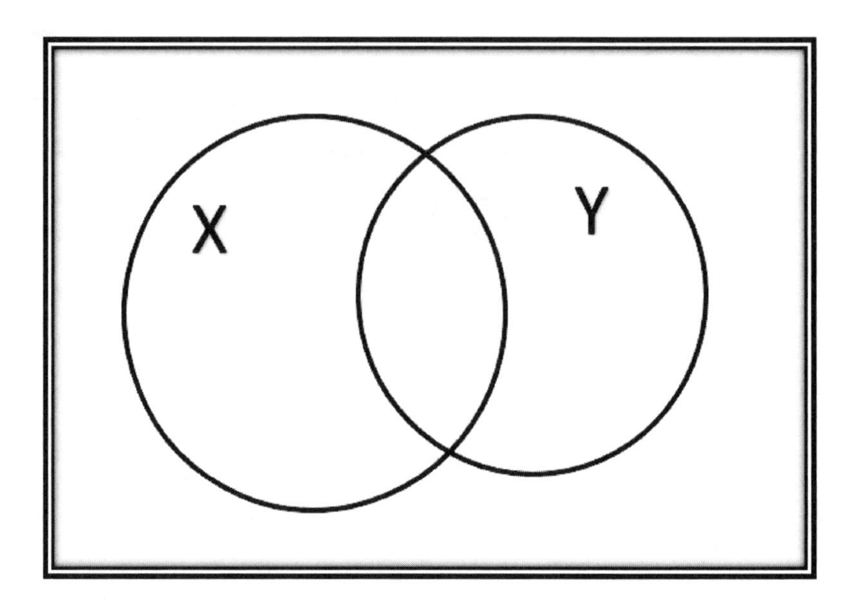

FIGURE 14.1 Illustration of *conditional* probability and *joint* probability.

14.1.3 Joint Probability

The rectangle in Figure 14.1 represents a *universe*. This universe contains two subsets, X and Y, each in the form of a circle. To establish the probability that a randomly selected point in the universe belongs to X, we divide the area of the X-circle by the area of the rectangle. The probability that a randomly selected point belongs to Y is calculated similarly.

The intersection of the two circles, $X \cap Y$, represents the case where a randomly picked point belongs to both X and Y. The *joint probability* of X and Y is established by dividing the intersection's area by the size of the rectangle. We denote this probability by $P(X, Y)$ and read, "the probability that Y and X occur at the same time."

14.1.4 Numeric Example

For the sake of illustration, Table 14.1 shows examples of joint probabilities. At the top, the probabilities of all possible combinations of the two events (including their negations), are given. The table also reminds us how to use these joint probabilities when calculating probabilities of x and y. For instance, x occurs either in combination with y or in combination with ¬y (there is no other possibility). This means that $P(\mathrm{x}) = P(\mathrm{x}, \mathrm{y}) + P(\mathrm{x}, \neg\mathrm{y}) = 0.13 + 0.47 = 0.60$.

14.1.5 Conditional Probability

If we divide the size of the intersection by the size of the X-circle, we obtain the *conditional probability* that a point belonging to X also belongs to Y. We denote this probability

The following table shows the probabilities of the joint occurrences of two events, x and y, and their negations, occurring at the same time.

	y	¬y
x	0.13	0.47
¬x	0.22	0.18

For instance, the probability of x and y occurring at the same time is $P(x, y) = 0.13$, and the probability of ¬y and x occurring at the same time is $P(x, \neg y) = 0.47$.

Recall how to calculate the probabilities of the events x and y:

$P(x) = P(x, y) + P(x, \neg y) = 0.13 + 0.47 = 0.60$

$P(y) = P(x, y) + P(\neg x, y) = 0.13 + 0.22 = 0.37$

Finally, note that the entries in the table sum to 1: $0.13 + 0.47 + 0.22 + 0.18 = 1.00$

Table 14.1 Examples of Basic Probabilities

by $P(Y|X)$ and read, "the probability of Y, given X." Conditional probability and joint probability are related according to the following formula:

$$P(X, Y) = P(X|Y) \cdot P(Y) = P(Y|X) \cdot P(X) \tag{14.1}$$

Joint probability has the same value regardless of the order of X and Y in the argument. Both formulations in Equation 14.1 are thus equivalent.

14.1.6 More General Formula

In the case of three sets, X, Y, and Z, Equation 14.1 is generalized as follows:

$$P(X, Y, Z) = P(X|Y, Z) \cdot P(Y|Z) \cdot P(Z) \tag{14.2}$$

The reader will find it easy to formulate an even more general equation that addresses the situation with four or more sets.

14.1.7 Rare Events: *m*-estimate

Suppose that we have conducted an experiment 10 times, $N_{all} = 10$. Suppose that, in these 10 trials, event x was observed three times; formally, $N_x = 3$. Such small numbers clearly do not support strong conclusions about probabilities.

However, we may have a reason to suspect that, in the absence of experimental verification, the probability of x is 20%, which we denote by $\pi_x = 0.2$. Let us call this our *prior expectation*. Using parameter m (its role will be explained later), we introduce a so-called m-estimate, a formula that combines our prior expectation with experimental evidence:

$$P(x) = \frac{N_x + m\pi_x}{N_{all} + m} \tag{14.3}$$

In the case of $N_x = N_{all} = 0$, note that the m-estimate is the same as the prior expectation, $P(x) = \pi_x$. Conversely, with a great many trials, the values of N_x and N_{all}

will dominate, making it possible to neglect m and $m\pi_x$, in which case the m-estimate approaches the relative frequency of x, which of course is N_x/N_{all}.

14.1.8 Quantifying Confidence by *m*

The role for parameter m is to control the formula's sensitivity to experimental evidence. High values of m indicate a situation where the prior expectation, π_x, is trusted, in which case a lot of evidence (i.e., high values of N_x and N_{all}) would be needed to override it. Conversely, small values of m indicate a situation where the prior expectation is deemed somewhat dubious, in which case it should be easy to override with just a few experimental observations.

14.1.9 Numeric Example

When flipping a coin, it is reasonable to expect that the coin will come up *tails* 50% of the time, $\pi_{tails} = 0.5$. Suppose we decide to put this prior expectation to test. We flip a coin four times and observe that it came up *tails* only once. The classical approach based on relative frequency will give $P(tails) = 1/4 = 0.25$. On the other hand, m-estimate with $m = 2$ will result in the following:

$$P(tails) = \frac{1 + 2 \cdot 0.5}{4 + 2} = 0.33$$

The reader will agree that these 33% are closer to reality than the 25% suggested by relative frequency.

High confidence in the prior expectation, $\pi_{tails} = 0.5$, is indicated by a higher value of m, say, $m = 50$. In this event, the calculation of the m-estimate will result in the following:

$$P(tails) = \frac{1 + 50 \cdot 0.5}{4 + 50} \doteq 0.48$$

We observe that this high-confidence case (suggested by the a high value of m) results in a probability estimate that is much closer to prior expectation.

Control Questions

To make sure you understand this topic, try to answer the following questions. If you have problems, return to the corresponding place in the preceding text.

- Summarize the classical approaches to probability estimates.
- Explain the difference between joint probability and conditional probability. Write down the equation that shows how the two are related.
- Explain the principle of m-estimate. How can the user control the formula's confidence in the reliability of the prior expectation, π?

14.2 Probability and Reasoning

Having reviewed the fundamental concepts, we are ready to proceed to the mechanisms that employ them in automated reasoning.

14.2.1 Examples from the Family-Relations Domain

The simple rules from Chapter 9 are always valid: If `bill` is `eve`'s parent, then `eve` is `bill`'s offspring, and no exceptions are permitted. In reality, though, many rules only capture certain general tendencies, rather than "eternal truths." And yet they make perfect sense! This is the case of the following two:

if `old(x)` *then* `has_children(x)`
if `young(x)` *then* `single(x)`

The first is *true* most of the time; the second, very often. Imperfections of this kind are typical of many rules in realistic knowledge bases. Common sense lacks the solidity of classical mathematics, but people find such rules plausible and have no difficulties drawing from them reliable conclusions. We want to accomplish the same in automated reasoning.

14.2.2 Rules and Conditional Probabilities

Suppose that among all individuals that are known to be old, 90% have children. This is expressed by the conditional probability $P(\texttt{has\_children}|\texttt{old}) = 0.9$.

Likewise, we may have established that the probability of a young person being single is $P(\texttt{single}|\texttt{young}) = 0.3$. Such number may have been obtained from a relative frequency or by m-estimate (see Section 14.1).

14.2.3 Dependent and Independent Events

The two *if-then* rules from the previous paragraphs reflect dependencies between related events. Thus the first rule says that a person's having children *depends* on his or her `age`.

Dependencies of this kind are easily expressed graphically. Figure 14.2 illustrates a domain where five events, `A` through `E`, are interrelated by the rules (dependencies) represented here by arrows. For instance, `D` depends on `A` and `B`, but does not depend on `C` because there is no arrow from `C` to `D`. Likewise `D` can be seen to be independent of `E` (again, no arrow).

The fact that some Y is *independent* of X means that the likelihood of Y has nothing to do with whether X has been observed. Put another way, the presence of X does not affect Y's probability. In the world of probabilities, this independence is captured by the following formula:

$$P(Y|X) = P(Y) \tag{14.4}$$

The arrows in the graph from Figure 14.2 can be assigned numbers that quantify conditional probabilities. The absence of an arrow leading from X to Y is telling us that Y is independent from X. Returning to Figure 14.2, we can state, for instance, that $P(D|E) = P(D)$.[1]

[1] This convention will be used in Section 14.3.

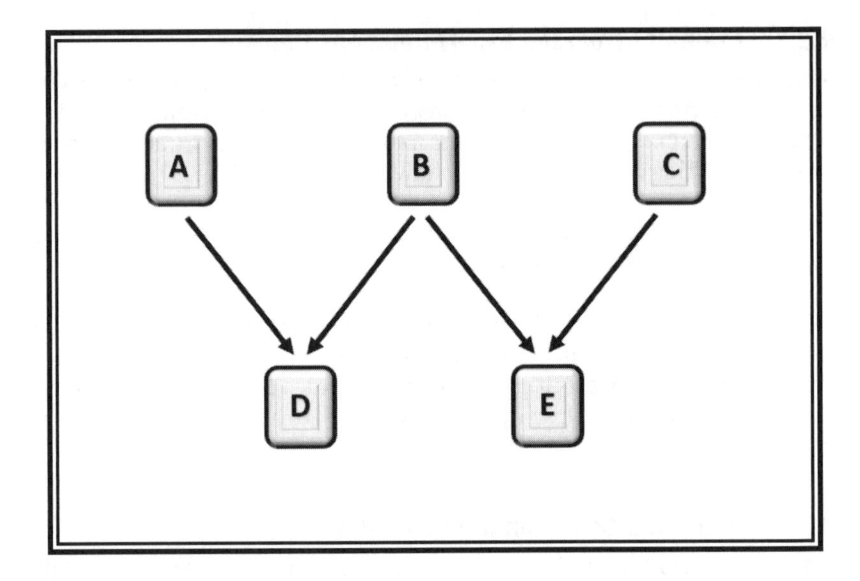

Figure 14.2 Illustration of the dependencies among five events. For instance, D depends on A and B, but it does not depend on C or E.

14.2.4 Bayes Formula

The most valuable tool for handling the probabilities needed in automated reasoning is the famous *Bayes formula*. This is easily derived from Equation 14.1 where we realized that

$$P(Y|X) \cdot P(X) = P(X|Y) \cdot P(Y)$$

Bayes formula is obtained by dividing both sides by $P(X)$:

$$P(Y|X) = \frac{P(X|Y) \cdot P(Y)}{P(X)} \tag{14.5}$$

14.2.5 Bayes Formula and Probabilistic Reasoning

Let us denote by H a hypothesis, say, that a patient suffers from a certain medical issue, and let E denote available evidence, perhaps the patient's symptoms. Finally, let us denote by $P(H|E)$ the probability that hypothesis H is *true* in the presence of evidence E. With these denotations, the Bayes formula acquires the following form:

$$P(H|E) = \frac{P(E|H) \cdot P(H)}{P(E)} \tag{14.6}$$

In plain English, the probability that H is the correct hypothesis provided that E has been observed is proportional to $P(E|H)$, which denotes the probability of E among the cases where H was correct. This number is then multiplied by the probability of H and divided by the probability of E. The probabilities can be obtained by relative frequency or by m-estimate.

14.2.6 Choosing the Most Likely Hypothesis

Let us assume the existence of a set of hypotheses, $H_1, \ldots, H_n$. Observing evidence E, we want to decide which of these hypotheses is the one most likely to be correct. To this end, we apply Equation 14.6 separately to each H_i:

$$P(H_i|E) = \frac{P(E|H_i) \cdot P(H_i)}{P(E)} \tag{14.7}$$

Once we know all these values, we simply choose the hypothesis with the highest probability $P(H_i|E)$.

Practically speaking, however, we do not need to calculate the probabilities. Seeing that the denominator, $P(E)$, is the same for all hypotheses, we can just as well ignore it and simply choose the hypothesis with the highest value of the formula's numerator, $P(E|H_i) \cdot P(H_i)$.

Control Questions

To make sure you understand this topic, try to answer the following questions. If you have problems, return to the corresponding place in the preceding text.

- How are *if-then* rules related to the concept of conditional probability?
- What are *dependent* and *independent* events?
- Derive the Bayesian formula and discuss the fundamental possibilities it offers to automated reasoning.

14.3 Belief Networks

Consider a knowledge base consisting of a set of facts and rules, each quantified probabilistically along the lines discussed in Section 14.2. In the process of reasoning, these probabilities need to be propagated throughout the reasoning process so that the certainty of the system's response can be quantified. A popular mechanism for doing so in the context of probabilities is known as the *belief network*.

14.3.1 Belief Network

Let us take another look at Figure 14.2. We see a Bayesian *belief network* that involves five Boolean variables, some of them mutually dependent (e.g., D depends on A and B) others mutually independent (e.g., D does not depend on C). In reality, the belief network may represent facts A, B, and C, plus a set of rules such as $A \wedge B \rightarrow D$. Any combination of the values (*true* or *false*) of the five variables defines one concrete situation.

In the context of this chapter, the rules are probabilistic. Knowing the probabilities of the facts and rules, and using the Bayes formula, we are able to calculate the probability of any situation.

The links in the graph are associated with the probabilities of the related rules. For instance, D depends on A and B, which defines rule $A \wedge B \rightarrow D$. Each of the two variables in the antecedent, A and B, can be either *true* or *false*. For the information to be complete, we thus need four probabilities, each for a different combination of those values.

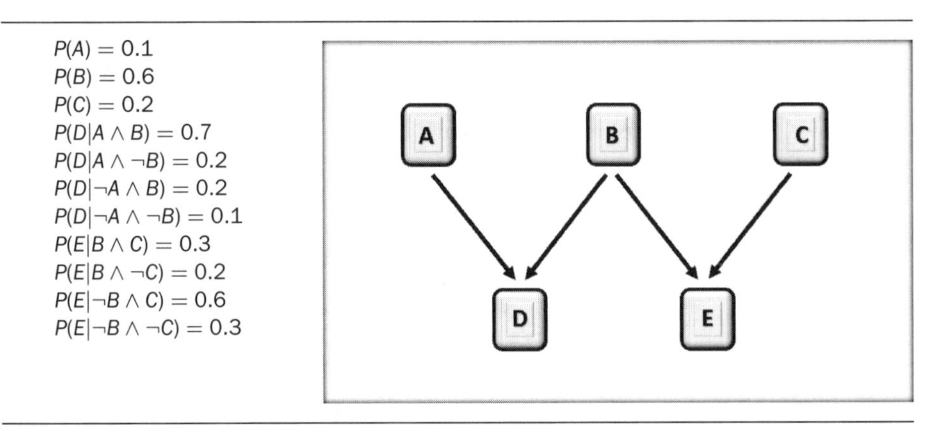

$P(A) = 0.1$
$P(B) = 0.6$
$P(C) = 0.2$
$P(D|A \wedge B) = 0.7$
$P(D|A \wedge \neg B) = 0.2$
$P(D|\neg A \wedge B) = 0.2$
$P(D|\neg A \wedge \neg B) = 0.1$
$P(E|B \wedge C) = 0.3$
$P(E|B \wedge \neg C) = 0.2$
$P(E|\neg B \wedge C) = 0.6$
$P(E|\neg B \wedge \neg C) = 0.3$

TABLE 14.2 A Complete Set of Probabilities for the Belief Network from the Previous Picture (An Example)

14.3.2 Numeric Example

Table 14.2 provides example values of the probabilities of the facts and rules represented by the given belief network. Note that the table contains only the probabilities of the "positive" facts and conclusions, such as $P(A)$ or $P(D|A \wedge B)$. The "negative" ones, such as $P(\neg A)$ or $P(\neg D|A \wedge B)$, are obtained as follows:

$$P(\neg A) = 1 - P(A) = 1 - 0.1 = 0.9$$

$$P(\neg D|A \wedge B) = 1 - P(D|A \wedge B) = 1 - 0.7 = 0.3$$

Probabilities $P(D|A \wedge B), P(D|A \wedge \neg B), P(D|\neg A \wedge B)$, and $P(D|\neg A \wedge \neg B)$ do *not* have to sum to 1, even though they all have the same consequent.[2] A beginner may find it unexpected.

14.3.3 Probability of a Concrete Situation

We define a *situation* as any concrete combination of the values of the belief network's Boolean variables. Let us use the numbers from Table 14.2 to calculate the probability of a situation where all the variables are *true*.

In view of what Equation 14.2 had to say about joint probability of multiple variables, we realize that the corresponding formula will look as follows (to "shorten" the formulas, we will represent conjunction, '$\wedge$.' by a comma, ','):

$$P(A, B, C, D, E) = P(E|A, B, C, D) \cdot P(D|A, B, C) \cdot P(C|A, B) \cdot P(B|A) \cdot P(A)$$

Some of the terms in this formula can be simplified. For instance, since E is independent of A and D, we can write $P(E|A, B, C, D) = P(E|B, C)$. Simplifying the other terms

[2] For illustration, suppose that A stands for "hard working," B stands for "brilliant," and D stands for "succeeds." It is conceivable that, in this case, $P(D|A \wedge B) = 0.9$ and $P(D|\neg A \wedge B) = 0.5$. The sum of these two numbers, $0.9 + 0.5 = 1.4$ already exceeds 1.

accordingly, we obtain the following:

$$P(A,B,C,D,E) = P(E|B,C) \cdot P(D|A,B) \cdot P(C) \cdot P(B) \cdot P(A)$$

The values of the terms on the right-hand side are available in Table 14.2. Specifically, we obtain the following:

$$P(A,B,C,D,E) = 0.3 \cdot 0.7 \cdot 0.2 \cdot 0.6 \cdot 0.1$$

14.3.4 Probability of a Conclusion

Suppose we know that A is *false*, and that B, C, and E are all *true*. Using the same belief network, we want to know whether D is *true* or *false*. The easiest way to find out is by comparing the probabilities of D and $\neg D$.

The belief network from Table 14.2 indicates that D depends only on $A \wedge B$. This leads us to the following formula:

$$P(D|\neg A,B,C,E) = P(D|A,B) = 0.7$$

To obtain the opposite case, namely that D is *false*, it is enough to calculate it as the complement of the probability that D is *true*:

$$P\neg(D|\neg A,B,C,E) = P(\neg D|A,B) = 1 - P(D|A,B) = 1 - 0.7 = 0.3$$

We see that, given the values of A, B, C, and E, the probability of D being *true* is higher than the probability of D being *false*.

14.3.5 Is *B true*?

Similarly, we can investigate the probability of any of the variables in the given belief network. The variable in question does not have to be found in any consequent; it can just as well be found in some of the antecedents. For instance, given that A, D, and E are *true*, and C is *false*, we may want to know whether B is more likely to be *true* than *false*.

The procedure is in principle the same. We simply compare $P(A,B,\neg C,D,E)$ with $P(A,\neg B,\neg C,D,E)$. If the former has higher value, then B is more likely to be *true* than *false*.

Control Questions

To make sure you understand this topic, try to answer the following questions. If you have problems, return to the corresponding place in the preceding text.

- Explain the principle of a *belief network*. How are belief networks related to the facts and rules in knowledge bases? How are they assigned probabilities?
- Explain the mechanism that calculates the probability of a given situation in a belief network.

14.4 Dealing with More Realistic Domains

Simple toy domains are good for the explanations of basic principles. On the other hand, they tend to ignore certain difficulties encountered in realistic applications. Let us take a look at some of them.

14.4.1 Larger Belief Networks

Figure 14.3 shows a belief network that is only slightly larger than the previous one. While still fairly simple, it already gives us a more plastic view of the nature of the calculations of all those conditional probabilities. For instance, suppose that we know that C and D are *true* whereas A is *false*. The probability of G being *true* is then obtained by the following formula that is easily derived from the graph of the network:

$$P(G|\ C,D,\neg A) = P(G|\ C,D) \cdot P(C|\ \neg A) \cdot P(D|\ \neg A)$$

Employing an appropriate data structure, the programmer will easily automate the construction of the formula for any reasonable probability, not only this one.

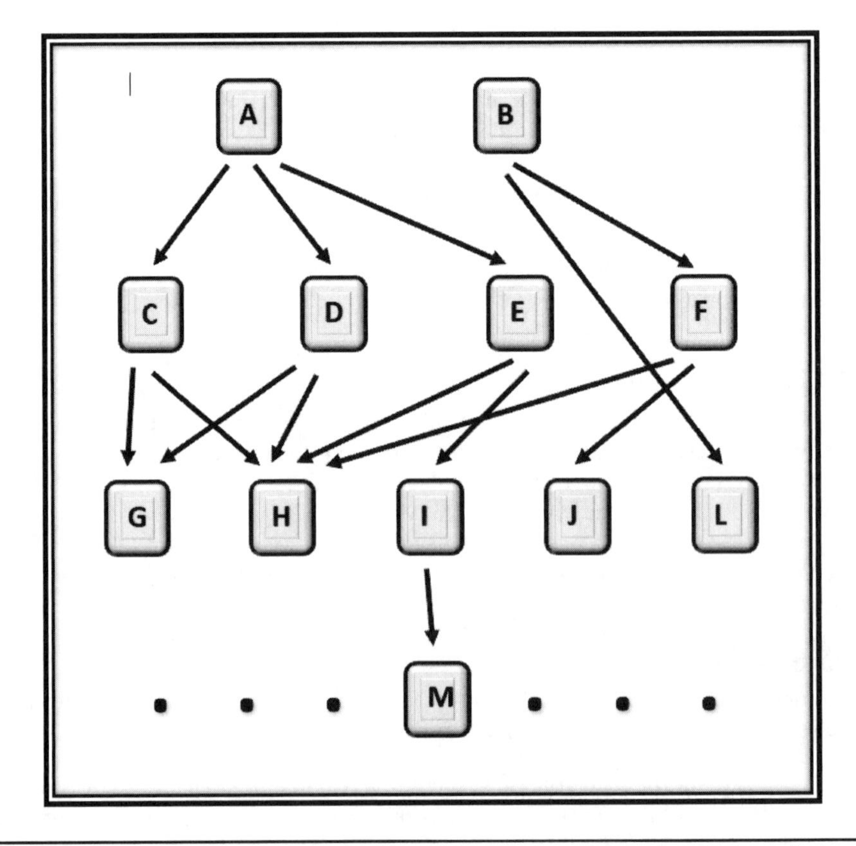

FIGURE 14.3 A somewhat larger network with a higher number of mutually interdependent Boolean variables.

Very often, the values (*true* or *false*) of some variables involved in the formula are unknown. In this event, the program may initiate a dialog during which it may ask the user to provide them. If the user does not know the values, one possibility is to consider all alternatives. For instance, if in the previous formula the value of A is not known, and the user is unable to provide it, the program may calculate separately $P(G)$ for the case with A being *true* and for the case with A being *false*. The difference may turn out to be only marginal.

14.4.2 Invisible Causes and *Leak* Nodes

Another problem is that the belief network is likely to be incomplete; it may ignore some important relations. In realistic domains, it is impossible to know all the dependencies among so many variables. Returning to Figure 14.3, it is conceivable that G depends not only on C and D, but also on other variables that are not included here. The engineer perhaps suspects the existence of other causes of G, but is unable to identify them explicitly. This is sometimes referred to as the problem of *invisible causes*.

One practical solution to the problem of invisible causes is the introduction of so-called *leak nodes*. The engineer simply combines all unknown causes into one single "artificially created" cause, denoted, say, by X. This X is the leak node. Probability $P(G|\ C, D)$ is then replaced with $P(G|\ C, D, X)$.

14.4.3 Too Many Probabilities Are Needed

In Figure 14.3, we can see that C depends on the single variable A. To enable the probabilistic reasoning outlined in the previous section, the user has to provide the conditional probabilities for both values of A: $P(C|A)$ and $P(C|\neg A)$.

The same graph shows that G depends on two variables, C and D. Since any of them can, again, have two values, the user has to provide the probabilities of four combinations: $P(G|C \wedge D), P(G|C \wedge \neg D), P(G|\neg C \wedge D), P(G|\neg C \wedge \neg D)$. Finally, we can see that H depends on four variables, C, D, E, and F. In this case, the user has to provide 16 probabilities, one for each combination of the *true/false* values of the four variables.

We can see that even in the still-relatively-small belief network from Figure 14.3, quite a few values need to be provided. Realistic belief networks, however, are much larger than this textbook graph. To provide the reasoning system with all the requisite information, a great many numbers would have to be entered. This may not always be possible. At the very least, it may not be practical because so many values simply cannot be known.

14.4.4 Naïve Bayes

The last paragraph has reminded us that, if some X depends on N other variables, we need to know 2^N conditional probabilities—one for each combination of the *true/false* values. Not only is this impractical, it may be downright impossible. A commonly used solution relies on the assumption that the individual variables are mutually independent.

Suppose that our system involves N events, $X_1, \ldots, X_N$, and that for each of them we know its probability: $P(X_i)$. If all of these events are pairwise independent, then the probability of all of them occurring at the same time is calculated as follows:

$$P(X_i \wedge \ldots \wedge X_N) = P(X_1) \cdot P(X_2) \cdot \ldots \cdot P(X_N)$$

In our context, we are dealing with Boolean variables, each of them acquiring one out of two possible values. For each variable, X_i, we thus need two probabilities, $P(X_i)$ and $P(\neg X_i)$. This means that, for N variables, we need $2N$ probabilities, which is much less than the 2^N probabilities required in the general case.

14.4.5 Is the Naïve Bayes Assumption Harmful?

Many variables *are* mutually related. For instance, `weight` is related to `size`, and `salary` is related to `age` and `education`. Generally speaking, therefore, the assumption of mutual independence can at best be an approximation of reality. Accepting that the assumption can rarely be justified without reservation, we call it naïve—hence the term, *naïve Bayes assumption*.

This said, advantages may outweigh shortcomings. After all, is there an alternative to resort to? If we do not assume mutual independence of variables, we have to include, in our calculations, a great many conditional and joint probabilities whose values can at best only be estimated, sometimes very inaccurately. Consequently, we will only carry out precise calculations with imprecise (sometimes even dubious) numbers. In that event, the results will be even less trustworthy than those obtained by *naïve Bayes*. The validity of this argument has been confirmed by many realistic experiments.

Besides, *naïve Bayes* is computationally much cheaper.

14.4.6 Probability of Negation (Reminder)

Suppose we have been provided with the following conditional probabilities: $P(X|A) = 0.8$ and $P(X|B) = 0.6$. Probabilistic reasoning often needs to know the conditional probabilities of the negations of the consequents, $\neg X$. In our case, these are calculated as follows:

$$P(\neg X|A) = 1 - P(X|A) = 1 - 0.8 = 0.2$$

$$P(\neg X|B) = 1 - P(X|B) = 1 - 0.6 = 0.4$$

14.4.7 What Is the Probability of $P(X|A_1 \vee A_2 \vee \ldots A_n)$?

Here, the truth of X may result from the truth of one out of the n possible causes, A_1 through A_n. Assuming that the causes are mutually independent, the theory of probability obtains the probability of X as follows:

$$P(X|A_1 \vee A_2 \vee \ldots \vee A_n) = 1 - P(\neg X|A_1) \cdot P(\neg X|A_2) \cdot \ldots \cdot P(\neg X|A_n) \tag{14.8}$$

The thing to remember is that the equation's right-hand side subtracts from 1 the product of the probabilities of the negations of all causes.

14.4.8 Probability of a Concrete Event

The main ambition of this section is to show how to calculate the probability of a certain event, X, that may potentially result from multiple causes. Calculating it, we take advantage of Equation 14.8.

The right-hand side of the equation from the last paragraph subtracted from 1 the product of all the probabilities of $\neg X$ due to the individual causes. Not all of these probabilities are used, however. We need only those where the corresponding cause, A_i, is *true*.

14.4.9 Numeric Example

Again, suppose we have been provided with the following conditional probabilities: $P(X|A) = 0.8$ and $P(X|B) = 0.6$. Besides the two known causes, A and B, all other (unknown) causes have been summarized in a leak node, L, with the conditional probability $P(X|L) = 0.7$. Here are the conditional probabilities of X's negation:

$$P(\neg X|A) = 1 - 0.8 = 0.2$$

$$P(\neg X|B) = 1 - 0.6 = 0.4$$

$$P(\neg X|L) = 1 - 0.7 = 0.3$$

In the case where A and L are *true* and B is *false*, we obtain the following:

$$P(X) = 1 - P(\neg X|A) \cdot P(\neg X|L) = 1 - 0.2 \cdot 0.3 = 0.96$$

Note that the right-hand side ignored the term $P(\neg X|B)$. This is in line with what was said: We ignore those terms whose antecedent (B, in our case) is *false*.

14.4.10 Where Do the Probabilities Come From?

So far, we have ignored the question of the source of all those probabilities. We tacitly assumed that their values could be estimated by the relative frequencies observed in real-world situations, or perhaps by m-estimates. This is indeed sometimes the case.

Beyond the scope of this textbook, however, let us mention in passing that the probabilities are often obtained by machine-learning techniques, most typically those that rely on the technology of *neural networks* and *deep learning*.

Control Questions

To make sure you understand this topic, try to answer the following questions. If you have problems, return to the corresponding place in the preceding text.

- Discuss the challenges faced when dealing with realistically sized domains.
- Explain the terms *invisible causes* and *leak nodes*. For each, discuss also the motivation behind their introduction.
- How is the *naïve Bayes assumption* used here? Summarize its strengths and weaknesses.

14.5 Demspter-Shafer Approach: Masses Instead of Probabilities

Theory of probability, a well-established field of mathematics, has been employed to great advantage in many AI applications. This said, certain aspects of automated reasoning are only inadequately modeled by classical probabilities. This is why an alternative, the Dempster-Shafer theory, is sometimes advocated. Let us introduce here at least some of its basic ideas.

14.5.1 Motivation

Suppose you are shown a landscape photograph and asked in which season it has been taken. The answer not being immediately obvious, you may respond with something like, "this looks like spring." When pressed for how much sure you are, you may clarify, "I am 70% sure this is a spring scenery."

What about the remaining 30%? A mechanical approach will divide them equally among the three remaining seasons, `summer`, `fall`, and `winter`, allocating 10% to each. But is this what you had in mind? Not really. You did *not* believe each of them equally likely. You gave 70% to `spring`, essentially relegating the remaining 30% to "I do not know." You cannot become more confident unless you receive additional information such as the scene's geographical location. For instance, what looks like `spring` in the south may actually be `summer` in the north.

Allocating 10% to each of the three remaining seasons is misleading because this is not what you said. The uniform division of the 30% represents here an artificial imputation of information that has not been there.

14.5.2 Mass Instead of Probability

The Dempster-Shafer theory, DST, was invented with this very situation in mind. In this approach, the probabilities are allocated to *sets* rather than to single items. Thus in our snapshot example, the 70% are allocated to the single-element set {spring}, and the 30% are allocated to the entire set of the four seasons: {spring, summer, fall, winter}. To prevent confusion, these "probabilities" (70% and 30%) are called *masses*.

These masses have to sum to 1—just as the probabilities in the classical probabilistic framework.

14.5.3 Frame of Discernment

Let us assume the existence of a set of n mutually exclusive and exhaustive propositions:

$$\Theta = \{\theta_1, \dots \theta_n\} \tag{14.9}$$

In the snapshot example, the frame of discernment consists of four seasons. These seasons are *mutually exclusive* in the sense that any day of the year belongs to one and only one season. They are also *exhaustive* because the year has only these four seasons, and none other. The masses mentioned in the previous paragraphs are allocated to subsets of the given frame of discernment.

14.5.4 Singletons and Composites

Any member, θ_i, of the frame of discernment is called a *singleton*. Any subset consisting of more than one singleton (e.g., $\{\theta_2, \theta_4\}$) is called a *composite*.

In the world of logical statements and rules, propositions apply to singletons and composites. For instance, one such composite proposition can state that "the snapshot was taken either during summer or fall." The number of all propositions is determined by the number of the frame of discernment's subsets: if it consists of $|\Theta|$ singletons, then the total number of all possible propositions is $2^{|\Theta|} - 1$ (we ignore the empty set, here).

Control Questions

To make sure you understand this topic, try to answer the following questions. If you have problems, return to the corresponding place in the preceding text.

- Why is the theory of probability sometimes inadequate when we want to quantify our uncertainties?
- What is the main difference between DTS's masses and classical probabilities?
- Explain the terms, *frame of discernment*, *singleton*, and *composite*.

14.6 From Masses to Belief and Plausibility

Knowing the meaning of masses, we are ready to proceed to the mechanism that DST uses to quantify uncertainty and ignorance: the notions of *belief* and *plausibility*.

14.6.1 Basic Belief Assignment

Given a concrete frame of discernment, Θ, DST assumes that each of the frame's subsets, $A \subseteq \Theta$, has been allocated a mass, $m(A)$. For some subsets (or perhaps for most of them), the mass can be zero, $m(A) = 0$, but no mass can ever be negative (just like probabilities cannot be negative).

For the set of the masses allocated within the given frame of discernment, we will use the term, *basic belief assignment*, BBA.

14.6.2 Elementary Properties of Any BBA

The mass of an empty subset is zero, $m(\emptyset) = 0$. Further on, the masses within the BBA must sum to 1, similarly as in the theory of probability:

$$\Sigma_{A \subseteq \Theta} m(A) = 1 \tag{14.10}$$

Let us denote by $\overline{A}$ the set-complement of A (so that $\overline{A}$ contains all singletons from Θ that are outside A). The sum of the masses of $\overline{A}$ and A must not exceed 1:

$$m(A) + m(\overline{A}) \leq 1 \tag{14.11}$$

For instance, $m(\{\texttt{spring}, \texttt{summer}\}) + m(\{\texttt{fall}, \texttt{winter}\}) \leq 1$. The reason why the sum is allowed to be less than 1 is that some masses may have to be assigned to other subsets of Θ.

14.6.3 Belief in a Proposition

Suppose we have a certain frame of discernment, such as the four seasons, and suppose we have a BBA with masses allocated to the frame's subsets. Let us investigate composite propositions such as the one defined by the set $A = \{\texttt{spring},\ \texttt{summer}\}$. How much confidence (how much *belief*) do we have in the statement that our snapshot has been taken during either of the two seasons?

The answer will depend not only on the mass assigned to {`spring`, `summer`}, but also on the masses assigned to {`spring`} and {`summer`}. Expressed formally, the belief in proposition A is defined as follows:

$$Bel(A) = \Sigma_{B \subseteq A}\, m(A)$$

14.6.4 Plausibility of a Proposition

Apart from belief, DST operates also with *plausibility*. Again, let $\overline{A}$ be the set-complement of A. Suppose we have already calculated the belief in $\overline{A}$. Plausibility of A is then defined as follows:

$$Pl(A) = 1 - Bel(\overline{A})$$

14.6.5 Uncertainty Is Quantified by the Two Values

Unlike the theory of probability, DST expresses uncertainty by a pair of values. Specifically, for each set A, the uncertainty is captured by the interval $[Bel(A), Pl(A)]$.

14.6.6 Numeric Example

Consider the following frame of discernment:

$$\Theta = \{a, b, c\}$$

Suppose the BBA has allotted the following masses (let us ignore the interpretation of these values):

$$m\{a\} = 0.6$$

$$m\{a, b\} = 0.3$$

$$m(\Theta) = 0.1$$

Suppose the task is to establish our uncertainty in $A = \{a, b\}$. We already know that the belief is the sum of the masses of all subsets of A. In our case, A has two subsets with non-zero masses, namely $\{a\}$ and $\{a, b\}$. We also see that $\overline{A} = \{c\}$. The beliefs in A and $\overline{A}$ are thus calculated as follows:

$$Bel(A) = m\{a\} + m\{a, b\} = 0.6 + 0.3 = 0.9$$

$$Bel(\overline{A}) = Bel(\{c\}) = m\{c\} = 0$$

From here, the plausibility of A is $Pl(A) = 1 - Bel(\overline{A}) = 1$.

Within the given frame of discernment and the given BBA, the two values, belief and plausibility, define an interval $[0.9, 1]$ that characterizes our confidence in A.

Control Questions

To make sure you understand this topic, try to answer the following questions. If you have problems, return to the corresponding place in the preceding text.

- Explain the meaning of BBA. What do you know about its basic properties?
- Write down the formulas that DST uses for the calculations of *belief* and *plausibility*.

14.7 DST Rule of Evidence Combination

Perhaps the most interesting feature of the Dempster-Shafer theory is the way it combines evidence that has been delivered by different sources of information.

14.7.1 Multiple Sources of Mass Assignments

Suppose there are two frames of discernment, Θ_1 and Θ_2, each with its own BBA (masses assigned to subsets). We would like to know how much different these two sources of knowledge are, whether they are compatible, of whether they contradict each other.

14.7.2 Level of Conflict

Let us denote by B_i the subsets of the first frame of discernment, $B_i \subseteq \Theta_1$, and by C_j the subsets of the second frame of discernment, $C_j \subseteq \Theta_2$. *Conflict* between the two frames of discernment (and the accompanying BBAs) is established by the following formula:

$$K_{1,2} = \Sigma_{B_i \cap C_j = \emptyset}\ m(B_i)m(C_j) \tag{14.12}$$

Note that the summation is taken over all such pairs of sets, $[B_i, C_j]$ (one from Θ_1 and the other from Θ_2) that have empty intersections.

$K_{1,2} < 1$ indicates that, overall, the masses of such non-intersecting pairs of sets are low, and we can thus assume that the two sources do not contradict each other, and as such can be combined (see below).

If the conflict is higher, $K_{1,2} \geq 1$, the two sources are conflicting to a degree that disqualifies attempts to combine them.

14.7.3 Rule of Combination

Consider two frames of discernment, Θ_1 and Θ_2 and their BBAs. Suppose their conflict has been found to be $K_{1,2} < 1$, which allows us to combine the evidence. We want to establish the masses of individual sets from the masses coming from the two sources.

The mass set A is calculated by the following formula:

$$m(A) = (m_1 \oplus m_2)(A) = \frac{\Sigma_{B_i \cap C_j = A}\ m_1(B_i)\ m_2(C_j)}{1 - K_{1,2}} \tag{14.13}$$

Here, the summation is over such pairs of sets (one from Θ_1, the other from Θ_2) whose intersection is A. The term "$(m_1 \oplus m_2)(A)$" is read, "mass of A as obtained by combining the evidence from the two sources."

Note that the denominator's value depends on the level of conflict between the two sources. The greater the conflict, $K_{1,2}$, the smaller the value of the denominator, $1 - K_{1,2}$, and the higher the mass.

Importantly, if $K_{1,2} > 1$, the conflict is so strong that the two sources cannot be combined. The reader is sure to have noticed that, in this event, the denominator would be negative, which means that the assigned masses would be negative, too, and this is not permitted (masses must always be positive).

14.7.4 Numeric Example

The upper part of Table 14.3 introduces two sources of knowledge. For simplicity, both operate with the same frame of discernment but they differ in the way they assign masses to the individual subsets. Seeing that the conflict between the two BBAs is small, $K_{1,2} = 0.18 < 1$, we realize it is possible to combine them.

Suppose that two sources of knowledge rely on the same frame of discernment: $\Theta = \{a, b, c\}$. The two sources differ in the way they allocate the masses:

$$m_1\{a\} = 0.6 \qquad m_2\{a\} = 0.7$$
$$m_1\{a,b\} = 0.3 \qquad m_2\{b,c\} = 0.3$$
$$m_1(\Theta) = 0.1$$

There is only one pair of non-intersecting mass-allocated sets such that one is from the first source and the other is from the second: $\{a\}$ from the first and $\{b, c\}$ from the second. The level of conflict of the two sources is thus established as follows:

$$K_{1,2} = m_1\{a\}m_2\{b,c\} = 0.6 \cdot 0.3 = 0.18$$

Seeing that $K_{1,2} < 1$, we conclude that the two sources *can* be combined.

Suppose we want to combine the two sources to establish the mass of the set $\{a\}$. To this end, we have to find all such pairs whose intersection is $B_i \cap C_j = \{a\}$. There are three such cases—see the numerator of this formula:

$$m\{a\} = \frac{m_1\{a\}\, m_2\{a\} + m_1\{a,b\}m_2\{a\} + m_1(\Theta)m_2\{a\}}{1 - K_{1,2}}$$

Using the concrete numbers from the two BBAs, we obtain for $\{a\}$ the following mass:

$$m\{a\} = \frac{0.42 + 0.21 + 0.07}{1 - 0.18} = 0.853$$

For other sets, such as $\{b\}$, the calculation is analogous.

TABLE 14.3 Numeric Example: Conflict Between Two Frames of Discernment, and the Application of DST's Rule of Evidence Combination

The bottom part of Table 14.3 shows how to obtain the "combined" mass of set $\{a\}$. Note that this mass is higher than either $m_1\{a\}$ or $m_2\{a\}$. This is because $m_1\{a\}$ and $m_2\{a\}$ tend to reinforce each other. Also $m_1\{a,b\}$ and $m_1(\Theta)$ add to the overall mass.

14.7.5 More Than Two Sources

Very often, we need to combine not two, but three or more BBAs. In that event, common practice proceeds stepwise. For instance, suppose we want to combine four sources, with indices $1, 2, 3$, and 4. First we combine m_1 with m_2, obtaining $m_{1,2}$. In the next step, we combine $m_{1,2}$ with m_3, obtaining $m_{1,2,3}$. Finally, we combine $m_{1,2,3}$ with m_4, obtaining $m_{1,2,3,4}$. Admittedly, the approach can be computationally demanding.

14.7.6 What the BBAs Typically Look Like

Let us return to the example of the landscape snapshot where we said, "I am 70% sure this is a `spring` scenery." This means the following mass assignment:

$$m_1\{\texttt{spring}\} = 0.7$$
$$m_1(\Theta) = 0.3$$

Another expert can then say, "I am 90% sure this is not `winter`." This means the following mass assignment:

$$m_2\{\texttt{spring}, \texttt{summer}, \texttt{fall}\} = 0.9$$
$$m_2(\Theta) = 0.1.$$

And yet another person will suggest yet another mass assignment. This is what the BBAs typically look like.

Control Questions

To make sure you understand this topic, try to answer the following questions. If you have problems, return to the corresponding place in the preceding text.

- How does DST quantify the degree of conflict between two BBAs?
- What is the DST rule of combination? Under what condition can it be applied? Write down the formula and explain its individual terms.
- How would you combine the evidence that comes from three or more sources?

14.8 Practice Makes Perfect

To improve your understanding, take a chance with the following exercises, thought experiments, and computer assignments.

- Suppose that, in Figure 14.3, A is *false* whereas C and D are *true*. Write the formula calculating the conditional probability of $\neg G$.
- Using the same graph, write the formula to calculate the probability that C and B are *false*, whereas all other variables are *true*.

- Using the same graph, how will you answer the question "Is D more likely than $\neg D$?" if we know that all other variables are *true*?
- Suggest an application domain of your own and create a belief network describing the inherent dependencies among its variables.
- What is the motivation behind the use of *leak nodes*? Illustrate their use on an example different from the one in this chapter.
- Explain the principle of the naïve Bayes assumption. When is it used? What are its advantages and what are its shortcomings?
- Write a two-page essay comparing the principles of *Mycin* with those of belief networks. What, in your view, are the individual advantages and disadvantages of these two paradigms?
- Return to Table 14.3. Calculate the mass of $\{b\}$ obtained by combining the two sources. Calculate the belief and plausibility of $\{a, b\}$ based on the first source (the one on the left at the top of the table).
- Write a one-page essay discussing the advantages of DST as compared to classical probabilities. In the essay, suggest also situations where you believe classical probabilities are to be preferred.

14.9 Concluding Remarks

At the core of probabilistic reasoning with belief networks is an eponymous theorem discovered by Thomas Bayes, a prominent English mathematician and philosopher of the 18th century. The concept of belief networks (also known as Bayesian networks, or Bayesian belief networks) was developed by Judea Pearl who introduced his ideas in the now-legendary paper, Pearl (1986). Since then, the paradigm has established itself as one of the most influential approaches to uncertainty processing.

The criticism that a belief network requires the programmer to enter just too many conditional probabilities (most of them subjective or speculative) is not altogether fair. The number of certainty factors that had to be entered into *Mycin*-based systems was big, too, and no less subjective. Further on, the requisite probabilities can be obtained in an automated manner by the use of techniques from the field of *neural networks*. These, of course, belong rather to a textbook of machine learning.

The idea of masses assigned to sets of items instead of probabilities assigned to individual items was first formulated by Dempster (1967), but it was developed into a full-scale theory only by Shafer (1976). The paradigm is commonly referred to as the Dempster-Shafer theory to give credit to both scholars. In comparison to classical probability, DST is less widespread, but this can change in the future. It is good to know about its existence.

CHAPTER 15
Fuzzy Sets

Whereas classical logic prefers to operate with rigorously defined concepts, our daily discourse is different. Most of the time, we rely on words whose meaning is less than clear. Indeed, what exactly do we have in mind when we say that *Bill is tall*? And is each student either *clever* or *not clever*, with nothing in between? These concepts are poorly defined, and yet we rarely experience major difficulties dealing with them. And if humans can do it, why not computers? These, then, are the considerations that once stood at the cradle of the *fuzzy set theory*.

This chapter explains the motivation behind the fuzzy-set framework, then explains its basic principles, and finally shows how to exploit these principles in automated-reasoning programs. Attention is paid to the differences between fuzzy sets and the theory of probability. The basic concepts are illustrated by simple examples.

15.1 Fuzziness of Real-World Concepts

Let us begin by clarifying what we mean by fuzzy concepts and by pointing out why artificial intelligence finds them so attractive.

15.1.1 Crisp Concepts and Fuzzy Concepts

Mathematicians and scientists prefer to work with clearly defined concepts such as *integer, hydrogen, tiger,* or *Monday*. In the terminology of AI, these terms are *crisp*. For a crisp concept, it is always possible to decide whether or not a given object is its representative; whether a given atom is hydrogen, and whether tomorrow is Tuesday.

In our daily discourse, however, this is no so common. We seem predominantly to deal with concepts whose meaning is intuitively clear, yet far from unambiguous. Consider such terms as *sunny day, smart politician, talented student,* or *boring movie*. Such concepts are vague, *fuzzy*. Many a student is regarded as talented to *some extent*, smaller or greater, but rarely talented without qualification. The same can be observed in most of what we talk about—and yet we rarely experience major difficulties!

It seems reasonable to expect an intelligent computer program to reason with similarly fuzzy terms without causing confusion, either.

15.1.2 Paradox of Heap

Ancient philosophers were aware of the problem. In a reaction to the advent of Aristotelian logic (with its limitation to crisp concepts), the following counterexample was raised. If you put one thousand stones in the same place, you have created a heap. Suppose you remove a single stone. Is what remains still a heap? It sure is, but suppose that you then remove another stone, and yet another, and so on—until, at a certain moment, there are only two or three stones left. Most certainly, this is no longer a heap. Here is the conundrum: When exactly did the heap cease to be a heap?

The reader will easily find other examples: bald man, expensive house, rich woman, and so on. Some words, like *little, many,* or *hot* inevitably evoke the heap paradox. How many hairs do I have to remove from a rock star's head to make him arguably bald? How much will the seller have to reduce the price to make us agree the house is no longer expensive?

All of these examples illustrate the notion of fuzziness.

15.1.3 Visual Example

Figure 15.1 contains four squares. One of them is black, another white, and the remaining two are gray, which means that they are neither white nor black—and yet they possess some measure of both extremes. Importantly, each possesses the property to a different *degree*. The bottom-left square is surely more black than the bottom-right one.

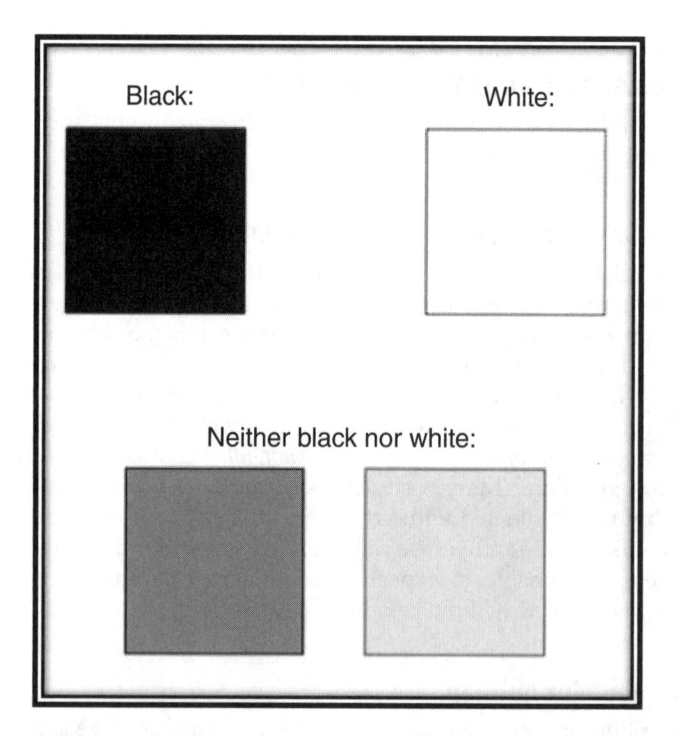

FIGURE 15.1 One square is black and one is white. The remaining two squares are black or white only to a smaller or greater degree.

15.1.4 Yet Another Example

Suppose an engineer has defined *warm room* as one whose (Fahrenheit) temperature falls into the interval $[70, 80]$. Everything below the lower bound is regarded as cold, and everything above the upper bound is hot.

This is what a classical programmer might do, but for the rest of us, definitions of this kind are too rigid, inflexible. We are not so dogmatic as to accept that 69.5 is cold whereas 70 is already warm. Our practical minds would prefer some transitory region along the range of temperatures. In that case we would be able to say, quite reasonably, "69.5 is on the borderline between cold and warm."

Control Questions

To make sure you understand this topic, try to answer the following questions. If you have problems, return to the corresponding place in the preceding text.

- Suggest a few examples of crisp concepts and a few examples of concepts that, by their very nature, are fuzzy.
- What is the *paradox of heap*, and how does it relate to the notion of fuzziness? Can you offer some other examples that are similar in nature?

15.2 Fuzzy Set Membership

Now that we have grasped the nature of fuzziness, we are ready take a look at how to address the phenomenon computationally.

15.2.1 Degree of Membership

In classical mathematics, each object either is or is not a member of a set. Fuzzy set theory is more flexible. In line with the observations from the previous section, each object belongs to the set *to a certain degree.* Practically speaking, object x is assigned a number, $\mu_A(x)$, that quantifies the degree of this object's membership in set A. By convention, the degree of membership is bound by the unit interval:

$$0 \leq \mu_A(x) \leq 1$$

Concrete values are interpreted in the following way:

$\mu_A(x) = 1$ means that x clearly belongs to A
$\mu_A(x) = 0$ means that x definitely does not belong to A
$0 < \mu_A(x) < 1$ means that x belongs to A to some degree
The higher the value of $\mu_A(x)$ the higher the degree of membership of x to A.

15.2.2 Black Squares

Let us illustrate the degrees of membership using the four squares from Figure 15.1. Suppose that A is defined as the set of black squares. For the square in the upper-left corner, $\mu_A(x) = 1$, and for the one in the upper-right corner, $\mu_A(x) = 0$. As for the two remaining

In the following table, each individual is assigned his or her degree of membership in the set of smart students.

x	bob	jim	eve	jane	fred	pete	...	jill	...
$\mu_{smart}(x)$	0.8	1.0	1.0	0.7	0.1	0.0	...	0.9	...

Table 15.1 The Simplest Way of Assigning the Degrees of Membership Is by Doing so Manually, by Entering the Value Separately for Each Single Object

ones (the bottom row), we may suppose that the one on the left has, say, $\mu_A(x) = 0.7$ and the one on the right, $\mu_A(x) = 0.5$ because it is "less black."

The example also illustrates one specific feature of fuzzy sets. To wit, the degrees of membership tend to be subjective. In practical applications, this is rarely a problem. What *is* important, however, is that the mutual relations of the values should reflect the underlying reality. In our case, the darker the square, the higher the degree of membership.

Of course, we could just as well have chosen to define the degrees of membership in the set B of *white* squares. In this event, the whiter the square, the higher the degree of membership, $\mu_B(x)$.

15.2.3 Talented Student

The simplest way of assigning the degrees of membership is by doing so manually as in the case shown in Table 15.1 where each individual is associated with one concrete value, perhaps one entered by a professor who knows the students personally.[1]

The table informs us that `jim` and `eve` are definitely smart, but `pete` most certainly is not. As for the others, their intelligence is graded: $\mu_{smart}(\texttt{fred}) = 0.1$ indicates that `fred`'s belonging to the set of smart students is questionable. As for `jill`, she is much smarter than `fred`, but only a little smarter than `jane`.

15.2.4 Tall Person

Another way of defining fuzziness is illustrated by the concept *tall person* whose degrees of membership are shown in the left part of Figure 15.2. Here, the engineer used a piecewise linear function that divides the domain of *height* (the x-axis) into three regions. The one on the left is marked by $\mu_{tall}(x) = 0$ (definitely *not* tall), the one on the right by $\mu_{tall}(x) = 1$ (definitely tall), whereas in the middle, the degrees of membership grow linearly with the growing height.

It is easy to define the membership function for the opposite concept: *short person*. The function will start with the $\mu_{small}(x) = 1$ region, which will then give way to the linearly descending part, and finally end up with the $\mu_{small}(x) = 0$ region.

[1] Again, the concrete degrees are in this case subjective. Another professor may have different views, and may therefore suggest other values.

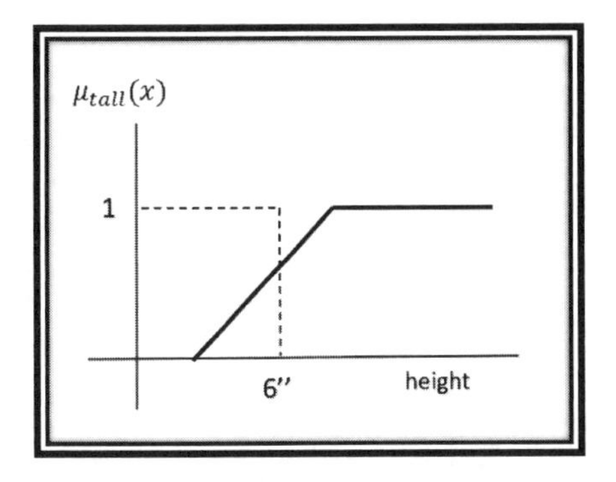

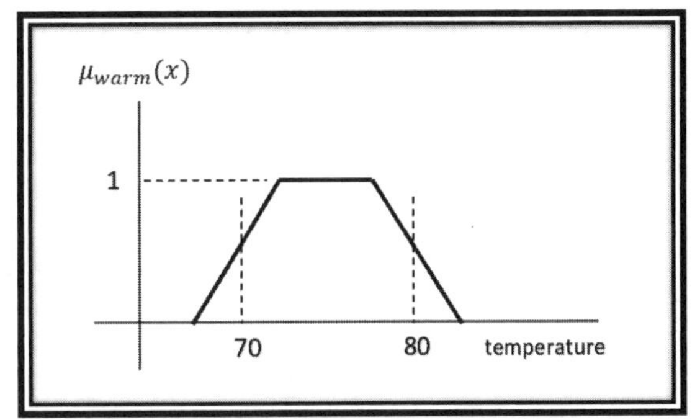

FIGURE 15.2 For each person (left) and for each room (right), the degree of membership into the set of tall people and warm rooms, respectively, is defined by a piecewise linear function.

15.2.5 Warm Room

The next concept is, again, slightly more complicated. The room is warm if it is not cold, but also if it is not hot. The situation is represented by the piecewise linear function in the right part of Figure 15.2.

Here we can see five different regions. Two of them assign zero membership to temperatures that are either too low or too high. In the middle is the region whose $\mu_{warm}(x) = 1$ indicates a clearly warm room. Finally, the function has two transitory regions where the degrees of membership increase or decrease linearly with the growing temperature.

15.2.6 Other Popular Shapes of the $\mu_A(x)$ Function

Membership functions do not have to be piecewise linear. Some popular non-linear alternatives are shown in Figure 15.3. Interpretation is similar to the piecewise linear function. Some experts find these continuous functions intuitively more appealing.

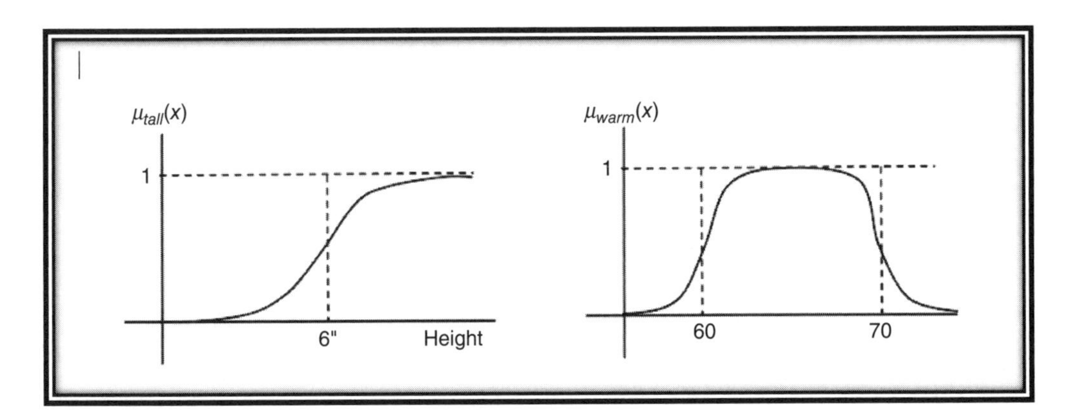

FIGURE 15.3 The functions to define degrees of membership do not have to be piece-wise linear. Among popular alternatives we find sigmoid and Gaussian functions and their various combinations and modifications.

15.2.7 Sources of the Values of $\mu_A(x)$

Where do the concrete values for the degrees of membership (or the functions to calculate them) come from? The simplest approach is to interview an expert who will suggest his or her subjective view, as in the case of the *smart student* from Table 15.1. To reduce subjectivity, one may interview a group of experts and take the averages of their individual opinions.

More sophisticated methods will construct $\mu_A(x)$ automatically by extracting and processing information available in a database. For instance, the student's intelligence can be derived from his or her grades, from extracurricular activities, and so on.

Finally, real-world applications of the fuzzy-set theory in control systems sometimes optimize their behaviors based on the feedback they receive from their environment.

Control Questions

To make sure you understand this topic, try to answer the following questions. If you have problems, return to the corresponding place in the preceding text.

- Explain the notion of the degrees of membership in the fuzzy-set theory. How are they interpreted?
- Discuss the possibilities of defining the degrees of membership by mathematical functions. What typical functions are usually employed?
- Comment on the possible sources of the degrees of membership and of the functions defining them.

15.3 Fuzziness versus Other Paradigms

First encounter with the fuzzy-set theory often raises the question of this paradigm's relation to the theory of probability. Does the idea of fuzziness really address problems that cannot be handled by traditional means? The question is legitimate, and surely deserves our attention.

15.3.1 Probability of a Crisp Event

Probabilities are inherently related to crisp events and variables. For instance, we may ask, "what is the likelihood that a flipped coin comes up heads?" The result here is clearly defined: either *heads* or *tails*, nothing in between. This enables us to count experimental outcomes, and then determine their probabilities by relative frequencies.

15.3.2 Extent of a Feature

In some domains, the outcomes are not as clearly delineated. Thus the examples we saw in the previous section dealt with the *extent* of a certain property, not its likelihood. Classical theory of probability cannot easily say, "what is the probability that a student in this class will be smart?" because each individual is smart to a certain degree. This means that smart students cannot be counted in the way that we can count the outcomes of coin flipping.

15.3.3 Probability of a Fuzzy Value

This said, attempts have been made to quantify probabilities of fuzzy variables. For instance, we may ask, "is the occurrence of a smart student of nuclear physics more likely that the occurrence of a smart student in a history class?" To answer a question of this kind, we need a mechanism of counting fuzzy values (in analogy to counting crisp values). Section 15.5 will explain one possible solution.

This does not mean that the fuzzy set theory can be replaced with the theory of probability. Far from it. Mechanisms for calculating the probabilities of fuzzy variables are still nothing but special aspects of the fuzzy set theory.

15.3.4 Fuzzy Probabilities

Classical theory of probability deals with numbers. For instance, we may say that the probability of a flipped coin coming up *heads* is 50%. In our daily conversations, however, we rarely rely on concrete numbers. We may say that "Bill's failing at this exam is unlikely" or that "it is highly probable that John will get married soon." The terms *unlikely* or *highly probable* are *linguistic variables* that are by their very nature fuzzy. They represent fuzziness of probabilistic statements.

Operations that allow us to deal with linguistic variables and with their counting will be discussed in Section 15.5.

Control Questions

To make sure you understand this topic, try to answer the following questions. If you have problems, return to the corresponding place in the preceding text.

- Discuss the difference between the probability of a crisp event and the extent of a fuzzy property.
- What do we mean by the term, "probability of a fuzzy value"? What, by contrast, is *fuzzy probability*?

15.4 Fuzzy Set Operations

Section 15.2 introduced the notion of an object's degree of membership in a set. This helped us model such fuzzy propositions as "Bill is smart." Suppose we decide to combine such propositions by way of logical operations. How do we quantify the fuzziness of the resulting compound statements?

15.4.1 Fuzzy Logic

Everybody knows how classical logic evaluates compound logical statements such as "*A* and *B*" or "*C* and (*A* or *B*)." Their truth or falsity is easily established by simple rules or by truth tables.

Suppose, however, that the atoms, *A*, *B*, and *C*, are fuzzy propositions such as "John is tall" and "Bill is quick." John's belonging to the set of tall persons and Bill's belonging to the set of quick persons are specified by degrees of membership: μ_{tall}(`john`) and μ_{quick}(`bill`). How do we evaluate the truth and falsity of compound statements in *this* context?

15.4.2 Conjunction

Let us begin with conjunction. If we denote by $\mu_A(x)$ the degree of membership of x in A, and by $\mu_B(x)$ the degree of membership of x in B, then the degree of membership of x in $A \cap B$ (the intersection of the two sets) is the smaller of the two values:

$$\mu_{A \cap B}(x) = \min\{\mu_A(x), \mu_B(x)\} \tag{15.1}$$

The formula is easy to generalize to more than two sets. Specifically, the degree of membership of x in the intersection of N sets is the smallest of the individual degrees of membership.

Instead of talking about intersections, we can express the same by way of logical conjunctions: "x belongs to A *and* x belongs to B."

15.4.3 Disjunction

The other important operation is disjunction. If we denote by $\mu_A(x)$ the degree of membership of x in A and by $\mu_B(x)$ the degree of membership of x in B, then the degree of membership of x in $A \cup B$ (the union of the two sets) is the greater of the two values:

$$\mu_{A \cap B}(x) = \max\{\mu_A(x), \mu_B(x)\} \tag{15.2}$$

The formula is easy to generalize to more than two sets. Specifically, the degree of membership of x in the union of N sets is the greatest of the individual degrees of membership.

Instead of talking about unions, we can express the same by way of logical disjunctions: "x belongs to A *or* x belongs to B."

15.4.4 Negation

In classical set theory, to say that x does not belong to A is the same as to say that x belongs to $\overline{A}$, the complement of A. In fuzzy-set theory, the degree of membership of x in the complement of A is obtained by the following formula:

$$\mu_{\overline{A}}(x) = 1 - \mu_A(x) \tag{15.3}$$

The reader will have noticed that, in fuzzy-set theory, x can belong to a certain degree both to A and to $\overline{A}$. In classical set theory, this is of course impossible.

15.4.5 Graphical Illustration

Figure 15.4 illustrates the concepts introduced in the previous paragraphs: set intersection, which models logical AND, and set union, which models logical OR. The values of x are plotted along the horizontal axis, and the degrees of membership along the vertical axis. The function for A is a triangle, and the function for B is a trapezoid.

On the left, we see how fuzzy set theory handles intersection: For each value of x, always the smaller of the two degrees of membership is considered, such as in the case of x_1 whose degree of membership in the intersection is highlighted by a black bullet.

On the right, we see how fuzzy set theory handles union: For each value of x, always the greater of the two degrees is considered, such as in the case of x_1 whose degree of membership in the union is highlighted by a black bullet.

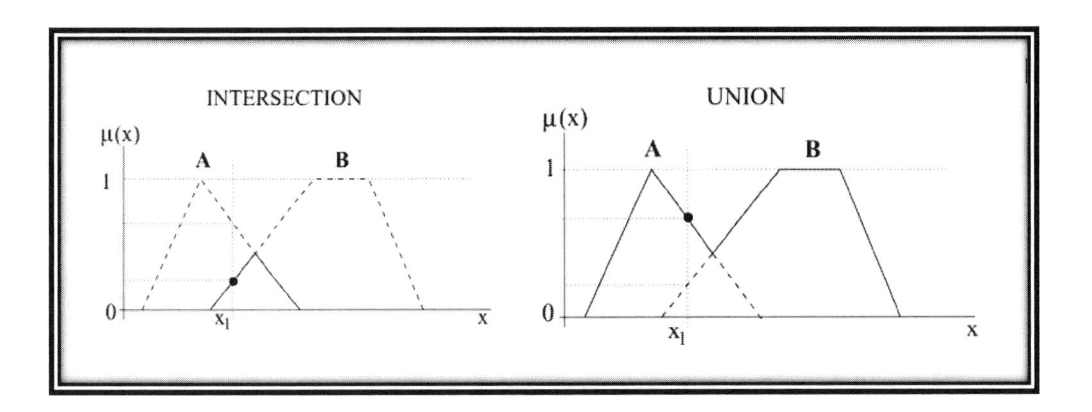

FIGURE 15.4 The graphs show how to determine the degrees of membership of *x* in sets *A* and *B*, and how to determine the degrees of membership of *x* in the intersection, and in the union, of the two sets.

15.4.6 Numeric Examples

Consider the fuzzy set of `smart` students from Table 15.1 and the fuzzy set of `tall` persons from the left part of Figure 15.2. For the latter, if x is height, then $\mu_{tall}(x) = 0$ if $x < 5.5$ feet, and $\mu_{tall}(x) = 1$ if $x > 6.5$ feet; for values between the two thresholds, we will use $\mu_{tall}(x) = x - 5.5$, so that, for instance, we get $\mu_{tall}(6) = 0.5$.

Suppose that `bob` is 5.8 feet tall. This means that the degree of his membership in the set of tall persons is $\mu_{tall}(\texttt{bob}) = 5.8 - 5.5 = 0.3$. Table 15.1 gives the degree of his membership in the set of smart students as $\mu_{smart}(\texttt{bob}) = 0.8$. The degree of his membership in the set of students that are *smart* AND *tall* is $\min(0.8, 0.3) = 0.3$, and the degree of his membership in the set of students that are *smart* OR *tall* is $\max(0.8, 0.3) = 0.8$. The degree of his membership in the set of students that are NOT tall is $1 - \mu_{tall}(5.8) = 1 - 0.3 - 0.7$.

15.4.7 Complex Expressions

Evaluation of the degrees of membership of more complex expressions is reduced to simple mathematical operations and does not cause any difficulties. For instance, suppose that `bob`'s degree of membership in the set of `diligent` students is $\mu_{diligent}(\texttt{bob}) = 0.9$, and suppose we want to know his degree of membership in the set defined by the following expression (he is either not diligent or he is smart and tall):

$$(\texttt{smart} \land \texttt{tall}) \lor \neg\texttt{diligent}$$

Using the values from the previous paragraphs, we obtain the value as follows:

$$\max\{\min(0.8, 0.3), (1 - 0.9)\} = \max(0.3, 0.1) = 0.3$$

Control Questions

To make sure you understand this topic, try to answer the following questions. If you have problems, return to the corresponding place in the preceding text.

- Write down the formulas that determine the degrees of membership in a disjunction of sets, union of sets, and the complement of a set.
- In what way does logical AND correspond to set intersection, logical OR to set union, and negation to set complement?

15.5 Counting Linguistic Variables

Scientists and mathematicians work with numbers. In our daily discourse, however, numbers are rare. Instead of saying that Bob is 6′2″, we will say that he is "rather tall," and everybody has an idea what we have in mind. In other words, we use *linguistic variables*. These are easily modeled by fuzzy sets.

15.5.1 Examples of Linguistic Variables

The way to handle linguistic variables in the context of fuzzy sets is to define for each of them the function that determines the degrees of membership. These functions may have the shape of triangles, trapezoids, or the curves from Figure 15.3. Their domains may overlap: With his 6′2″, bob may be characterized by the following values:

$$\mu_{very-tall}(\texttt{bob}) = 0.9$$
$$\mu_{tall}(\texttt{bob}) = 1.0$$
$$\mu_{average}(\texttt{bob}) = 0.7$$
$$\mu_{rather-short}(\texttt{bob}) = 0.0$$

15.5.2 Subjectivity of Linguistic Variables

Degrees of membership tend to be subjective. Fortunately, this subjectivity is rarely a problem in practical applications where the difference between, say, $\mu_{tall}(\texttt{bob}) = 1.0$ and $\mu_{tall}(\texttt{bob}) = 0.9$, rarely plays a major role. Of course, mutual relations between the values have to be reasonable. It would be weird to give high values to $\mu_{tall}(\texttt{bob})$ and $\mu_{short}(\texttt{bob})$, while giving a small value to $\mu_{average}(\texttt{bob})$.

15.5.3 Context Dependence

Degrees of membership usually depend on context. The term high_income has one interpretation in the community of physicians and trial lawyers, and another among construction workers. Its degrees of membership will be different in rich countries and in poor. Their values may even vary in time: In the nineteenth century United States, $2,000 represented a handsome annual income whereas today it does not.

15.5.4 Counting Fuzzy Objects

In classical set theory, the question "how big is set A" is answered by counting the set's members. In the context of fuzzy sets, we sum the degrees of membership. If we denote by $S(A)$ the size of set A, the following formula calculates its value:

$$S(A) = \Sigma_i \mu_A(x_i) \tag{15.4}$$

Here, $\mu_A(x_i)$ is the degree of membership of the i-th object, x_i, to set A.

15.5.5 Numeric Example

Suppose that five students, `bob`, `jim`, `eve`, `jane`, and `fred` are sitting in a classroom. In an earlier section, Table 15.1 listed for them the following degrees of membership to the set of smart students: $0.8, 1.0, 1.0, 0.7$, and 0.1. Using Equation 15.4, we establish that the number of smart students in the classroom is $0.8 + 1.0 + 1.0 + 0.7 + 0.1 = 3.6$.

Note that counting fuzzy objects does not have to result in integers.

15.5.6 More Advanced Example

Consider an AI program for a travel agency. Pondering the pluses and minuses of a vacation resort, the customer wants to know, "are there many high-rise buildings close to the beach?"

The statement involves two linguistic variables related to the buildings in the area: `high_rise` and `close` (i.e., close to the beach). Since both requirements are to be satisfied at the same time (logical AND), the individual degrees of membership are subject to the min function. For instance, suppose that building A belongs to high-rise buildings with degree 0.7, and its degree of membership to objects close to the beach is 1.0. In this case, the building satisfies the two conditions to the degree $\min(0.7, 1.0) = 0.7$.

Table 15.2 summarizes the procedure. The calculations begin with a table that summarizes for each building its degrees of membership in the set of high-rise buildings, and in the set of buildings close to the beach. These values may have been obtained from user-specified functions such as those suggested in Figures 15.2 and 15.3 (see Section 15.2).

For each building, the program then calculates the degree of membership in both sets at the same time (using the min function) and then sums the degrees of all buildings in the resort to obtain the count. This count is then subjected to the function that determines the degree of membership of this number into *many*, one possibility being shown at the bottom of Table 15.2. Note that this last function may have to be more sophisticated, to reflect the meanings of `many` in diverse contexts. What is `many` in one area may be `few` in another.

Control Questions

To make sure you understand this topic, try to answer the following questions. If you have problems, return to the corresponding place in the preceding text.

- Suggest a few examples of *linguistic variables*. Why do we need to consider them in our AI programs?
- Explain the mechanism that counts objects described by linguistic variables.

Each building in the resort is assigned the degree of membership to the two sets.

x	# stories	$\mu_{high}(x)$	distance	$\mu_{close}(x)$
A	6	0.7	100	1.0
B	9	1.0	300	0.7
...	...	...	...	...
Z	5	0.4	500	0.4

Here is the count of the buildings that satisfy both conditions at the same time:

$$N = \Sigma_i \min\left[\mu_{high}(x),\ \mu_{close}(x)\right]$$
$$= \min[0.7, 1.0] + \min[1.0, 0.7] + \ldots + \min[0.4, 0.4] \ldots$$

The resulting value of N is then submitted to a function that calculates the degree of membership of N in `many`. Below is one possibility.

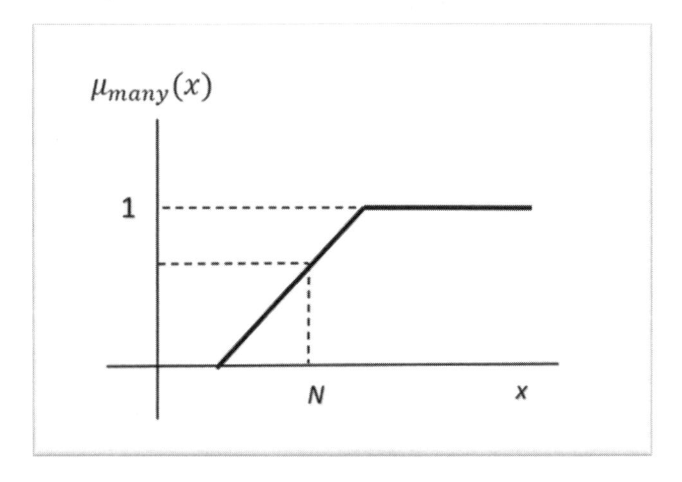

The degree of membership is read from the graph's vertical axis.

Table 15.2 Degrees of Membership for Two Concepts That Characterize the Buildings in the Resort: `high_rise` and `close` to the Beach. Are There `many` Such Buildings?

15.6 Fuzzy Reasoning

Now that we have mastered the preliminaries, we are ready to proceed to the question of how to use fuzzy logic in automated reasoning. In a sense, there is nothing new about it. We use the same *if-then* rules as before. The only problem is how to propagate fuzzy uncertainty through the rules.

15.6.1 Fuzzy Rules

The rules are essentially the same as before, only that now they involve the fuzzy and linguistic variables from the previous sections. For illustration, consider the following:

if `smart(x)` *then* `successful(x)`

The rule tells us that smart students tend to be successful. For any x, the degree of membership in the set successful is the same as the degree of membership in the set smart: $\mu_{\texttt{successful}}(\mathrm{x}) = \mu_{\texttt{smart}}(\mathrm{x})$.

if very_tall{(x)} *then* good_in_basketball(x)

Here, too, the degree of membership of x in the set of good players of basketball is the same as x's degree of membership in the set of very_tall persons.

15.6.2 More Realistic Rules

If the antecedent has two or more conjuncted conditions, the antecedent's degree of membership is the smallest degree of membership among the conditions. For illustration, consider the following rule:

if few_rebellious_students(c) and many_smart_students(c)
then class_moderately_good(c)

Here is how we establish the degree of membership of class c in the set of moderately_good classes. First, we decide for each student in the class his or her degree of membership in the set rebellious. In the next step, we establish how many such students there are (see Section 15.5, Counting Linguistic Variables), and the number is then passed through the function that models few. Let us denote the result by $\mu_{\texttt{FRS}}(\mathrm{c})$ where c is the class, and the acronym FRS stands for few rebellions students.

In like manner, we evaluate also the second condition. Any student belongs to a certain degree to the set smart. We count these students, and then pass the result through the function that models many. Let us denote the result by $\mu_{\mathrm{MSS}}(c)$, where c is the class and the acronym MSS stands for many smart students.

The degree of membership of the antecedent is then $\min\{\mu_{\texttt{FRS}}(\mathrm{c}),\ \mu_{\mathrm{MSS}}(\mathrm{c})\}$. This is then also the degree of membership of c into the set of moderately good classes.

15.6.3 Reasoning with Fuzzy Rules

Suppose that the knowledge base contains also the following rule

if interesting_topic(c) *then* class_moderately_good(c)

Evaluating the antecedent, we also get the class's membership in moderately good classes. Since a rule with the same consequent appeared also in the previous paragraph, we need a way to combine the two values.

Let us denote the degree of membership obtained in the previous paragraph by $\mu^{(1)}_{\mathrm{MG}}(\mathrm{c})$ and the degree of membership obtained in this paragraph by $\mu^{(2)}_{\mathrm{MG}}(\mathrm{c})$. Since each rule gives different reasons for the class being moderately good, it makes sense to sum the two degrees:

$$\mu_{\mathrm{MSS}}(\mathrm{c}) = \mu^{(1)}_{\mathrm{MG}}(\mathrm{c}) + \mu^{(2)}_{\mathrm{MG}}(\mathrm{c}) \tag{15.5}$$

Of course, degrees of membership have to fall in the unit interval, $[0, 1]$. If the sum in Equation 15.5 exceeds the upper bound, we set $\mu_{\mathrm{MG}}(\mathrm{c}) = 1.0$.

15.6.4 Propagating Degrees of Membership

A realistic knowledge base will have at least hundreds of such rules. Consequents of some rules appear in the antecedents of other rules. In the end, the knowledge base can be seen as a graph similar to those we used for the representation of belief networks in Chapter 14. The difference is that, instead of probabilities, we deal here with fuzziness. The reader has noticed that propagation of the degrees of membership is easier.

15.6.5 Fuzzy Control

The principles outlined in this chapter are employed in what is known as *fuzzy control*. In this context, the task is not to answer a user's query, but rather to suggest the value of some output quantity. For instance, based on diverse measurements made throughout a technological process, the control system is to decide whether to increase temperature and/or pressure—and how much to increase them.

The input variables are then *fuzzified*, by which we mean conversion of numeric values into linguistic values. This makes it possible to employ fuzzy rules that are similar to those subconsciously used by human operators. The output then takes the form of another linguistic variable, such as `increase temperature sharply`. This output is then converted into a concrete number by a process known as *defuzzification*.

Further details would exceed the needs of an introductory textbook.

Control Questions

To make sure you understand this topic, try to answer the following questions. If you have problems, return to the corresponding place in the preceding text.

- Explain the principle of fuzzy rules. How are they interpreted and how are they evaluated?
- Discuss briefly the principle of a knowledge base consisting of fuzzy rules. How are the degrees of membership combined and propagated?

15.7 Practice Makes Perfect

To improve your understanding, take a chance with the following exercises, thought experiments, and computer assignments.

- Explain the conceptual differences between the classical set theory, fuzzy set theory, and the theory of probability. Why are the three paradigms not interchangeable? What, by contrast, do they have in common?
- Briefly discuss how the popular saying, "the last straw that broke the camel's back" relates to the *paradox of heap* mentioned in Section 15.4.
- Section 15.3 mentioned that attempts have been made to find mathematical models for fuzzy probabilities. Practically, this means fuzzy models of such linguistic probabilities as `highly probable` or `rather unlikely`. Write a two-page essay in which you describe a concrete mechanism to do so. Provide also numeric examples to illustrate the point.

- Using fuzzy set theory, analyze the following expression with linguistic variables: "Are smart and hard-working students more common in engineering or in the sciences?" Identify the relevant linguistic variables and suggest the mechanism for evaluating them.
- Summarize the principles of fuzzy reasoning. Suggest a small fuzzy knowledge base illustrating the main points. How are the degrees of membership propagated throughout the network of these rules?
- Section 15.6 showed two rules that had the same consequent. Since these consequents have been obtained by different arguments, the section proposed that the two degrees of membership thus obtained should be added. Suggest a situation where taking the minimum or the maximum of the two values would be more appropriate.

15.8 Concluding Remarks

Fuzzy-set theory was born with the publication of the legendary paper by Lotfi Zadeh (1965). At first, his ideas did not attract much attention. In the 1970s, however, experts on knowledge-based systems and automated reasoning discovered the limitations of classical logic, especially when it came to *linguistic variables*. Zadeh (1975) showed that his fuzzy set theory could treat these variables in a very natural way. From then on, interest in this paradigm started to grow—and to grow very fast.

Another impetus was provided by Mamdani (1977) who proposed a simple mechanism for reasoning with fuzzy rules. Soon after, scholars found ways of employing fuzzy reasoning in automated control. Reports of successful industrial applications were only a matter of time. Famous are the control systems developed in the 1980s for some Japanese subway systems where fuzzy logic was used to control acceleration, braking, and stopping. Some textbooks also like to mention applications in Japanese cameras in the 1990s.

Today, fuzzy sets are no longer as hot as they used to be. In the last decades of the twentieth century, the theory was an epoch-making invention; now it is well-established, and in a sense even commonplace. PhD students in search of attractive research topics look elsewhere, mathematicians no longer anticipate groundbreaking discoveries, and funding agencies are less inclined to pour money into it. Perhaps it is inevitable. Some of the early claims and predictions were so audacious that anything short of miracles was bound to cause disappointment. As every so often, overheating gave way to a period of cooling.

This said, the fuzzy-set paradigm *is* useful. Particularly impressive are the applications of fuzzy control. Also the ability to model linguistic variables is attractive. The simplicity and intuitive appeal of the underlying concepts are likely to make sure that the paradigm is not forgotten any time soon.

CHAPTER 16

Highs and Lows of Expert Systems

The main beneficiary of the studies of automated reasoning and uncertainty processing is the technology of *expert systems*, an early success story of artificial intelligence. The idea is to emulate human decision-making in domains that depend on non-trivial background knowledge.

Typical solutions combine a well-designed knowledge base with techniques for reasoning so as to answer users' queries. Typically, the knowledge has the form of *if-then* rules, although semantic networks and frame-based representation deserve to be mentioned as significant alternatives. The imperfections of available knowledge are quantified by some of the paradigms from the previous chapters: certainty factors, probabilities, and fuzzy sets. An important part of the product is its ability to explain the expert system's reasoning and to lead a quasi-intelligent conversation with the user.

This chapter briefly outlines some critical moments in the history of expert systems, and a few lessons gleaned from experiments with early products.

16.1 Early Pioneer: *Mycin*

Mycin is the oldest expert system. The presentation of its early version in 1969 became something of a bombshell, and quickly inspired the work of many pioneers of AI. In the 1970s, several research groups experimented with a wide range of application domains, introduced improvements, and accumulated experience that helped establish the reputation of these promising new tools.

16.1.1 Implementation

The program was implemented in *Lisp*, programming language that had been developed with the specific needs of artificial intelligence in mind. An early version of *Mycin*'s knowledge base consisted of hundreds of *if-then* rules whose imperfections were addressed by the certainty factors discussed in this book's Chapter 13. Soon, however, much larger, and more ambitious, knowledge bases were developed.

16.1.2 Intended Field of Application

Extensive experimentation soon demonstrated *Mycin*'s ability successfully to diagnose specific diseases, among them *bacteremia* and *meningitis*. For these, the program could also recommend appropriate choices of antibiotics. Its diagnostic performance was said to compare favorably with that of human experts. In spite of its good reputation, however, the tool was not employed in regular practice.

16.1.3 Early Concerns

Why did the success fail to encourage physicians to start using the new tool regularly? An early impediment was concerns of ethical and legal nature. Here is the opponents' main line of argument. Suppose a physician prescribes drugs based on the software's recommendation. Who will be responsible if the decision proves incorrect, perhaps even having some unpleasant consequences? Will it be the physician or the programmer? The prospects of possible lawsuits were deemed unattractive.

Today, we see things differently. If *Mycin* were first introduced now, rather than in the atmosphere that was typical of the early days of AI, it would be perceived as an advice-giving tool that may offer arguments and inspiration—and nothing more. The ultimate decisions would be the responsibility of the physician. In the middle of the heated discussions that followed the software's first presentation, this did not seem obvious.

16.1.4 Early Hopes

Some scholars argued that expert systems should not serve highly qualified specialists who, after all, still performed better than computer programs. Rather, the argument went, the new technology might prove useful in regions where doctors were rare, or perhaps not available at all. Large areas of third-world countries were plagued by natural disasters, epidemics, and civil wars. At a time of acute need, trained physicians were often unavailable—in which event, a nurse armed with a small computer might be better than no help at all.

To the best knowledge of this author, however, this idea did not venture far beyond coffee-break discussions of scientific conferences.

Control Questions

To make sure you understand this topic, try to answer the following questions. If you have problems, return to the corresponding place in the preceding text.

- When was the first expert system introduced, and what was its primary field of application? What kind of knowledge base did it employ and how big was this knowledge base? How did it handle uncertainty?
- Discuss some of the early hopes and concerns related to the new technology.

16.2 Later Developments

The success of *Mycin*, and the passionate discussions that it inspired, led to an outburst of activities. Within a few years, dozens of expert systems were introduced, many of them experimenting with diverse new mechanisms of knowledge representation and reasoning, many exploring new application domains. The experience proved invaluable. Step by step, artificial intelligence established itself as a field likely to enhance the potential of computer machinery.

16.2.1 Another Medical System

Medicine remained one of the primary fields of interest. Perhaps the most impressive expert system of the 1980s was *Caduceus*. Its authors intended to build on the successes of *Mycin*, but to move one step further and overcome one serious limitation. To wit, they found it unnatural that the range of the tool's abilities was so narrow: just the recognition of a handful of bacteria. Human experts are never so extremely specialized. The more they know, the more easily they add to their knowledge, and the more easily they move on to new challenges. A true expert system should exhibit similar versatility.

Caduceus could deal not just with a few, but with hundreds of diseases. This broad scope made it perhaps the most knowledge-intensive expert system of those times, one of the peaks scaled by the new technology. Where *Mycin* pointed the way, *Caduceus* demonstrated expert systems' real potential.

16.2.2 Prospector

Many scholars felt uncomfortable about the *ad hoc* nature of the uncertainty factors employed by *Mycin*. By way of offering an alternative, the authors of *Prospector* preferred to implement in their system the probabilistic reasoning principles we now know from Chapter 14. The tool was reported to have helped discover rich deposits of molybdenum ore. This was a real triumph!

This success, however, was not followed by reports of comparable achievements, and this rendered the molybdenum discovery less convincing than expected. While the authors of *Prospector* no doubt did a great job, it is conceivable they had inadvertently entered in the knowledge base some information that was known to them, but should not have been made explicit to the program. Something like, "I wonder if you ever find the wristwatch I have hidden in the left pocket of my pants." Such suspicions, however, were never proved, and reservations of this kind should not detract from *Prospector*'s great achievement. By pioneering the probabilistic approach to reasoning, this expert system established itself as a valuable milestone.

16.2.3 Hundreds of Expert Systems

At a certain period, in the 1980s and 1990s, literally hundreds of expert systems were developed by diverse research groups, each with a different application domain in mind, each with different inference engine at its disposal. *Dendral* addressed molecular chemistry, *R1* helped engineers configure complex computer systems, and *Mistral* monitored safety of dams. Other systems offered incremental solutions to advanced engineering problems, diagnosed student behavior, dealt with voice-recognition challenges, even offered advice to robots facing unexpected situations.

16.2.4 Dangers of Premature Excitement

The field became fashionable, and expectations ran high. Many scholars believed that artificial intelligence had reached a stage where machines would become truly intelligent and capable of offering advice in every walk of life, that the world had stepped on the threshold of a new technological era. As every so often, in similar circumstances, the excitement spawned predictions that today's engineer—armed with the wisdom of hindsight—is tempted to call naïve, if not ridiculous.

Some of it was of course opportunistic. To attract government funding for their research, scientists sometimes inflate their expectations, and this is what happened here, too. But let us be fair. The breakthroughs *were* impressive, and many people honestly believed something grandiose was afoot.

16.2.5 Skepticism

The old cliché has it right: Action causes reaction. The name of the penalty for excessive excitement is disappointment. This is what happened to expert systems, too. Uncritical optimism proved counterproductive, and in the end it even backfired. The turn of the century's interest in the field was lukewarm at best.

Those who believed that *Mycin*, *Caduceus*, or *Prospector* were only first steps of a triumphant long march, waited in vain for advancements of comparable weight. Pessimists started to suspect that these systems were the maximum that would ever be achieved. They also identified the fundamental bottleneck: the difficulties related to the need to create, and then fine-tune, very large knowledge bases. Beyond a certain size and complexity, difficulties mushroomed.

16.2.6 Today's Situation

But the technology did not disappear, it only adopted a new guise. Rather than standalong packages, expert systems—or programs based on their principles—became incorporated in broader systems. Examples are easy to find. For instance, many lecturers who prepare *PowerPoint* presentations for their classes have become acquainted with the function known as "design." This offers formatting suggestions that help them convert their text into graphs and other means of visualization. These suggestions are obtained by rules akin to those employed in expert systems. Something similar has been in use in health-care facilities where a computer program navigates a nurse through diagnoses and treatment recommendations that previously were the responsibility of physicians.

Today, no one thinks of marketing such software as cutting-edge application of expert systems. The world has changed. The technology that was once so hot and sexy no longer constitutes a credible selling point.

Control Questions

To make sure you understand this topic, try to answer the following questions. If you have problems, return to the corresponding place in the preceding text.

- What expert systems do you know and what can you say about their most popular application domains?
- Summarize the main reasons underlying the popularity of expert systems in the '80s and '90s. Why did this popularity later fade off?

16.3 Some Experience

In the days when expert systems were all the rage, many scientists experimented with them, and gleaned lot of practical experience. Let us mention some of the most important lessons.

16.3.1 The 5-Minutes-to-5-Hours Rule

When choosing an appropriate application domain, experience shows that two extremes should be avoided: problems that are too easy, and problems that are too difficult. As a guideline, the *5-minutes-to-5-hours rule* used to be mentioned.

Problems that a human can solve in less than 5 minutes are too simple to merit the attention of any expert-system developer. Why bother if it is so easy? At the other end of the range, problems that the human specialist cannot solve in a few hours are probably so difficult that an attempt to create an expert system to deal with them is likely to be doomed from the get-go. The program will be either too expensive to develop or too unreliable to use—or both.

16.3.2 Bottleneck: The Knowledge Base

Depending on the concrete application, perhaps as much as 90% of all the engineer's effort will go into the development of the knowledge base. This is where most of the difficulties are encountered. The first difficulty is how to obtain the requisite knowledge in the course of lengthy interviews with the field expert. Then comes the question of how to convert human understanding and intuition into *if-then* rules (or some other formalism). But the greatest challenge of them all is the fine-tuning of the probabilities, certainty factors or degrees of membership to make the software give correct answers to a wide range of the users' queries.

By comparison, implementation of the automated-reasoning mechanism is relatively easy, even if sophisticated ways of uncertainty processing are involved. Resolution or *modus ponens* can be implemented even by classical search techniques, enhanced by a few auxiliary functions to speed-up the process.

16.3.3 Communication Module

Early pioneers focused predominantly on knowledge representation and reasoning. Practitioners soon realized, however, that a lot of work has to be devoted to the development of the programming module that will communicate with the users. In the course of this communication, the system will ask the users to provide additional facts related to the given query so that the right rules in the knowledge base can be identified.

Of particular importance is the necessity to implement an *explanation module*. The thing is, the user is unlikely to trust the system's recommendations unless they are supported by convincing arguments. For this reason, the user should at any moment during the dialog have the opportunity to ask, *why?*—and the system should be capable of explaining everything.

In the last century, implementing the communication and explanation modules required a lot of programming effort. Today, this is easier. Not only can we rely on all sorts of high-level GUI software, but even fresh college graduates have much more experience writing such programs than their fathers and grandfathers had after years in the programming business. Interactive programs were rare.

16.3.4 Graceful Degradation

One weak spot of the first rule-based expert systems was that a tiny imprecision often had disastrous consequences. When you talk to a human specialist, and say something mildly inaccurate, the specialist easily corrects it; and even if he or she does not, the inaccuracy is unlikely to lead to a total collapse of all communication and to nonsensical decisions. By contrast, early expert systems appeared to be dogmatic.

This experience led to the formulation of what came to be known as the requirement of *graceful degradation*. To wit, the system's way of malfunctioning should be "graceful": A minor mistake should not lead to absurd behavior.

Control Questions

To make sure you understand this topic, try to answer the following questions. If you have problems, return to the corresponding place in the preceding text.

- Discuss the difficulties of implementing the main modules of a functional expert system.
- Explain the meaning of the term, *graceful degradation*.

16.4 Practice Makes Perfect

To improve your understanding, take a chance with the following exercises, thought experiments, and computer assignments.

- Suggest an application domain appropriate for the technology of expert systems. What will the knowledge base consist of? Where will the requisite knowledge come from? What uncertainty-processing mechanism will you prefer?
- Search Wikipedia for information about successful expert systems, the principles they employed, and the domains they addressed. Write a two-page essay on what you have thus learned.
- Write a one-page essay discussing the historical role of expert systems, the reasons behind their loss of popularity, and the chances of their future recovery.

16.5 Concluding Remarks

Early motivation for expert systems was provided by Edward Feigenbaum who is sometimes called the "father of expert systems." He started publishing his ideas on this technology in late 1960s. Later, he participated in the development of *Dendral*, and collaborated on the development of some other systems, including the famed *Mycin*. The latter, however, is usually associated with the names Shortliffe and Buchanan (1975), although other scholars collaborated on its development, too. The experience with *Caduceus* was discussed in depth by Bobrow, Mittal, and Stefik (1986).

The period of keen interest in this technology spanned a whole generation. It is unfortunate (though perhaps inevitable) that the failure to deliver on hyped promises resulted in such an anticlimax. Our twenty-first century is largely indifferent; other directions of AI research have gained prominence and dominate the field. The days of stand-along expert systems are over. However, the underlying principles have found their way into modern software packages without being marketed under the now-somewhat-obsolete label.

CHAPTER 17
Beyond Core AI

Intelligent behavior cannot be reduced to problem solving and reasoning. An agent will hardly be called intelligent, if it cannot interpret the environment in which it operates, if it is unable to communicate in something akin to natural language, and if it cannot learn from its mistakes.

The author of this book *does* share these sentiments. However, he believes that the studies of computer vision, natural language processing, and machine learning have by now followed their own specific directions, acquiring a measure of independence from the core AI. The same applies to another field that was popular in more recent textbooks: agent technology.

Still, it would be inappropriate to ignore them totally, as if they never existed. This chapter offers the reader a cursory look at what these disciplines seek to accomplish, and how they in the past employed the methods, algorithms, and techniques of classical artificial intelligence.

17.1 Computer Vision

From the early days of artificial intelligence, scientists agreed that an intelligent agent (e.g., a robot), should be capable of orienting itself in its environment. Very often, this referred to the ability to analyze and interpret visual information.

17.1.1 Image and Its Pixels

A digital image consists of a vast number of tiny dots called *pixels*.[1] Each pixel in a black-and-white image specifies the level of darkness. To this end, it uses an integer from the interval $[0, 255]$ (one byte), where 0 means totally white and 255 means totally black (it can also be the other way round).

Figure 17.1 illustrates the point. On the right is a picture of a human face. The area indicated by the small rectangle is represented by the matrix on the left. The reader can see that the pixels corresponding to the black curve are marked by higher numbers. Note that the numbers are not strictly 255 or 0. This is caused by the noise that is in realistic images all but unavoidable.

---

[1] The word *pixel* is an shorthand for *picture element*.

12	13	211	212	13	12	23	11	10
10	11	245	239	78	38	34	09	08
13	14	75	225	330	98	52	68	20
26	14	75	220	234	17	64	68	20
14	14	175	20	230	228	13	68	20
11	14	75	30	130	228	214	68	20
10	11	45	96	98	286	215	219	08
22	13	14	17	13	12	08	209	10

FIGURE 17.1 The digital version of a concrete image consists of a huge matrix of pixels, each characterized by an integer that quantifies the degree of blackness. The matrix on the left describes the small square shown in the picture on the right.

In reality, the whole picture can be represented by $1{,}000 \times 1{,}000 = 10^6$ pixels or even more.

17.1.2 Noise Removal

The first step in classical computer vision was noise removal. A simple version of a technique called *convolution* replaces each pixel's value with the average of its neighborhood. Thus in the second row, the current value of the second pixel from the right is 9. Convolution recalculates as follows:

$$\frac{23 + 11 + 10 + 34 + 9 + 8 + 52 + 68 + 20}{9} = 26$$

The same correction is applied to every single pixel in the entire image.

17.1.3 Edge Detection

The next step seeks to identify those locations where the values of neighboring pixels significantly differ. Thus in the first row, the values 12 and 13 are followed by 211, which indicates a steep increase of darkness; the almost-white area represented by 12 and 13 is followed by the almost-black pixels 211 and 212. This steep gradient betrays an *edge*—and indeed, in the picture on the right we see that this is indeed the case.

17.1.4 Connecting the Edges

Once those many minuscule edges have been identified, the next step seeks to connect them, thus creating lines. Many of the edges will be isolated, perhaps representing noise in the picture. Others do form lines that can then be combined into specific shapes. These shapes are then compared to models of typified objects in the computer memory.

This, then, is how the original matrix of apparently meaningless integers can gradually be interpreted ("a chair in front of a table"), and perhaps even reasoned about ("this is an obstacle to be avoided").

17.1.5 Texture

Looking at pictures of diverse materials, the observer will easily tell metal from, say, textile fabric. The reason they are so easily distinguished is that their surfaces have different *textures*. Scientists specializing in computer vision have developed methods of converting the integer matrix into advanced features that help them identify specific categories of textures.

Among the most popular such features are *wavelets* (e.g., *Gabor wavelets*). Other features are statistical; for instance, standard deviations of light intensity, gradients, and so on.

17.1.6 Color

In color images, each pixel is represented by three integers, one for each of the three basic colors: red, blue, and green. This multiplies the volume of analyzed data. Other than that, however, the same procedures are applied as in black-and-white images: edge detection, model comparison, texture analysis, and so on.

17.1.7 Segmentation

Another method of analysis seeks to identify in the image specific regions of interest. Thus in Figure 17.2, one area contains a house, another a tree, and yet another a flower. Certain smaller segments contain the house's windows and the door. To identify the segments, a combination of image processing and computer vision techniques is employed, including edge detection, texture analysis, and color.

17.1.8 Scene Interpretation

After these preliminary steps, the program combines all the information about the edges, objects, colors, textures, and regions in the picture's 3-D model. This is followed by an attempt to identify what the image really represents. For instance, analysis of the picture in Figure 17.2 may tell us that there is "a tree next to a house."

17.1.9 Modern Approach

Three generations of computer-vision experts brought to perfection the techniques described in the previous paragraphs—and on many occasions achieved impressive results. Nowadays, however, the mood has changed. Recent advances in the field of machine learning (specifically the so-called *deep learning* based on artificial neural networks) have been shown to be so powerful that they sometimes outperform human subjects in such tasks as face recognition.

Figure 17.2 In the process of segmentation, the computer program identifies specific regions of interest. In this particular picture, one region contains the house, another a tree, and yet another a flower.

The fact that such high-recognition performance has been achieved without the need for edge detection or texture analysis, has led some observers to predict (perhaps hastily) the demise of classical computer vision techniques. Only future will tell whether this view is justified.

Control Questions

To make sure you understand this topic, try to answer the following questions. If you have problems, return to the corresponding place in the preceding text.

- Summarize the principles of *edge detection*. What is it used for?
- What is *texture* and what role does it play in computer vision?
- What is *segmentation* and *scene interpretation*?

17.2 Natural Language Processing

Another aspect of an agent's intelligent behavior is the ability to communicate with the human user in the user's own language such as English, whether spoken or written. To develop software capable of such communication is a task for a field referred to by the acronym NLP (natural language processing).

17.2.1 Signal Processing

If the goal is to interpret *spoken language* (rather than text), the first task faced by any NLP system is to convert the analog signal to phonemes and words. The program has to detect the moment when the phoneme begins, and then describe the following brief period by features that may characterize the phoneme. The simplest features are the coefficients of harmonics, but much more sophisticated ones are typically used.

After this, one has to compare the signal thus described to the specific patterns stored in the computer memory. Each phoneme can be represented by one or more such patterns. Once the phonemes have been identified, they have to be combined into words. This can be accomplished by checking the given sequence of phonemes against a dictionary. If the sequence does not seem to represent any known word, then the most similar word is tried.

This signal-processing task is the most advanced aspect of NLP research.

17.2.2 Syntactic Analysis (Parsing)

Words are not enough. Most of the time, the real meaning of an utterance is conveyed by the way the individual words are related to each other. These relations are captured by grammar that is in our context called *syntax*. We know that a sentence often begins with a subject followed by a verb followed by an object. This sequence conveys meaning. The following two sentences illustrate the point.

"Fred called Jim."

"Jim called Fred."

Classical NLP has relied on many syntactic rules, representing them by what is known as *augmented transition networks*, ATN: special graphs that are used to parse sentences. Thus in the subject-verb-object structure mentioned above, the ATN may specify that the subject may or may not begin with a definite or indefinite article followed by a noun; alternatively, the first word can be a name, a pronoun, etc.

The analysis is also assisted by a specifically implemented *lexicon* that for each word specifies its grammatical circumstances. For instance, the lexicon will contain the information that the word *goes* is third person singular, present tense. Similarly, *went* is in past tense, and *houses* is plural.

17.2.3 Semantic Analysis

Whereas syntax represents grammar and structure, the term *semantics* refers to meaning. To figure out the precise meaning of each word may not be easy. After all, we know that the same word often acquires different meanings in different contexts.

17.2.4 Disambiguation

One of the most difficult challenges is the removal of ambiguities. The process is known as *disambiguation*. As we know, the same word may have two or more meanings.

Moreover, the same recorded signal may indicate one of a few similarly sounding words. Which of them is the right one was in the past established from the word's context.

For instance, *cardinal* can refer to an ecclesiastical official, a bird, or perhaps even to a rock band or a football club, not to speak of cardinal numbers in math. Which of these meanings is the correct one depends on the context. A newspaper article reporting recent events in Vatican is unlikely to refer to a bird. In the context of social networks, the meaning can also be narrowed by the history of an individual's recent web browsing. The ultimate decision can in many cases be assisted by background knowledge, including the *if-then* rules that this book used in automated reasoning.

17.2.5 Language Generation

Even more challenging than language understanding is generation of grammatically correct sentences. In one advanced application, the user presents to the computer a certain text (perhaps a newspaper article or a legal document) and expects the machine to return the text's brief summary, preferably written in decent English or some other language. In the context of expert systems, we want the software to be capable of explaining its conclusions. This means that the list of *if-then* rules that have been employed during the query-answering process have to be converted to something that can be communicated to the human user. Finally, language generation is needed in automated translation between natural languages.

Techniques capable of language generation do exist, but they are too advanced to be included in an introductory text.

17.2.6 Modern Approach: Machine Learning

Current efforts have largely departed from parsing and reasoning. Augmented transition networks and lexicons are much less fashionable than they used to be. Perhaps the most widespread use of NLP in these days are such applications as phone-call handling software. Here, correct identification of isolated words (date of birth, address, etc.) is often deemed sufficient. This essentially eliminates the need for parsing or for semantic analysis. Language generation is here very simple, too: just pre-recorded words or sentences ("What is your address?").

For more challenging applications, recent advances in *recurrent neural networks* proved invaluable—in particular the technology known as *long short-term memory*. These are sometimes assisted by a mathematical field called *hidden Markov models*. All of these, however, find themselves well outside the realm of core AI.

Control Questions

To make sure you understand this topic, try to answer the following questions. If you have problems, return to the corresponding place in the preceding text.

- Within the context of NLP, summarize the roles of signal processing, syntactic analysis, and semantic analysis. What is understood under the term *disambiguation*?
- What is *language generation* and where is it needed?

17.3 Machine Learning

A chess player that repeatedly makes the same blunder in the same position will hardly be called smart. Nor is a card player admired who always falls for the same simple trick. In other words, intelligence is unthinkable without the ability to learn. Indeed, some will go as far as to declare that, "there is no intelligence without learning!"

17.3.1 Knowledge Acquisition: The Bottleneck of AI

During the 1970s and early 1980s, the AI community gradually came to recognize the main difficulty of expert systems: Where was the requisite knowledge to come from? Many assumed that the knowledge base would be put together by engineers who had gained adequate level of understanding after extensive interviews with field experts. In reality, this was realistic only when the number of the *if-then* rules was manageable. For more challenging applications, attempts to formulate thousands of rules, and to fine-tune all those certainty factors, probabilities, or degrees of membership, proved daunting, to say the least.

To begin with, communication between engineers and field experts was not easy. For instance, physicians found logical formulations lamentably mechanical. In their world, a lot depended on intuition and on insight obtained in the course of long years of studies and a lot of practical experience. They were appalled that AI specialists kept confusing critical pieces of knowledge that any fresh medical-school graduate found elementary—while the engineer did not have a clue. The reader will recall, in this context, what Section 13.1 had to say about the tacit assumptions that appear so obvious that experts fail to convey them to the computer.

17.3.2 Learning from Examples

The observation that people are very good at learning from examples led to an idea that gained traction in the 1980s. To wit, if the knowledge base cannot be created manually, why not explore the possibilities of doing so indirectly, by way of well-chosen examples?

To illustrate the point, Table 17.1 shows a set of six training examples, each described by the values of three features, `a`, `b`, and `c`, and classified into two categories, `good` or `bad`. From these, the machine-learning technique is to induce the requisite knowledge, encoded in an appropriate data structure.

In reality, of course, the training sets will consist of thousands of examples that can easily be described at least by dozens of features.

17.3.3 Rules and Decision Trees

The knowledge induced from the examples can acquire various forms. Some of the most popular are shown in Table 17.1. In the upper-right corner is a decision tree and underneath are *if-then* rules. The reader will have noticed that this particular set of rules has been obtained as a direct translation of the decision tree where the `t` values are interpreted as *true* and the `f` values are interpreted as *false*.

Many techniques have been developed for efficient induction of decision trees and/or rules, and for their conversion to *Prolog* programs. Other techniques sought to

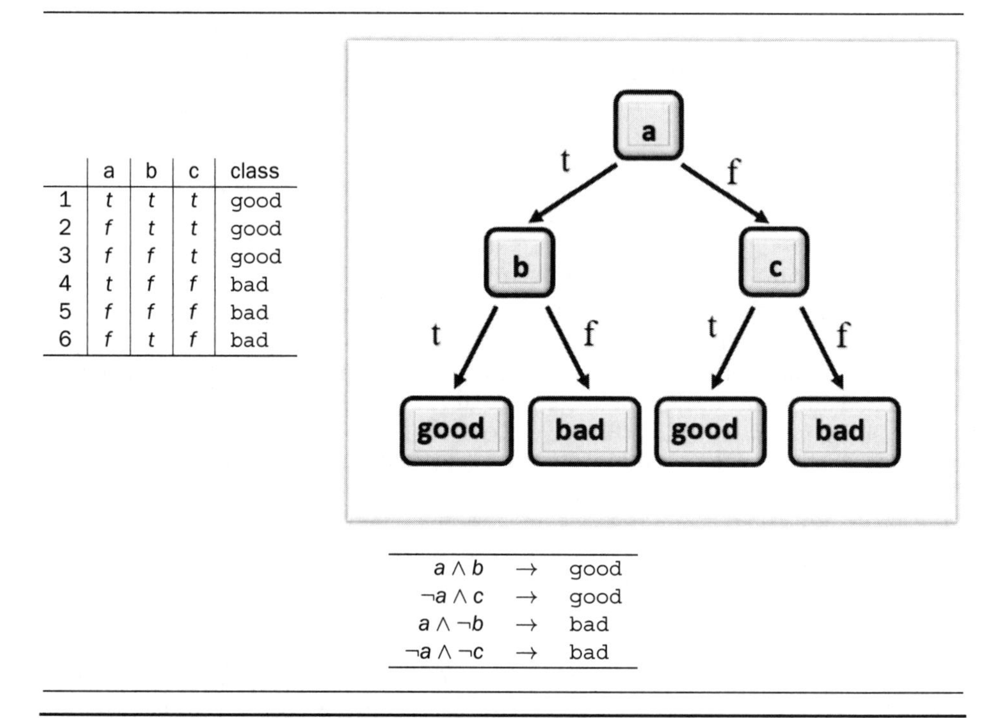

	a	b	c	class
1	*t*	*t*	*t*	good
2	*f*	*t*	*t*	good
3	*f*	*f*	*t*	good
4	*t*	*f*	*f*	bad
5	*f*	*f*	*f*	bad
6	*f*	*t*	*f*	bad

$a \wedge b$	$\rightarrow$	good
$\neg a \wedge c$	$\rightarrow$	good
$a \wedge \neg b$	$\rightarrow$	bad
$\neg a \wedge \neg c$	$\rightarrow$	bad

TABLE 17.1 Next to the Set of Six Training Examples Is a Decision Tree That Can Be Induced from Them. Below Are Four *if-then* Rules into Which the Tree Can Be Translated. Here, "t" Stands for *true* and "f" Stands for *false*.

optimize these programs, and perhaps even refine the knowledge base in response to new examples.

17.3.4 Other Approaches

Besides decision trees and rules, machine learning has studied other methods of knowledge representation. Among the paradigms that have been found most beneficial in AI, we can mention the diverse ways of storing typical prototypes of various concepts, probabilities to be employed in Bayesian networks, and artificial neural networks that have of late become so popular as almost to dominate the field.

17.3.5 Prevailing Philosophy of Old Machine Learning

In the 1990s, many scientists were convinced that the idea of machine learning is ideally approached as an exercise in classical AI search. This was shown to be the case when it came to induction of decision trees, rules, and neural networks. Also the benefits of relying on logic seemed all but self-evident—and even inevitable in induction of *Prolog* programs. In this sense, the field of machine learning seemed to fit the bigger picture of artificial intelligence.

17.3.6 Machine Learning Today

Later, however, this worldview faded, and gave way to techniques and mechanisms that no longer resonated with traditional AI. The most popular frameworks employed in the twenty-first century machine learning (hidden Markov models, neural networks, deep learning, long short-term memory, and other numeric approaches) no longer seemed to have much to do with the AI methods of symbol manipulation and logic. Theoretically speaking, they *can* be coached as examples of search, but attempts to do so are somewhat artificial.

More importantly, the field no longer cares about induction of knowledge for expert systems. That ambition stood at the cradle of machine learning. Since then, however, the world has changed. Today's machine learning pursues different goals.

Control Questions

To make sure you understand this topic, try to answer the following questions. If you have problems, return to the corresponding place in the preceding text.

- What was perceived as the main difficulty in the development of realistic expert systems in the 1980s? Where did scientists hope to find a solution?
- Draw an example decision tree and convert it into *if-then* rules.
- In what way did the field of machine learning depart from the realm of traditional artificial intelligence?

17.4 Agent Technology

Another topic that deserves to be mentioned here at least briefly is the field of *agent technology*. At a certain time, many AI specialists felt that intelligent agents are bound to share certain functional and architectural properties that should be studied and analyzed—and then employed in the implementation of functional AI programs.

17.4.1 Why Agents?

Throughout this book, the word *agent* has been used a great many times, mainly because the writer found the term so versatile. Indeed, the concept of an agent springs to mind whenever we seek to explain the techniques and mechanisms of AI within the context of problem solving.

From the programmer's perspective, however, the term has acquired a very specific meaning. It turns out that it helps to formalize many functions of intelligent programs in a unified way.

17.4.2 Architecture

Figure 17.3 illustrates the generic idea. The agent interacts with what is known as *environment* (which, broadly speaking, can even be a human user). Through *sensing*, the agent receives information about the environment's current state. The agent processes this information through what the picture calls *deliberation*. This results in the selection of an action by which the agent *acts* on the environment.

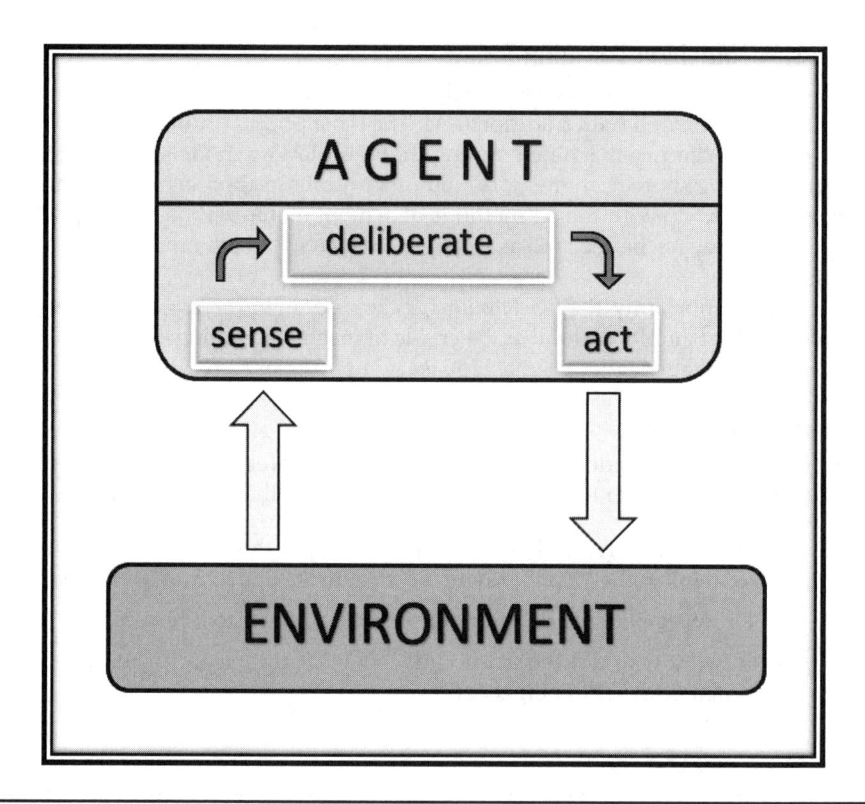

FIGURE 17.3 Essential architecture of an AI agent.

Deliberation can consist of just finding the appropriate action in a look-up table. More interestingly, however, the action is obtained through automated reasoning, search, or even some swarm-intelligence-based analysis.

Control Questions

To make sure you understand this topic, try to answer the following questions. If you have problems, return to the corresponding place in the preceding text.

- What is an agent? Why do we need the term?
- Explain the principle of the overall architecture of a typical agent employed in AI software.

17.5 Concluding Remarks

Older generations of AI scholars would be surprised if they heard that such fields as computer vision and natural language processing would one day be excluded from an introductory AI textbook. For them, these skills were inseparable from any intelligent behavior. Younger scientists seem to adhere to different views. For them, the classical approaches based on search and reasoning are virtually dead, having been largely replaced by the *deep learning* technology. The evolution appears well-nigh inexorable. From the perspective of the developments that could be observed during the last decade or two, there is no doubt that the scales have been tilted toward the novel techniques.

In machine learning, this seems to be true, also, but only up to a certain degree. The fact of the matter is that some classical tasks cannot be easily carried out by neural networks. This is most certainly the case if we want the machine to induce *if-then* rules from raw data. True, such applications are outside of today's mainstream, and certainly out of fashion. However, at least part of the once-so-vivid interest may yet return.

As for agent technology, this attracted a lot of attention at the turn of the century. The author of this book agrees that the technology is fascinating. This said, he sees it rather as a family of programming techniques that are useful, but do not really belong to core AI.

CHAPTER 18
Philosophical Musings

An engineer wants to develop computer programs capable of solving difficult problems. Philosopher digs deeper, wishing to know what it is that makes a certain behavior intelligent. Once the answer has crystallised, it is legitimate to ask: Will truly intelligent machines ever be developed?

This book deals with technology, not philosophy. Still, even the most practically minded reader may want to be acquainted with two major thoughts. The first, the Turing test, asks how to recognize intelligence. The second, Searle's Chinese Room, alerts us to the fact that apparently intelligent behavior is not yet a proof of consciousness.

18.1 Turing Test

As early as in the late 1940s, scholars asked themselves whether an intelligent machine would ever be created. To this end, it seemed necessary to know how to recognize when this happened. Claude Shannon suggested that the ultimate test would be passed when the world champion in chess had been beaten by a computer. More subtle, however, was the intriguing idea proposed by Allan Turing.

18.1.1 Turing's Basic Scenario

The principle is illustrated by Figure 18.1. You are sitting in Room 1 whereas in the other room is either an intelligent machine or simply another person. Your task is to figure out which of the two it is. To find out, you are allowed to send typed questions that the other side will answer and send back, again in the form of typed message.

Turing believed that the answers will enable you to tell the machine from a human. The stipulation was that there is something, in our thinking, that a computer program cannot emulate; and if it can, then it deserves to be called intelligent.

18.1.2 Additional Complications

The basic scenario from the previous paragraph is still too simple. Suppose you ask, "what is the fifth root of π." If you receive the correct answer within a second, you may conclude that only a machine could succeed so fast. If the response takes longer, or if it is incorrect, then you may be dealing with a human.[1]

[1] In those days, there were no calculators to help the human with the answer.

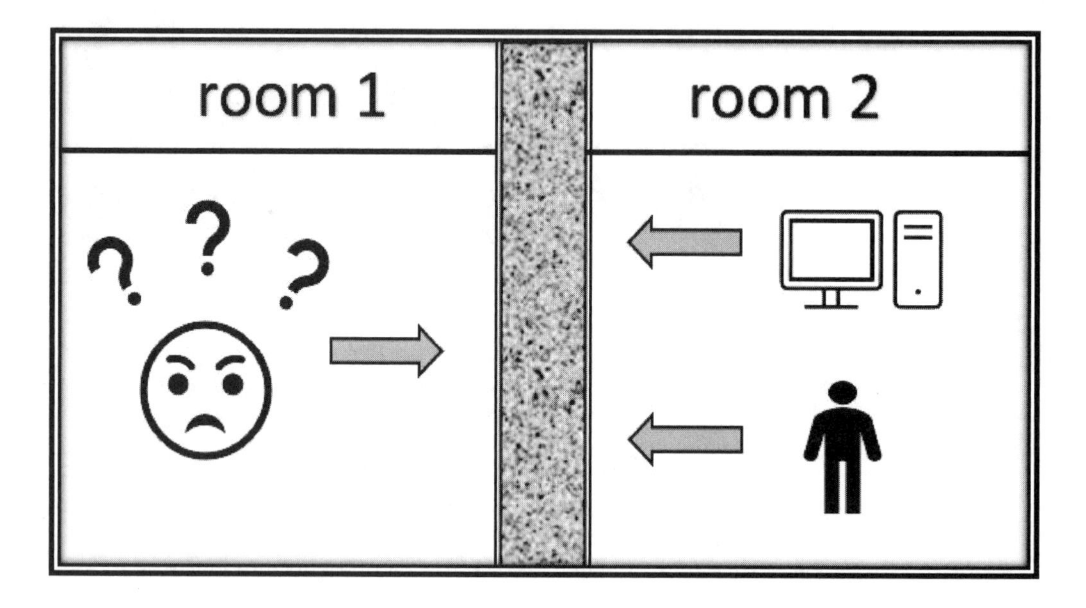

Figure 18.1 An intelligent agent should be able to recognize whether it communicates with a machine or with a person.

The situation becomes less trivial if we allow the occupant of Room 2 to cheat, to give answers that are intentionally incorrect to mislead us. For example, the computer may say that the fifth root of π is 0.5, assuming the dismissive attitude of a bored individual who does not want to be bothered.

Allan Turing believed that, even under such circumstances, the nature of the answers will sooner or later reveal the source of the answers to be the machine, no matter how many tricks the machine's programmer may have employed.

18.1.3 Beating the Turing Test

As early as in the mid-1960s, the computer program ELIZA seemed to pass the test in the sense that those who conversed with it believed they were interviewing a person. The idea was disarmingly simple. The user's query was compared to a set of keywords. If the keyword was found, in the query, the program invoked rules that helped it create an intelligent-looking response.

If the keywords failed, ELIZA dodged the question by a pre-programmed general answer such as, "give me a break, this is a topic I'd rather avoid." Or, if you asked, say, "what do you think about the liberal movement," the program would pick the last word, *movement*, and reply, "what exactly do you mean by movement?" Either response appeared very human, and many participants were indeed mislead.

Later, other such programs were introduced, some of them so powerful as to seem almost to invalidate Turing's idea. This only reminds us that what appears impossible today may be solved tomorrow, and perhaps even ridiculed soon afterwards. Still, the test enjoys unceasing popularity. Every now and then, competitions on this theme are organized, bearing witness to Turing's genius. More sophisticated programs are subjected to more sophisticated tests, and many specialists believe that no program has truly passed the test.

Control Questions

If you are unable to answer the following questions, return to the appropriate place in the preceding text.

- Describe the general scenario of the Turing test.
- Offer examples of questions you would ask. Explain how the answers might help you tell the difference between the human and the machine.
- What was the name of the first program that appeared to beat Turing's test, and what "tricks" did it employ?

18.2 Chinese Room and Other Reservations

Turing's approach was pragmatic in that it suggested a way to put the machine's intelligence to test. Whether the machine really *is* intelligent, or only appears so, is another question. Some thinkers expressed serious doubts, arguing that a computer program, or the machine executing it, will never possess true understanding of anything. Perhaps the most famous of their arguments was John Searle's Chinese room.

18.2.1 Searle's Basic Scenario

Consider a closed room with a person that does not know a single Chinese word. The room contains a huge store of index cards. Each card contains two Chinese sentences, one a question, the other an answer. As in Figure 18.2, someone slips under the door a piece of paper on which a Chinese question is written. The person inside the room finds the index card containing this very question, and copies from it the suggested answer. The paper with the answer is then slipped back under the door.

Theoretically speaking, it is conceivable that the file system is big enough to contain an index card for any question anybody can ever ask. The man will thus always provide the correct answer—and nobody outside the room will doubt his command of the language. A real person many not be able to find the right index card quickly enough, but a computer program will do so in a second.

18.2.2 Does the Person Understand Chinese?

The man does not understand Chinese. He does not even attempt to figure out the nature of the questions presented to him. His only activity is to identify the card where the characters from the question match those on the sheet of paper slipped under the door. This, of course, does not constitute understanding.

The scenario neatly illustrates the limits of AI. A computer capable of millions of operations per second can quickly return the solution of a difficult problem, but will do so without any real insight into the problem's nature, without any trace of understanding.

Figure 18.2 A man in a closed room can possibly find, in a huge file system, a written answer to any question written in Chinese. He may then return this answer without understanding a single Chinese word.

18.2.3 Philosopher's View

Philosophers tend to associate intelligence with consciousness (engineers do not do that). They point out that our human condition cannot be separated from our desires and wishes, from our sense of pain and happiness, from countless instincts acquired throughout the eons of Darwinian evolution. The machine has none of these. The computer's actions are unlikely to be driven by emotions or sexuality, nor do we ever expect it to be embarrassed or angry or contrite.

Many philosophers believe that the computer can only become human-like if it is endowed with typically human traits.

18.2.4 What Chess-Playing Programs Have Taught Us?

The concluding remarks of Chapter 4 mentioned an important historical milestone: In 1997, the program *Deep Deep Blue*[2] beat world chess champion Garry Kasparov, thus passing Claude Shannon's test. For many scientists, however, the success only re-opened the old question: How to identify intelligence? Was problem-solving performance enough? The machine achieved its playing strength by evaluating billions of chessboard positions per each move; the grandmaster could scarcely consider more than a few dozen—and yet he lost only by a narrow margin. Who was more intelligent, then? A skeptic may regard the *Deep Deep Blue* experience as a variation of Searle's Chinese Room, each of those billions of positions investigated by the machine constituting one of the question/answer index cards.

[2] The two occurrences of "deep" are not a typo. First version of the program was *Deep Blue*. Later it was improved and presented under the modified name, *Deep Deep Blue*.

One might even argue that the prodigious amount of wasteful number-crunching provided evidence that the machine was pretty dumb. In that event, Shannon's test was not any stronger than Turing's test.[3] And if we accept the notion that sheer power can make it, it is legitimate to ask: Where is the end of mechanical number-crunching, and where does true intelligence begin?

18.2.5 Turing's Response to Theological Reservations

Not surprisingly, the conundrum of an intelligent machine has been scrutinized even from the religious point of view. Theologians argued that it was the God's decision that only human beings should be endowed with soul and reasoning. To this Turing offered his famous tongue-in-cheek rejoinder that this argument would unfairly limit the Almighty's powers. Would the theologians question the God's ability to put soul in, say, an elephant if He chose so? And if in an elephant, why not in a machine?

18.2.6 Weak AI versus Strong AI

The gist of this section can be condensed to two pieces of terminology. The notion that a machine might *act as if* it were intelligent (while not being so in reality) is known as the *weak-AI* hypothesis. This is what matters to an engineer, and this is what satisfies the ambitions of practical-minded AI scientists.

The notion that one might one day develop a machine that would *really be thinking* is known as the *strong-AI* hypothesis. The reader already senses that explorations of the strong-AI hypothesis belong more to the realm of philosophers, theologians, and perhaps science-fiction writers.

Control Questions

If you are unable to answer the following questions, return to the appropriate place in the preceding text.

- Summarize the essence of the Chinese Room metaphor. What was John Searle's point in proposing it?
- When a computer program beat world champion in chess, some scholars believed the feat did not really prove the machine to be intelligent. What was their point?
- What do you know about reservations raised by philosophers and theologians? What is meant by the terms, strong-AI hypothesis and weak-AI hypothesis?

18.3 Engineer's Perspective

The question of whether AI products are indeed intelligent and whether they might develop their own consciousness seems to resurface every few years. Not long ago, some

[3] To be fair, recent advances in game programming are getting closer to the human approach in that they rely extensively on the recognition of patterns induced by techniques from the relatively new fields of *reinforcement learning* and by *deep learning*.

media even organized discussions on the topic of AI-related ethical issues. Participants raised arguments and counterarguments regarding the urgency of granting the machines something analogous to animal rights. Some went even further, and predicted that an intelligent robot of the future would deserve the right to vote at presidential elections. All of this, of course, was based on the assumption that the strong-AI hypothesis from the previous section is realistic.

18.3.1 Practicality

In the eyes of an engineer seeking to develop a useful tool, all these discussions are irrelevant. What we want is to develop computer programs capable of solving problem for which algorithms are not known. This does not have much to do with attempts to create an intelligent being, no matter how attractive the idea may appear. This author hopes that this chapter has conveyed to the reader the essence of the main difficulty: We are not sure *what* constitutes intelligence; until we know it, we are unlikely to bring it about.

18.3.2 Should People Worry?

The growing power of AI programs raises labor-related questions. Over the last generations, machines kept taking over tedious and non-qualified jobs. Will they now start replacing even engineers, physicians, and highly skilled professionals in general?

This author is not so pessimistic. Recall the early stages of the industrial revolution. Many jobs were indeed lost; still, we no longer see the mass unemployment that the invention of machines was supposed to cause. Rather, workers have been freed from backbreaking drudgery, and could move on to safer jobs. As a matter of fact, new types of jobs were created!

During the 1960s and '70s computers started to take over a lot of administrative work, and yet they did not generate massive unemployment of bureaucrats. Latest technology will have similar consequences. AI will not replace human experts; it will assist them, increase their efficiency, and most likely even give rise to new types of jobs.

18.3.3 Augmenting Human Intelligence

AI scientists and engineers do not want to replace human intelligence; far from it. Their main focus is on solutions to difficult problems. In this, they often rely on mechanisms that are very different from those of the human brain. This difference can actually become beneficial. The reader will recall one important lesson gleaned from chess programming: A lot might be gained by combining the two approaches, and from endeavors to *augment* human intelligence by powerful computers that run AI programs.

18.3.4 Limitations of Existing AI

Creativity is likely to remain the domain of human superiority for a long time to come. In spite of the recent breakthroughs in the field of AI, the computer is still easily taken aback.

Can a crocodile ride a bicycle? If not, explain why. Any giggling 8 years old will tell you that the reptile's legs are too short, that the animal is unlikely to keep its balance, that its body is poorly adjusted to the bicycle's seat, and so on and so forth. Even the most sophisticated computer program is bound to fail the same test miserably.

Finding other examples is easy. Can a locomotive read newspapers? Can an airplane wave its wings? Can a sumo fighter dance the leading part in Tchaikovski's Nutcracker? Any schoolchild will happily educate you, but the computer will struggle. The machine may give you a correct answer by virtue of the closed-world assumption we know from the chapters on automated reasoning. Even so, it will be unable to support the verdict with convincing arguments, let alone with a sense of humor.

Control Questions

If you are unable to answer the following questions, return to the appropriate place in the preceding text.

- Should we be concerned about the goals of the proponents of the strong-AI hypothesis? Will smart computer program take over the jobs of educated professionals just like steam-powered machines took over the jobs of nineteenth century workers?
- What did this section have to say about the possibilities of *augmenting* human intelligence?
- Suggest tasks that an AI program is likely to find difficult. Use examples different from those mentioned in this section.

18.4 Concluding Remarks

Turing (1950) introduced his legendary test under the name "imitation game." Its popularity soon gave rise to all sorts of modified versions which, however, are in our context unimportant. The first program to beat the original Turing test, ELIZA, was presented by Weizenbaum (1966). The Chinese Room argument is attributed to Searle (1980), though a similar idea is said to have been independently formulated somewhat earlier by the Russian thinker Dneprov.[4]

As the accomplishments of artificial intelligence grow more and more impressive, they inevitably attract the attention of media and the public in general. The faddish excitement gives rise to excessively optimistic predictions alongside raised index fingers and dire warnings. The engineer better be prepared to face the arguments of alarmists and dreamers alike.

[4] This is what they say. However, the author of this book is not aware of any concrete paper or book where Dneprov published the said ideas.

Bibliography

[1] Applegate, D.L., Bixby, R.E., Chvatal, V., and Cook, W.J. (2007). *The Traveling Salesman Problem*. Princeton University Press, New Jersey.

[2] Axelrod, R. (1984). *The Evolution of Cooperation*. Basic Books, New York.

[3] Bobrow, D.G., Mittal, S., and Stefik, M.J. (1986). Expert systems: perils and promise. *Communications of the ACM*, Vol. 29 (9), pp. 880–894.

[4] Brachman, R.J. and Schmolze, J.G. (1989). An Overview of the KL-ONE Knowledge Representation System. In: *Readings in Artificial Intelligence and Databases*, pp. 207–230, Morgan Kaufmann, Massachusetts.

[5] Bratko, I. (2001). *Prolog Programming for Artificial Intelligence.* Pearson Education, London.

[6] Chakraborty (eds). (2009). *Computational Intelligence in Flow Shop and Job Shop Scheduling*. Springer Nature, Germany.

[7] Colmerauer, A. and Roussel, P. (1996). The birth of Prolog. *ACM Digital Library*, Vol. 38, pp. 331–367.

[8] Dempster, A.P. (1967). Upper and lower probabilities induced by a multivalued mapping. *Annals of Mathematical Statistics*, Vol. 38 (2), pp. 325–339.

[9] Dorigo, M. (1992). *Optimization, Learning and Natural Algorithms*, PhD thesis, Politecnico di Milano, Italy.

[10] Fikes, R.E. and Nilsson, N.J. (1971). STRIPS: A new approach to the application of theorem proving to problem solving. *Artificial Intelligence*, Vol. 2 (3–4), pp. 189–208.

[11] Gambardella, L.M. and Dorigo, M. (1995). Ant-Q: a reinforcement learning approach to the traveling salesman problem, *Proceedings of ML-95, Twelfth International Conference on Machine Learning*, Prieditis, A. and Russell, S. (eds.), Morgan Kaufmann, pp. 252–260.

[12] Gardner, M. (1970). The fantastic combinations of John Conway's new solitaire game "life." *Scientific American*, Vol. 223 (4), pp. 120–123.

[13] Ginsberg, M. (1987). *Readings in non-monotonic reasoning.* Morgan Kaufmann, San Mateo, California.

[14] Ginsberg, M. (1993). *Essentials of Artificial Intelligence.* Morgan Kaufmann Publishers, San Mateo, California.

[15] Hart, P., Nilsson, N.J., and Raphael, B. (1968). A formal basis for the heuristic determination of minimum cost paths. *IEEE Transactions on Systems Science and Cybernetics*, Vol. 4 (2), pp. 100–107.

[16] Hayes, P. (1973). *The Frame Problem and Related Problems in Artificial Intelligence.* University of Edinburgh Press, UK.

[17] Holland, J.H. (1975). *Adaptation in Natural and Artificial Systems*. University of Michigan Press, Ann Arbor, Michigan.

[18] Karaboga, D., and Baturk, B. (2007). A powerful and efficient algorithm for numerical function optimization: Artificial Bee Colony (ABC) Algorithm. *Journal of Global Optimization*, Vol. 39, pp. 459–471.

[19] Kennedy, J., and Eberhart, R. (1995). Particle Swarm Optimization. *Proceedings of IEEE International Conference on Neural Networks, ICNN-95*, Perth, Australia, Nov. 27–Dec. 1, pp. 1942–1948.

[20] Kirkpatrick, S., Gelatt Jr, C.D., and Vecchi, M.P. (1983). Optimization by simulated annealing. *Science*, Vol. 220, pp. 671–680.

[21] Kowalski, R.A. (1988). The early years of logic programming. *Communications of the ACM*, Vol. 3, pp. 38–43.

[22] Lindenmayer, A. (1968). Mathematical models for cellular interaction in development. *Journal of Theoretical Biology*, Vol. 18, pp. 280–315.

[23] Mamdani, E.H. (1977). Application of Fuzzy Logic to Approximate Reasoning Using Linguistic Synthesis. *IEEE Trans, Computers*, Vol. 26, pp. 1182–1191.

[24] Martello, S., and Toth, P. (1990). *Knapsack Problems: Algorithms and Computer Implementations*. Wiley-Interscience, New Jersey.

[25] Minsky, M. (1961). Steps toward artificial intelligence. In: Luger, G.F. (ed.) *Computation and Intelligence–Collected Readings*, MIT Press, Massachusetts.

[26] Pearl, J. (1986). Fusion, propagation and structuring in belief networks. *Artificial Intelligence*, Vol. 29, pp. 241–288.

[27] Rechenberg, I. (1973). *Evolutionsstrategie: Optimierung technischer Systeme nach Principien der biologischen Evolution*. Frommann-Holzboog, Stuttgart, Germany.

[28] Robinson, J.A. (1965). A machine oriented logic based on the resolution principle. *Journal of the ACM*, Vol. 12, pp. 23–41.

[29] Rozsypal, A., and Kubat, M. (2001). Using the genetic algorithm to reduce the size of a nearest-neighbor classifier and to select relevant attributes. *Proceedings of the 18th international conference on machine learning*, Williamstown, Australia, pp. 449–456.

[30] Searle, J. (1980). Minds, brains and programs. *Behavioral and Brain Sciences*, Vol. 3 (3), pp. 417–457.

[31] Shafer, G. (1976). *A Mathematical Theory of Evidence*. Princeton University Press, New Jersey.

[32] Shannon, C. (1950). Programming a computer for playing chess. *Philosophical Magazine*, Vol. 7 (pages unknown).

[33] Shortliffe, E.H., Buchanan, B.G. (1975). A model of inexact reasoning in medicine. *Mathematical Biosciences*, Vol. 23 (3–4), pp. 351–379.

[34] Turing, A. (1950). Computing machinery and intelligence. *Mind*, Vol. 236, pp. 433–460.

[35] Ulam, S.M. (1962). On some mathematical problems connected with patterns of growth of figures. *Mathematical Problems in Biological Sciences*, Vol. 14, pp. 215–224.

[36] Weizenbaum, J. (1966). ELIZA – A computer program for the study of natural language communication between man and machine. *Communications of the ACM*, 9 (1), pp. 36–45.

[37] Zadeh, L.A. (1965). Fuzzy Sets. *Information and Control*, Vol. 8, pp. 338–353.

[38] Zadeh, L.A. (1975). The concept of a linguistic variable and its application to approximate reasoning. *Information Sciences*, Vol. 8 (3), pp. 199–249.